D1716167

Waterborne Diseases in the United States

Editor

Gunther F. Craun
Health Effects Research Laboratory
U.S. Environmental Protection Agency
Cincinnati, Ohio

CRC Press, Inc.
Boca Raton, Florida

Library of Congress Cataloging in Publication Data
Main entry under title:

Waterborne diseases in the United States.

Includes bibliographies and index.
1. Waterborne infection—United States.
2. Water quality management—United States.
I. Craun, Gunther F.
RA642.W3W38 1986   614.4'3   85-9617
ISBN 0-8493-5937-6

This book represents information obtained from authentic and highly regarded sources. Reprinted material is quoted with permission, and sources are indicated. A wide variety of references are listed. Every reasonable effort has been made to give reliable data and information, but the author and the publisher cannot assume responsibility for the validity of all materials or for the consequences of their use.

All rights reserved. This book, or any parts thereof, may not be reproduced in any form without written consent from the publisher.

Direct all inquiries to CRC Press, Inc., 2000 Corporate Blvd., N.W., Boca Raton, Florida, 33431.

© 1986 by CRC Press, Inc.

International Standard Book Number 0-8493-5937-6
Library of Congress Card Number 85-9617
Printed in the United States

To Mary, Michael, Matthew, and Emily

PREFACE

Nostalgia for the "good old days" is a common phenomenon of survivors. This book provides a landmark of the bygone hazardous life of most people in the U.S. or in most of the industrialized world. In the first half of the 20th century, the family of man was an endangered species, not normally listed in legislative protection. Life expectancy at birth was still no more than 50 years. Infant mortality frequently reached or exceeded 50/1000 births. Infectious diseases were rampant. Prevention was still elementary, even though specific causes of transmission were gradually being understood.

During that public health gestation period, bases for literal miracles of disease prevention were implemented. The present volume records those experiences from 1920 to 1983 — an heroic accomplishment by Mr. Craun. No small part of the lists of waterborne diseases are the result of his own continuing identification of epidemic episodes throughout that period. They were the responsible indicators upon which sanitary engineers constructed and operated the myriads of public water supply systems of the U.S. These sophisticated devices represented one of the means by which the war against infectious diseases was largely won in this country and in the developed countries elsewhere.

As evidence accumulated throughout the decades as to the means of transmission via water, the management of human wastes and potable waters made tremendous progress in "building out" biological and physical causes of illness and death. The number of waterborne outbreaks still occurring is infinitesimal. They are predominantly in smaller noncommunity systems inadequately disinfected, in most instances by interruption of chlorination. Incidentally, chlorine should be noted as perhaps saving more lives throughout the world than any other chemical.

The pedestrian rehearsal in the volume conceals this dramatic extension of the lives of people to a present expectancy of some 75 years. The book deserves a high place on the book shelf with Von Clausewitz on the logistics of war and on the detection performances of Sherlock Holmes, James Bond, and Maigret. Their record is no better than that of Craun.

All of this does not suggest that vigilance of water quality can now be relaxed. Much too often we get vivid reminders that the familiar salmonellas, dysenteries, and the less familiar giardia are disturbing us. Those who think the task is finished are unaware that the bacterial-viral insults have a strange and long half-life.

But what of the future, only briefly noted by Craun? We entered a new and complex world in the last 2 decades. We call it the "Chemical Era". Manmade and natural chemicals total in the millions and thousands of new ones are synthetically produced every year. Too many find their way into the ambient waters of the country and into sources of drinking water. Their actual, potential, and alleged toxicities are known only for a relatively small percentage of the total number. The challenge ahead of us in the water management field is particularly difficult, because the manifestation of insults is 10, 20, and 30 years deferred — some perhaps genetically affecting our great grandchildren. We live in a world of chemicals, many of which are necessary and salutary for health, safety, and welfare. We wait upon major scientific clarifications. In the meantime, we should follow our past fruitful experience by taking them out of our potable waters as we continue to do with the familiar biological ones.

In all of this, we cannot ignore the remaining billions of people in the developing world. They are still the victims of those infectious diseases no longer major threats to us. Their misfortunes are now being multiplied as well by the chemical threats.

I close by recalling the wide reminder by the late great public health statesman, Prof. C.E.-A. Winslow. He emphasized that our world is a battlefield, not a nursery.

Abel Wolman
September 1984

THE EDITOR

Gunther F. Craun* has served in various capacities over the past 20 years as an environmental engineer and epidemiologist with the U.S. Public Health Service and U.S. Environmental Protection Agency (EPA). Since 1970 he has been associated with the drinking water and health research activities of the EPA. In addition to infectious waterborne diseases, his current research interests include relationships between drinking water contaminants and cardiovascular disease and cancer. He received his education in civil engineering (B.S.) and sanitary engineering (M.S.) at Virginia Polytechnic Institute, Blacksburg, Va. and public health (M.P.H.) and epidemiology (S.M.) at Harvard University, Cambridge, Mass.

He has authored and coauthored numerous articles in the international scientific, public health, and engineering literature. The American Water Works Association and the New England Water Works Association have recognized Mr. Craun for his work on waterborne disease outbreaks and trace metals in the drinking water of the Boston metropolitan area. The EPA awarded Mr. Craun a meritorious performance citation for his participation in the Community Water Supply Study, which identified deficiencies in the public water supplies of the Nation.

He is registered as a professional engineer in the Commonwealth of Virginia and is a member of the American Water Works Association Committee on the Status of Waterborne Disease Outbreaks in the U.S. and Canada (chairman, 1977 to 1982) and the International Association of Milk, Food, and Environmental Sanitarians Committee on Communicable Diseases Affecting Man. He served as liasion representative to the Safe Drinking Water Committee of the National Research Council from 1974 to 1977 and as a member of the World Health Organization Working Group of Sodium, Chloride, and Conductivity in Drinking Water in 1978. He has also served as a member of the Water Pollution Control Federation Research Committee and the International Association on Water Pollution Research Study Group on Water Virology.

Mr. Craun is currently Coordinator of the Environmental Epidemiology Program in the Health Effects Research Laboratory of the EPA, Cincinnati. In his present capacity, he works with a number of other governmental agencies, including the National Cancer Institute, Bethesda, Md. and Oak Ridge National Laboratory, Oak Ridge, Tenn. on epidemiological studies of drinking water contaminants. He is also involved in projects with the National Academy of Sciences, Washington, D.C. and the Center for Environmental Epidemiology, University of Pittsburgh, Pittsburgh, Pa. to identify new research areas and methodologies for environmental epidemiology.

---

* This book was edited by Gunther F. Craun in his private capacity. No official support or endorsement by the Environmental Protection Agency or any other agency of the Federal Government is intended or should be inferred.

CONTRIBUTORS

Frank L. Bryan
Food Microbiologist
Food Safety Consultation and Training
Tucker, Georgia

Gunther F. Craun
Epidemiologist and Sanitary Engineer
Health Effects Research Laboratory
U.S. Environmental Protection Agency
Cincinnati, Ohio

Alfred P. Dufour
Chief, Bacteriology Group
Health Effect Research Laboratory
U.S. Environmental Protection Agency
Cincinnati, Ohio

T. H. Ericksen
Microbiologist
Health Effects Research Laboratory
U.S. Environmental Protection Agency
Cincinnati, Ohio

Jeffrey R. Harris
Medical Epidemiologist
Enteric Diseases Branch
Division of Bacterial Diseases
Centers for Disease Control
Atlanta, Georgia

John C. Hoff
Research Microbiologist
Drinking Water Research Division
U.S. Environmental Protection Agency
Cincinnati, Ohio

Peter C. Karalekas, Jr.
Sanitary Engineer
Water Supply Branch
U.S. Environmental Protection Agency
Boston, Massachusetts

Edwin C. Lippy
Sanitary Engineer
Health Effects Research Laboratory
U.S. Environmental Protection Agency
Cincinnati, Ohio

Gary S. Logsdon
Research Sanitary Engineer
Drinking Water Research Division
U.S. Environmental Protection Agency
Cincinnati, Ohio

Floyd B. Taylor
Executive Director
New England Water Works Association
Dedham, Massachusetts

Abel Wolman
Professor Emeritus
Department of Sanitary Engineering and
　Water Resources
The Johns Hopkins University
Baltimore, Maryland

TABLE OF CONTENTS

SECTION I: WATERBORNE DISEASES

Chapter 1
Introduction 3
Gunther F. Craun

Chapter 2
Clinical and Epidemiological Characteristics of Common Infectious
Diseases and Chemical Poisonings Caused by Ingestion of
Contaminated Drinking Water 11
Jeffrey R. Harris

Chapter 3
Diseases Caused by Water Contact 23
Alfred P. Dufour

Chapter 4
Chemical Drinking Water Contaminants and Disease 43
Gunther F. Craun

SECTION II: WATERBORNE OUTBREAK STATISTICS

Chapter 5
Statistics of Waterborne Outbreaks in the U.S. (1920—1980) 73
Gunther F. Craun

Chapter 6
Recent Statistics of Waterborne Disease Outbreaks (1981—1983) 161
Gunther F. Craun

SECTION III: INVESTIGATION OF WATERBORNE OUTBREAKS

Chapter 7
Epidemiologic Procedures for Investigation of Waterborne Disease Outbreaks 171
Frank L. Bryan

Chapter 8
Methods to Identify Waterborne Pathogens and Indicator Organisms 195
T. H. Ericksen and Alfred P. Dufour

Chapter 9
Engineering Aspects of Waterborne Outbreak Investigation 215
Edwin C. Lippy

SECTION IV: PREVENTION OF WATERBORNE OUTBREAKS

Chapter 10
Regulations and Surveillance 233
Peter C. Karalekas, Jr. and Floyd B. Taylor

Chapter 11
Barriers to the Transmission of Waterborne Disease 255
Gary S. Logsdon and John C. Hoff

Epilogue..275
Gunther F. Craun

Index...279

Section I: Waterborne Diseases

Chapter 1

INTRODUCTION

Gunther F. Craun*

TABLE OF CONTENTS

I. Content of Book 4

II. Water-Related Illness 4
    A. Waterborne Diseases 5
    B. Water-Washed Diseases 5
    C. Water-Based Diseases 5
    D. Water-Vectored Diseases 6

III. Miscellaneous Causes of Water-Associated Illness 6
    A. Shellfish 6
    B. Airborne 7
    C. Algae 8

IV. Summary 9

References 9

---

* This chapter was written by Gunther F. Craun in his private capacity. No official support or endorsement by the Environmental Protection Agency or any other agency of the Federal Government is intended or should be inferred.

I. CONTENT OF BOOK

This book examines, in both a current and historical context, water-related illness in the U.S. Emphasis is placed upon the transmission of infectious diseases through contaminated drinking water supplies and those deficiencies in water supply systems which allow waterborne outbreaks to occur. Chapters have been included on the important etiologic agents responsible for waterborne outbreaks in the U.S., surveillance activities, regulations, water treatment to prevent the occurrence of waterborne outbreaks, and procedures for investigating waterborne outbreaks. For completeness, discussions have been included on illnesses contracted by ingestion of or contact with waters for bathing, swimming, or wading and chronic ingestion of low levels of chemical contaminants in drinking water; however, because of space limitations these are necessarily brief, and the reader is directed toward the provided references, which discuss these subjects in more depth.

Waterborne outbreaks of infectious disease and acute chemical poisonings are usually readily apparent, whereas the sporadic occurrence of illness, subclinical illness, and adverse effects associated with chronic exposures are often difficult to detect. However, the reporting of most water-related illness is voluntary in the U.S., and many waterborne outbreaks are not recognized, investigated, or reported. This should be kept in mind when evaluating the statistics presented in this book.

The manifestations of waterborne illnesses range from asymptomatic infection and slight discomfort to severe reactions which may result in death, depending upon the etiologic agent and host response. Among the etiologic agents responsible for water-related illness throughout the world, some are important causes of waterborne outbreaks and illness in the U.S. while others are not. Foreign travel exposes persons to water that sometimes contains pathogens not usually found in this country, and whether these pathogens become important causes of waterborne illness in the U.S. depends upon many factors, including the infectious dose, excretion patterns of the organism, environmental conditions which allow survival of the pathogen, level of protection against contamination of water sources, and the adequacy of treatment of water supplies to remove or inactivate the pathogen. For infections which are transmitted by the fecal-oral route and which require a large number of organisms to infect, the organisms may never reach infective-dose levels in large surface water sources, but in the developing countries and in small water systems in the U.S., it may be possible to receive a sufficient contamination to cause illness.

II. WATER-RELATED ILLNESS

In the past, infectious diseases were frequently transmitted through contaminated drinking water, but with improvements in waste water disposal practices and the development, protection, and treatment of water supplies, the prevalence of infectious diseases has been reduced in the U.S. and other developed countries. The impact of water supply development and improvements in water quality on the health status of populations in developing countries, however, has varied because of the geographical, cultural, climatic, and socioeconomic differences which exist among the various countries of the world. In the developed countries the concern for prevention of water-related illness is primarily with the quality and quantity of drinking water, but these are only two of the components necessary to reduce water-related illness in other countries. To consider the problem of water-related illness and its prevention in a world-wide context, it is important to understand Bradley's classifications[1-3] of water-related illness which are based upon epidemiologic considerations and permit generalizations about the likely effect of environmental changes and other actions on their incidence (Table 1).

Table 1
BRADLEY'S
CLASSIFICATION OF
WATER-RELATED
ILLNESSES[1]

Waterborne
Water-washed
Water-based
Water-vectored

A. Waterborne Diseases

Waterborne diseases are those transmitted through the ingestion of contaminated water, and water acts as the passive carrier of the infectious or chemical agent. The classic waterborne diseases, cholera and typhoid fever, so frequently observed in the past in densely populated areas of the world, have been effectively controlled by the protection of water sources and treatment of contaminated water supplies. These classic diseases gave water supply its reputation as an important factor in the reduction of infectious diseases.[2] Diseases caused by other bacteria, viruses, protozoa, and helminths may also be transmitted by contaminated drinking water; however, it is important to remember that these diseases are transmitted through the fecal-oral route from human to human or animal to human with drinking water being only one of several possible sources of infection. Depending upon the infectious dose and the number of excretors in a population, some infectious diseases are likely to continue to occur in the developing countries regardless of steps taken to improve water supplies. In addition to infectious disease, chemical poisonings and methemoglobinemia have also been caused by contaminated water supply systems throughout the world. The chronic ingestion of low levels of chemical contaminants in drinking water has been associated with adverse human health effects, but with a few exceptions these associations are not completely understood at present.

B. Water-Washed Diseases

Water-washed diseases are closely related to poor hygienic habits and sanitation, and the availability of a sufficient quantity of water is generally felt to be more important than the quality of the water. The unavailability of water for washing and bathing contributes to diseases that affect the eye and skin, such as trachoma, infectious conjunctivitis, and scabies, and to diarrheal illnesses which, in the developing countries, are an important cause of infant mortality and morbidity. The diarrheal diseases are transmitted person to person and through contaminated foods and utensils by persons whose hands are fecally contaminated. The availability of water for hand washing has been shown to decrease the incidence of diarrheal diseases and prevalence of excretion of enteric pathogens, especially *Shigella* in various countries, including areas of the U.S.[4]

C. Water-Based Diseases

Water-based diseases are those in which the pathogen spends an essential part of its life in water or is dependent upon aquatic organisms for the completion of its life cycle. Schistosomiasis and dracontiasis are examples. Water quality and cultural-social behavior play important roles in the transmission of these diseases. Dracontiasis, or guinea worm disease, is the only disease which I know to be transmitted exclusively through contaminated drinking water.

Dracontiasis, an infection to the subcutaneous and deeper tissues with a large nematode, *Dracunculus medinensis*, is widely distributed in India and west Africa, but also occurs in northeastern Africa, the Middle East, the West Indies, and northeastern South America.

Dracontiasis is primarily associated with poverty, particularly with rural communities with inadequate water supplies. The intermediate host, a crustaces of the genus *Cyclops,* is required for transmission. This disease is manifested by a vesicle that usually appears on a lower extremity, especially the foot. Within the vesicle lies the female worm which can discharge larvae when the vesicle is immersed in water. Larvae discharged into the water are ingested by *Cyclops* and develop into the infective stage in about 2 weeks. Man swallows the infected copepods by drinking water from contaminated step-wells and ponds. Larvae are liberated in man's stomach or duodenum, migrate through the viscera, become adults, and reach the subcutaneous tissues. The cycle is continued when infected individuals wade in step-wells and ponds used for water supplies where *Cyclops* are present. The disease is transmitted only in areas where drinking water is obtained by persons wading in step-wells or ponds and is not likely to occur in the U.S.

Schistosomiasis is not indigenous to the continental U.S., but schistosome dermatitis[5] has been documented in the U.S. The three major species of schistosome that develop to maturity in humans are *Schistosoma japonicum, S. haematobium,* and *S. mansoni.* Each has a different snail host and a different geographic distribution. Warren[6] estimated that more than 200 million people in Asia, Africa, South America, and the Caribbean are presently infected with one, or perhaps two, of these species of schistosomes. Immigrants to the U.S. have been found to be infected with schistosomiasis, and it is estimated that some 300,000 persons in Puerto Rico are infected. The economic effects of schistosomiasis were estimated by Wright[7] to amount to some $642 million annually, and this includes only the resource loss attributable to reduced productivity and not the costs of public health programs, medical care, or compensation of illness.

The infections transmitted through contact with contaminated water in the U.S. and discussed in Chapter 3 may be included within this classification of water-related illnesses, but when compared with the occurrence of dracontiasis and schistosomiasis, their occurrence on a world-wide basis is of relatively minor importance.

D. Water-Vectored Diseases

Water-vectored diseases, such as yellow fever, dengue, filiariasis, malaria, onchocerciasis, and sleeping sickness, are transmitted by insects which breed in water (like malaria-carrying mosquitoes) or insects which bite near water (like insects which transmit the filarial infection onchocerciasis). These diseases are controlled through means other than water supply considerations (destruction of breeding grounds, pesticides, etc.) and are not considered in this book.

III. MISCELLANEOUS CAUSES OF WATER-ASSOCIATED ILLNESS

A. Shellfish

Outbreaks of gastroenteritis and other infectious diseases have been caused by the consumption of contaminated shellfish, and although these illnesses are considered to be foodborne rather than water-related, several key references[8-12] are provided for those readers interested in these or other illnesses transmitted by foods contaminated by waste waters. Statistics recently compiled by Verber[13] on outbreaks caused by the consumption of contaminated oysters, clams, and mussels show that over 11,600 cases of shellfish-associated illness have been reported in the U.S. since 1900 (Table 2). The last reported cases of shellfish-associated typhoid fever in the U.S. occurred in 1951, 1952, and 1954 after the consumption of contaminated clams in New York and New Jersey. Some 1400 cases of shellfish-associated hepatitis A were reported from 1961 through 1980 along the East and Gulf Coasts from Connecticut to Mississippi, and in recent years shellfish-associated gastroenteritis caused by non-0 group 1*Vibrio cholerae*[11] and Norwalk virus[12] have been reported.

Table 2
SHELLFISH-ASSOCIATED ILLNESS REPORTED IN THE U.S.[13]

Time period	Cases of illness					
	Unspecified gastroenteritis	Typhoid fever	Hepatitis A	Norwalk agent	Cholera	Other[a]
1900—1904	—	150				
1905—1909	52	160				
1910—1914	—	18				
1915—1919	79	110				
1920—1924	—	1591				
1925—1929	31	539				
1930—1934	11	202				
1935—1939	85	269				
1940—1944	980	156				
1945—1949	536	67				100
1950—1954	100	4				—
1955—1959	—					
1960—1964	7		1013			—
1965—1969	226		36			99
1970—1974	15		304			—
1975—1979	317		35		12	74
1980—1984	4337		11	6	12	28

[a] Includes shigellosis, salmonellosis, *Vibrio parahaemolyticus* AGI, *Escherichia coli* AGI, and AGI caused by several other agents.

B. Airborne

The potential for transmission of illness through inhalation of aerosols, gases, or vapors from contaminated water and waste water should not be overlooked even though these do not fall within Bradley's classifications of water-related illness. Outbreaks of legionellosis, an acute respiratory disease, have been attributed to the airborne dissemination of *Legionella pneumophila* through cooling towers and evaporative condensers of hotels, hospitals, a shopping mall, and a golf clubhouse.[13-17] More recently, legionellosis has been epidemiologically associated with hot and cold potable water systems in hospitals and hotels, and *Legionella* have been isolated in water samples from showerheads, faucets, and water storage tanks. *Legionella* appears to be able to colonize and grow in certain parts of water piping systems within buildings, especially hot water systems.[17] It is suspected that aerosolization of potable water through showerheads is important for the transmission of illness. The reader is directed to three excellent reviews[17-19] which discuss the nature of the illness, epidemiology, and potential means of transmission. An illness with symptoms consistent with hypersensitivity pneumonitis has also been suggested as being associated with contaminated water in humidifiers,[20] and the aerosolization of nontuberculous mycobacteria from aquatic environments may be a pathway for human infection in the eastern U.S.[21,22]

Atterholm et al.[23] described an outbreak of repeated chills, fever, leukocytosis, respiratory tract symptoms, and muscle pain (beginning 4 hr after a hot bath and lasting for 6 to 15 hr) occurring in 1975 among 56 persons who lived in an area of Sweden supplied with water from Lake Vombsjon. Bacteriological and chemical studies and investigation of the water for endotoxins and algae revealed nothing unusual, and no causal agent could be identified. In a series of experiments with volunteers, it was found that inhalation for about 10 min with the face just above water was enough to provoke a reaction. There was no reaction if a gas mask was worn. Susceptible individuals could take a bath outside the area supplied by lake water without being affected. Similar cases had occurred in 11 different places in southern Sweden between 1952 and 1975. Aro et al.[24] reported an outbreak of a similar respiratory illness among more than half the adult population in an area near Tampere, Finland, from August to December 1978. Symptoms included cough, dyspnea, chills, fever, headaches, muscle pain, and aching of joints, and appeared to be associated with exposure to vapor from tap water during bathing and washing dishes. The etiologic agent could not be specifically identified; however, endotoxin concentrations produced by blue-green algae were high in both the lake water source and tap water and might have been a possible cause. Neither massive chlorination of the water nor changing the sand filter had any effect, but the symptoms disappeared after the source of the water supply was changed.

The volatilization of organic compounds and release of radon gas from drinking water supplies contribute contaminants to the indoor air environment, and in some instances the inhalation of these contaminants may result in substantial individual doses when compared with the ingestion route of exposure. In a plant whose well was contaminated with 9.1 mg/ℓ benzene, Lucas[25] reported airborne benzene concentrations of 23.4 mg/m^3 in the shower room and 1.2 mg/m^3 near a wash basin with running water and calculated an approximate daily dose of 1 to 10 mg benzene attributed to inhalation exposure compared with 1 to 4 mg from ingestion. Air concentrations of trichloroethylene (TCE) in homes using individual wells from an aquifer containing about 40 mg/ℓ of TCE were reported to be as high as 81 mg/m^3 in bathrooms with the shower running.[26] This concentration of TCE is approximately one third of the occupational threshold limit value of the American Conference of Governmental and Industrial Hygienists,[27] emphasizing the need to consider inhalation exposures when assessing the possible adverse effects of chemical drinking water contaminants, especially the low molecular weight organic compounds. Skin absorption of chemical contaminants in bath water may also be important. Harris et al.[28] reported that the calculated skin absorption doses for a Tennessee infant bathing in well water contaminated by carbon tetrachloride, chloroform, chlorobenzene, and tetrachloroethylene was 25 to 40% higher than for ingestion.

Radon, a water-soluble inert gas, has been measured in the indoor air, originating from various sources including well waters, soils and rocks, and construction materials. Radon contained in ground waters drawn from soil or rock rich in radium is outgased into the home environment when water is used for showering, bathing, dishwashing, and washing clothes. In Maine,[29] an average concentration of 22,000 pCi/ℓ radon-222 was found in well waters from granite areas, but some wells contained up to 180,000 pCi/ℓ. The airborne radon-222 in homes ranged from 0.05 to 210 pCi/ℓ with airborne concentrations being high when water concentrations were high. Hess et al.[29] showed a geographic correlation between lung cancer mortality and the estimated waterborne radon exposure in Maine counties. The individual lifetime risk of dying from lung cancer due to an airborne radon concentration of 2 pCi/ℓ is estimated between 0.5 to 1.0%. Radon progeny tend to adhere to particulates and aerosols and to the surfaces of air passageways in the lungs, emitting α and β radiation during decay. The effects of α radiation from radon progeny are felt to be most important, particularly from those attached to the bronchial passageways, and radon progeny ingested with water are felt to have a minimal effect on the GI tract because of the short range of α radiation.[30]

C. Algae

The most common species of blue-green algae (Cyanobacteria) are also the most likely to produce toxins during algae blooms in reservoirs, lakes, and ponds, and if the cells and toxins become highly concentrated, the potential exists for illness or death in mammals, birds, and fish after ingestion of the toxic cells or extracellular toxin. Major losses to animals include cattle, sheep, pigs, birds, and fish, and minor losses are reported for dogs, horses, small wild mammals, amphibians, and invertebrates. Acute oral toxicity to humans has not been documented, but evidence is available which suggests the toxins may cause human gastroenteritis when ingested from municipal drinking water supplies.[31] Lipopolysaccharide endotoxin produced by certain Cyanobacteria, including *Schizothrix calcicola* and *Anabaena flos-aquae*, has been implicated in waterborne outbreaks of gastroenteritis in humans.[31] Contact dermatitis in humans has also been associated with the use of recreational water and may be due to an immunogenic response to the filaments or exotoxins secreted into the water.[32] For additional information on this subject, the reader is directed toward a recent publication of the proceedings of a conference on waterborne algae toxins and health.[33]

IV. SUMMARY

Water, an essential component of all organisms, has multiple roles in human societies. It is essential for drinking and is used for a wide variety of purposes including washing, bathing, cleaning, food preparation and processing, crop irrigation, aquaculture, industrial processes, sewage carriage and disposal, energy generation, and recreation. It may become a vehicle for the transmission of illness as well as a reservoir for the development of organisms that harbor or transmit illness. There are a wide variety of sources of water, each presenting a potential for water-related illness depending upon factors such as a country's general level of sanitation, cultural and socioeconomic characteristics of the population, sewage disposal practices, general technological development and degree of industrialization, endemic levels of disease, climate, and geological and hydrological conditions. The incidence of water-related illnesses in the U.S. and other developed countries is low compared with most of the rest of the world. Nevertheless, there is a residual occurrence of waterborne disease which can and should be eliminated.

This book is intended to serve as a guide and reference for those interested in investigating and preventing water-related illness. It is hoped that this book will stimulate interest in the documentation and reporting of waterborne outbreaks in the U.S. Only through good reporting can we be certain of the true extent of the problem, causes of water-related illness, and how we should approach solving the problem. Vigilance is essential if we hope to reduce or eliminate the residual occurrence of waterborne disease which occurs in the U.S. Knowledge of the currently important and potential new etiologic agents and causes of waterborne outbreaks is important if we are to design effective surveillance and regulatory programs for the protection and treatment of water supplies to prevent the transmission of illness.

REFERENCES

1. **Bradley, D. J.**, Health problems of water management, *J. Trop. Med. Hyg.*, 73, 286, 1970.
2. **Schneider, R. E., Shiffman, M., and Faigenblum, J.**, The potential effect of water on gastrointestinal infections prevalent in developing countries, *Am. J. Clin. Nutr.*, 31, 2089, 1978.
3. **Cutting, W. A. M. and Hawkins, P.**, The role of water in relation to diarrhoeal disease, *J. Trop. Med. Hyg.*, 85, 31, 1982.
4. **Blum, D. and Feachem, R. G.**, Measuring the impact of water supply and sanitation investments on diarrhoeal diseases: problems of methodology, *Int. J. Epidemiol.*, 12, 357, 1983.
5. **Wills, W., Fried, B., Carroll, D. F., and Jones, G. E.**, Schistosome dermatitis in Pennsylvania, *Public Health Rep.*, 91, 469, 1976.
6. **Warren, K. S.**, Precarious odyssey of an unconquered parasite, *Nat. Hist.*, 83, 46, 1974.
7. **Wright, W. H.**, A consideration of the economic impact of schistosomiasis, *Bull. WHO*, 47, 559, 1972.
8. **Bryan, F. L.**, Diseases transmitted by foods contaminated by wastewater, *J. Food Prot.*, 40, 45, 1977.
9. **Bryan, F. L.**, Diseases Transmitted by Foods (A Classification and Summary), Center for Disease Control, Publ. No. (CDC) 75-8237, U.S. Department of Health, Education and Welfare, Atlanta, Ga., 1975.
10. Foodborne Disease Surveillance Annual Summary 1981, Center for Disease Control, HHS Publ. No. (CDC) 83-8185, Atlanta, Ga., 1983.

11. **Rodrick, G. E., Lotz, M., Alexiou, N. G., and Ambrusko, J.**, A case of shellfish associated cholera in South Florida, *J. Fla. Med. Assoc.*, 68, 816, 1981.
12. **Gunn, R. A., Janowski, H. T., Lieb, S., Prather, E. C., and Greenburg, H. B.**, Norwalk virus gastroenteritis following raw oyster consumption, *Am. J. Epidemiol.*, 115, 348, 1982.
13. **Verber, J. L.**, Shellfish Borne Disease Outbreaks, Northeast Technical Services Unit, Public Health Service, Food and Drug Administration, Davisville, R. I., 1984.
14. **Dondero, T. J., Rendtorff, R. C., Mallison, G. F., Weeks, R. M., Levy, J. S., Wong, E. W., and Schaffner, W.**, An outbreak of legionnaires' disease associated with a contaminated air-conditioning cooling tower, *N. Engl. J. Med.*, 302, 365, 1980.
15. **Cordes, L. G., Frasher, D. W., Skalig, P., Perlino, C. A., Elsea, W. R., Mallison, G. F., and Hayes, P.**, Legionnaires' disease outbreak at an Atlanta, Georgia, country club: evidence for spread from an evaporative condensor, *Am. J. Epidemiol.*, 111, 425, 1980.
16. **Grist, N. R., Reid, D., and Najera, R.**, Legionnaires' disease and the traveler, *Ann. Intern. Med.*, 90, 563, 1979.
17. **Dufour, A. P. and Jakubowski, W.**, Drinking water and legionnaires' disease, *J. Am. Water Works Assoc.*, 74, 631, 1982.
18. **Thornsberry, C., Balows, A., Feeley, J. C., and Jakubowski, W., Eds.**, *Legionella: Proceedings of the 2nd International Symposium*, American Society for Microbiology, Washington, D. C., 1984.
19. **Broome, C. V. and Fraser, D. W.**, Epidemiologic aspects of legionellosis, *Epidemiol. Rev.*, 1, 1, 1979.
20. **Ganier, M., Lieberman, P., Fink, J., and Lockwood, D., G.**, Humidifier lung. An outbreak in office workers, *Chest*, 77, 183, 1980.
21. **Falkinham, J. O., Parker, B. C., and Graft, H.**, Epidemiology of infection by nontuberculous mycobacteria. I. Geographic distribution in the Eastern United States, *Am. Rev. Respir. Dis.*, 121, 931, 1980.
22. **Wendt, S. L., George, K. L., Parker, B. C., Graft, H., and Falkinham, J. O.**, Epidemiology of infection by nontuberculous mycobacteria. III. Isolation of potentially pathogenic mycobacteria from aerosols, *Am. Rev. Respir. Dis.*, 122, 259, 1980.
23. **Atterholm, I., Ganrot-Norlin, K., Hallberg, T., and Ringertz, O.**, Unexplained acute fever after a hot bath, *Lancet*, 2, 684, 1977.
24. **Aro, S., Muittari, A., and Virtanen, P.**, Bathing fever epidemic of unknown aetiology in Finland, *Int. J. Epidemiol.*, 9, 215, 1980.
25. **Lucas, J. B.**, Health effects of nonmicrobiologic contaminants, in *Wastewater Aerosols and Disease*, Pahren, H. and Jakubowski, W., Eds., U.S. Environmental Protection Agency, EPA-600/9-80-028, Cincinnati, Ohio, 1980.
26. **Couch, A. and Andelman, J.**, Unpublished data, 1984.
27. Documentation of Threshold Limit Value Criteria, American Conference of Governmental Industrial Hygienists, Cincinnati, Ohio, 1983.
28. **Harris, R. H., Rodricks, J. V., Rhamy, R. K., and Papadopulos, S. S.**, Adverse health effects at a Tennessee hazardous waste disposal site, in *Evaluation of Health Effects from Waste Disposal Sites*, Andelman, J. B., Ed., Princeton Scientific, Princeton, N.J., in press.
29. **Hess, C. T., Weiffenbach, C. V., and Norton, S. A.**, Environmental radon and cancer correlations in Maine, *Health Phys.*, 45, 339, 1983.
30. **Weiffenbach, C. V.**, *Radon, Water, and Air Pollution: Risks and Control*, University of Maine, Orono, 1982.
31. **Carmichael, W. W.**, Freshwater blue-green algae (cyano-bacteria) toxins — a review, in *The Water Environment: Algal Toxins and Health*, Carmichael, W. W., Ed., Plenum Press, New York, 1981, 1.
32. **Billings, W. H.**, Water-associated human illness in northeast Pennsylvania and its suspected association with blue-green algae blooms, in *The Water Environment: Algal Toxins and Health*, Carmichael, W. W., Ed., Plenum Press, New York, 1981, 243.
33. **Carmichael, W. W., Ed.**, *The Water Environment: Algal Toxins and Health*, Plenum Press, New York, 1981.

Chapter 2

CLINICAL AND EPIDEMIOLOGICAL CHARACTERISTICS OF COMMON INFECTIOUS DISEASES AND CHEMICAL POISONINGS CAUSED BY INGESTION OF CONTAMINATED DRINKING WATER

Jeffrey R. Harris

TABLE OF CONTENTS

I. Introduction 12

II. Bacterial Diseases 12
    A. Campylobacteriosis 12
    B. Cholera 12
    C. Enterotoxigenic *Escherichia coli* Gastroenteritis 14
    D. Salmonellosis 15
    E. Typhoid Fever 15
    F. Shigellosis 16
    G. *Yersinia enterocolitica* Infection 16

III. Viral Diseases 17
    A. Hepatitis A 17
    B. Norwalk Gastroenteritis 17
    C. Rotavirus Gastroenteritis 17

IV. Parasitic Diseases 18
    A. Amebiasis 18
    B. Giardiasis 18

V. Acute Chemical Poisonings 19

References 19

I. INTRODUCTION

This chapter is intended as a practical introduction to waterborne diseases which are caused by the ingestion of water and which are of public health importance in the U.S. The chapter is divided into sections for bacterial, viral, and parasitic diseases, and acute chemical poisonings. The epidemiologic, clinical, and diagnostic laboratory features of each disease are discussed. Although the relative importance of waterborne transmission is considered for each disease, more complete information on the incidence of waterborne outbreaks of these diseases can be found in Section II of this volume. For those readers wanting more complete clinical and diagnostic laboratory information, key references are provided. Laboratory procedures for the identification of pathogens from water are given in Chapter 10.

For the convenience of the reader faced with a possible waterborne disease outbreak, the pathogens and agents are grouped by incubation period (Table 1). The symptoms and incubation period of each disease are also given in Table 2 for quick reference. Additional information on the clinical syndrome and laboratory criteria for the confirmation of waterborne disease outbreaks is provided in Table 1, Chapter 8.

II. BACTERIAL DISEASES

A. Campylobacteriosis

Most reported human campylobacteriosis, and perhaps all waterborne campylobacteriosis, has been caused by *Campylobacter jejuni*. Within the last 10 years, *C. jejuni* has been recognized as one of the most common causes of human diarrhea in the U.S.[1] Although the organism is most frequently transmitted in outbreaks by foods (especially poultry and raw milk), waterborne transmission has been important, both as a cause of epidemics and as a cause of sporadic infections in backpackers. The first reported waterborne campylobacteriosis outbreak in the U.S. occurred in Bennington, Vt.,[2] in 1978. In this outbreak, 3000 townspeople became ill after contamination of a stream that was Bennington's main unfiltered water source. Taylor et al.[3] recently confirmed the role of waterborne *C. jejuni* in causing sporadic diarrhea among backpackers. In this investigation, campylobacteriosis was significantly associated with the consumption of untreated surface water in back-country areas, and *C. jejuni* was isolated from multiple specimens of surface water.

C. jejuni invades the colon, but the exact pathogenic mechanisms are unknown. Infection results in a gastroenteritis that cannot readily be distinguished from the illnesses caused by other enteric pathogens. Symptoms follow an incubation period of 2 to 5 days and include diarrhea, nausea, vomiting, abdominal cramps, fever, malaise, and constitutional symptoms. A history of blood in the stool is common. The duration of illness varies from less than 1 day to 1 week in most cases, but a chronic relapsing colitis can occur.

C. jejuni infection can be confirmed by isolation of the organism from stool or, in an outbreak, is suggested if the sera of ill persons contain IgM antibody to the epidemic organism.[4] Techniques for isolation have been thoroughly described[1] and include the use of selective media, microaerophilic conditions, and incubation at 42°C.

B. Cholera

Cholera, which is caused by *Vibrio cholerae* 01, is rare in the U.S. Although most infections are acquired abroad by travelers and imported into this country,[5] it has recently become apparent that cholera is endemic along the Gulf Coast.[6,7] As in outbreaks in other parts of the world, water and seafood have been important in the Gulf Coast outbreaks. A 1978 outbreak in Louisiana was caused by undercooked crab;[6] a 1981 outbreak on a floating oil rig off the coast of Texas was caused either by drinking water contaminated with *V. cholerae* 01 or by eating cooked rice which had been washed in the contaminated water.[7]

Table 1
INFECTIOUS AND CHEMICAL AGENTS CAUSING ACUTE WATERBORNE DISEASE IN THE U.S., BY INCUBATION PERIOD

Short (<6 hr)	Intermediate (7 hr to 6 days)	Long (≥ 1 week)
Fluoride	*Campylobacter jejuni*	*Salmonella typhi*
Heavy metals	Enterotoxigenic *Escherichia coli*	Hepatitis A virus
	Nontyphoidal *Salmonella*	*Entamoeba histolytica*
	Shigella	*Giardia lamblia*
	Vibrio cholerae 01	
	Yersinia enterocolitica	
	Norwalk virus	
	Rotavirus	

Table 2
CLINICAL SYNDROMES AND INCUBATION PERIODS OF INFECTIOUS AND CHEMICAL AGENTS CAUSING ACUTE WATERBORNE DISEASE IN THE U.S.

Agent	Incubation period	Clinical syndrome
Bacteria		
Campylobacter jejuni	2—5 days	Gastroenteritis, often with fever
Enterotoxigenic *Escherichia coli*	6—36 hr	Gastroenteritis
Salmonella	6—48 hr	Either gastroenteritis (often with fever), enteric fever, or extraintestinal infection
Salmonella typhi	10—14 days	Enteric fever — fever, anorexia, malaise, transient rash, splenomegaly, and leukopenia
Shigella	12—48 hr	Gastroenteritis, often with fever and bloody diarrhea
Vibrio cholerae 01	1—5 days	Gastroenteritis, often with significant dehydration
Yersinia enterocolitica	3—7 days	Either gastroenteritis, mesenteric lymphadenitis, or acute terminal ileitis; may mimic appendicitis
Viruses		
Hepatitis A	2—6 weeks	Hepatitis — nausea, anorexia, jaundice, dark urine
Norwalk virus	24—48 hr	Gastroenteritis, of short duration
Rotavirus	24—72 hr	Gastroenteritis, often with significant dehydration

Table 2 (continued)
CLINICAL SYNDROMES AND INCUBATION PERIODS OF INFECTIOUS AND CHEMICAL AGENTS CAUSING ACUTE WATERBORNE DISEASE IN THE U.S.

Agent	Incubation period	Clinical syndrome
Parasites		
Entamoeba histolytica	2—4 weeks	Varies from mild gastroenteritis to acute fulminating dysentery with fever and bloody diarrhea
Giardia lamblia	1—4 weeks	Chronic diarrhea, epigastric pain, bloating, malabsorption, and weight loss
Chemicals		
Fluoride	<1 hr	Nausea, vomiting, and abdominal cramps
Heavy metals		
Antimony		
Cadmium		
Copper		
Lead		
Tin		
Zinc, etc.	<1 hr	Nausea, vomiting, and abdominal cramps, often accompanied by a metallic taste
Others		
Pesticides		
Petroleum products, etc.	Variable	Variable

V. cholerae 01 is noninvasive and causes disease via its enterotoxin, which stimulates adenylate cyclase in small bowel epithelial cells to produce a secretory diarrhea. Asymptomatic infection is common, especially in endemic areas.[8] Following an incubation period of 1 to 5 days, symptomatic disease is variable, ranging from mild diarrhea and vomiting, usually without fever, to profuse painless diarrhea with rapid dehydration. If untreated, the dehydration can result in death. The illness is usually self-limited, lasting 2 to 7 days.

The diagnosis of cholera can be confirmed either by isolation of *V. cholerae* 01 from the stool or by serologic tests. Thiosulfate citrate bile salts sucrose (TCBS) agar, which is available commercially, should be used for isolating the organism from the stool.[9] In the U.S., although paired sera for antitoxic or vibriocidal antibodies are preferred for serologic diagnosis, a vibriocidal antibody titer $\geq$ 1:1280[10] and an antitoxic antibody titer $\geq$ 4 times the net optical density of the negative control[11] in a single convalescent-phase serum drawn 2 to 5 weeks after onset of illness is virtually diagnostic of infection.

C. Enterotoxigenic *Escherichia coli* Gastroenteritis

Enterotoxigenic *E. coli* (ETEC) infection is an uncommon cause of sporadic illness or outbreaks in the U.S.[12] ETEC, however, is the most common cause of diarrhea in American travelers to foreign countries.[13] Because of the large infectious dose required to cause ETEC infection, food- and waterborne ETEC transmission are thought to be important, while person-to-person transmission is not.[12] The role of water in transmitting ETEC traveler's diarrhea is unknown. One large waterborne ETEC outbreak[14] occurred in the U.S. at Crater

Lake National Park in Oregon in 1975. In this outbreak, 2000 park visitors and 200 park employees became ill after they consumed water which had been contaminated by sewage.

Like *V. cholerae*, ETEC causes a secretory diarrhea by the effects of enterotoxin on the small bowel. ETEC produces two kinds of toxin: a heat-stable toxin (ST) and a choleragen-like heat-labile toxin (LT). The ability of ETEC strains to cause diarrhea depends not only on the production of one or both of these toxins, but also on production of a colonization factor which allows *E. coli* to adhere to the samll bowel wall. Symptoms of ETEC infection follow an incubation period of 6 to 36 hr, are nonspecific, and include diarrhea, which can be severe, abdominal cramps, nausea, vomiting, and myalgias, often with low-grade fever. The duration of illness can be as brief as 1 day or as long as 2 weeks or more, but is usually 7 to 10 days.

Confirmation of suspected ETEC infection is difficult. *E. coli* is part of the normal bowel flora and can be isolated from MacConkey agar or other selective media.[15] Although certain *E. coli* serotypes are more commonly enterotoxigenic than others,[16] confirmation of enterotoxigenicity depends on LT and ST assays, which are expensive, time-consuming, and not available in most laboratories. In an outbreak setting, ETEC should be a suspected agent if most of the illnesses are compatible with the symptoms of ETEC infection and no other pathogen is found.

D. Salmonellosis

Nontyphoidal salmonellosis, caused by the more than 2000 *Salmonella* serotypes, is a common illness, with an estimated 2 million cases a year in the U.S.[17] It is primarily a disease of children, with over half of all isolates coming from persons under 5 years of age.[17] *Salmonella* transmission is usually foodborne, via mishandled poultry, meat, eggs, or dairy products; person-to-person transmission accounts for approximately 10% of cases.[17] Waterborne transmission is relatively uncommon; only 3% of waterborne disease outbreaks from 1971 to 1978 were caused by *Salmonella*.[18] The largest waterborne salmonellosis outbreak reported in the U.S.[19] affected over 16,000 persons in Riverside, Calif. in 1965. In this outbreak, contamination of the Riverside community water supply was demonstrated, but the exact source of the contamination was not determined.

Salmonella invades the colon, although the exact pathogenic mechanisms are unknown. Following an incubation period of 6 to 48 hr, infection with *Salmonella* results in various clinical syndromes ranging in severity from asymptomatic infection to enteric fever. Most common is an invasive gastroenteritis syndrome, with diarrhea, nausea, vomiting, fever, abdominal cramps, and malaise. *S. cholerae-suis* and *S. dublin* are more likely to cause severe illness, with bacteremia or extraintestinal infections such as meningitis, abscesses, or arthritis.[20,21] All *Salmonella* serotypes, but particularly *S. paratyphi* A, B, and C, can cause a typhoid fever-like enteric fever.[17] The duration of salmonellosis is usually 2 to 5 days.

Salmonella infection can be confirmed by isolating the organism from the stool. Isolation requires selective media and is sometimes aided by enrichment techniques.[15]

E. Typhoid Fever

Typhoid fever, which is caused by *S. typhi*, is now an uncommon illness in the U.S. Only about 500 cases are reported each year, and most of these are acquired during foreign travel.[22] Common-source outbreaks account for one fourth of domestically acquired cases.[22] Waterborne typhoid fever outbreaks are now rare,[18] although in the first half of this century, typhoid fever was the most commonly reported cause of waterborne disease outbreaks in the U.S. The most recently reported large waterborne typhoid fever outbreak affected 225 persons in a migrant labor camp in Dade County, Fla.[23] in 1973. In this outbreak, the water supply of the camp, a well, was contaminated by surface water.

S. typhi causes illness by invading the intestinal mucosa, replicating in lymph nodes, and entering the bloodstream, sometimes infecting secondary sites. Typhoid fever is a prolonged systemic illness. Symptoms and signs follow an incubation period that varies from 1 to several weeks, but is usually from 10 to 14 days. They include fever, headache, malaise, anorexia, epistaxis, a transient rash, splenomegaly, and leukopenia. Constipation is as common as diarrhea, and both may occur intermittently in the same patient. Mortality is high, ranging from 12 to 16% in untreated patients and 1 to 4% in antibiotic-treated patients.[24] The duration of untreated illness is from 1 to 8 weeks or more, and relapses occur frequently.

Confirmation of the diagnosis of typhoid fever rests on the isolation of *S. typhi*, with selective media,[15] from the bone marrow, blood, urine, or stool. Bone marrow and blood cultures are most likely to be positive in the 1st week of illness, whereas urine and stool cultures are more likely to be positive later in the course of the illness.[25] Serologic diagnosis by the Widal test is unreliable; among adults from endemic areas, an elevated titer is a common and nonspecific finding.[26]

F. Shigellosis

The genus *Shigella* contains four pathogenic species: *S. boydii*, *S. dysenteriae*, *S. flexneri*, and *S. sonnei*. Because of the extremely small infectious dose of *Shigella*,[27] person-to-person transmission is common, accounting for two thirds of the outbreaks in the U.S.[28] Day-care centers and institutions for the mentally retarded are particularly high-risk settings. Person-to-person, fecal-oral transmission presumably also accounts for the recent emergence of *Shigella* as an important diarrheal pathogen of homosexual men.[29] Common-source shigellosis outbreaks also occur, and waterborne transmission is responsible for a majority of cases in common-source outbreaks.[30] The largest reported waterborne shigellosis outbreak affected 3000 persons in Newton, Kan.[31] in 1942.

Shigella is an invasive organism, penetrating the colon and causing colitis with ulceration; the exact pathogenic mechanisms are unknown. Symptoms of *Shigella* infection follow an incubation period of 12 to 48 hr and are varied. Asymptomatic infections occur, as do mild, watery diarrhea and frank dysentery, with fever and grossly bloody stools. The illness is usually self-limited, ending within 1 week.

Shigella infection is confirmed by isolation of the organism from the stool. Direct plating of fecal swabs onto selective media is the optimal method of isolating the organism.[15] If the specimen must be stored on transport media, however, refrigeration or freezing of the specimen is better than storage at room temperature.[32]

G. *Yersinia enterocolitica* Infection

Much about the transmission of *Y. enterocolitica* remains unknown; the organism is isolated frequently from animals and the environment. Certain serotypes (particularly 0:3, 0:8, and 0:9), however, are well recognized as causes of human disease.[33] These serotypes frequently have unique geographic distributions, with 0:8 being the most commonly isolated serotype in the U.S. and 0:3 and 0:9 being more common in Canada and Europe.[33] For human illness, foodborne transmission appears to be important;[34,35] the role of water in transmitting the disease is uncertain. Although no true waterborne *Y. enterocolitica* outbreaks have been reported in the U.S., an outbreak in Washington state in 1981 and 1982 was caused by tofu which was packaged in untreated spring water contaminated by *Y. enterocolitica*.[36]

Y. enterocolitica invades the terminal ileum, but the exact pathogenic mechanisms are unknown. The clinical syndrome of *Y. enterocolitica* infection follows an incubation period of 3 to 7 days and is largely age dependent, but typically consists of fever and abdominal cramps, with or without diarrhea. Children under the age of 5 years have mild gastroenteritis, while those aged 5 to 15 years more commonly have mesenteric lymphadenitis, with pseudoappendicitis. Community outbreaks of *Y. enterocolitica* infection have been associated commonly with a marked increase in the occurrence of appendectomies. Adolescents with *Y. enterocolitica* infection most commonly have acute terminal ileitis, while adults more commonly have gastroenteritis or non-GI infection. The duration of illness varies from 1 day to several weeks, but is usually from 5 to 7 days.

Y. enterocolitica can be isolated with selective media[15] from stool, infected lymph nodes, or other infected sites. Cold enrichment aids isolation efforts.[37] Serologic diagnosis of infection is not reliable, particularly if the patient's isolate is unavailable for use as the test antigen.

III. VIRAL DISEASES

A. Hepatitis A

Hepatitis A (formerly infectious hepatitis), is caused by the 27-nm hepatitis A virus (HAV). Viral hepatitis is the second most commonly reported infectious disease in the U.S., and hepatitis A is the most commonly reported type of viral hepatitis.[38] More than half of the U.S. population over age 40 has serum IgG antibody to HAV.[39] While person-to-person transmission is by far the most important reported cause of HAV infection, common-source outbreaks occur frequently.[38] Recently, HAV has been the third most commonly reported cause of waterborne outbreaks in the U.S.[18]

HAV infects hepatocytes, causing cytologic damage, necrosis, and inflammation. Symptoms follow an incubation period of 2 to 6 weeks and include a prodrome of fever, nausea, anorexia, and malaise, often with mild diarrhea, followed shortly by jaundice. Asymptomatic infection is common in children. Illness usually lasts from 1 to 2 weeks, but may last several months.

HAV infection is diagnosed serologically: current or recent infection is indicated by the presence of IgM antibody to HAV; past infection is indicated by the presence of IgG antibody to HAV.[40]

B. Norwalk Gastroenteritis

Norwalk gastroenteritis is caused by the newly recognized 27-nm Norwalk virus.[41] Similar, but antigenically distinct, Norwalk-like viruses have also been shown recently to cause gastroenteritis. The Norwalk virus is an important cause of epidemic nonbacterial gastroenteritis in the U.S.[42] Approximately 40% of the 74 nonbacterial gastroenteritis outbreaks investigated by the Centers for Disease Control (CDC) from 1976 to 1980 were caused by the Norwalk virus.[42] Three quarters of the outbreaks in which a vehicle was identified were waterborne; the rest were foodborne. In these outbreaks, secondary person-to-person transmission was also common.

Although Norwalk and Norwalk-like viruses produce histologic abnormalities in the proximal small intestine, the exact pathogenic mechanisms are unknown. Abrupt onset of gastroenteritis follows an incubation period of 24 to 48 hr. Vomiting is the most common symptom in children, while diarrhea is more common in adults. Headache occurs in approximately half of those affected. Illness is usually brief, ending in 24 to 48 hr.

Norwalk virus infection can sometimes be confirmed early in the illness by visualization of the agent in the stool by immune electron microscopy. It is more frequently diagnosed serologically by demonstration of a fourfold rise in antiviral antibody in paired sera. Reagents for serologic diagnosis are scarce and not routinely available.[41]

C. Rotavirus Gastroenteritis

Rotaviruses are newly recognized 70-nm reovirus-like viruses. Unlike the Norwalk virus, rotavirus appears to be an important cause of nonepidemic nonbacterial gastroenteritis, especially in children. It is responsible for approximately half the hospitalized cases of acute diarrhea in children under the age of 2 years in the U.S.[43] Illness in adults is much less common.[41] Transmission is thought to be mostly person to person, but a waterborne rotavirus outbreak was reported in a Colorado ski resort[44] in 1981.

Rotavirus infection produces histologic changes in the proximal small intestine; the exact pathogenic mechanisms are unknown. Symptoms follow an incubation period of 24 to 72 hr and range from asymptomatic infection to severe gastroenteritis with significant dehydration requiring hospitalization.[43,45] The severity of illness decreases with age, apparently because of acquired immunity.[45,46] When patients are treated with rehydration, the illness usually lasts from 2 to 5 days.

Rotavirus infection can be confirmed by detection of the virus in stool or by serologic tests.[41] Virus particles in stool can be visualized by direct electron microscopy or can be detected by immunologic assays, such as the commercially available enzyme-linked immunosorbent assay (ELISA). Serologic diagnosis requires a fourfold rise in antiviral antibody in paired sera.

IV. PARASITIC DISEASES

A. Amebiasis

Amebiasis, which is caused by *Entamoeba histolytica,* is common in the U.S.; the stool prevalence in Americans is from 0.6 to 5%[47,48] Among sporadic cases, person-to-person transmission is thought to be important, as in male homosexuals, who are infected by fecal-oral contact.[49] Waterborne outbreaks are now rare in the U.S., none have been reported in recent years.[18]

E. histolytica trophozoites inhabit the colon, and only occasionally invade the colonic mucosa and cause colonic ulcerations. The symptoms associated with *E. histolytica* infection follow an incubation period of 2 to 4 weeks and are varied. Most infections are asymptomatic. Mild gastroenteritis can result, as can frank dysentery, with fever and grossly bloody stools. Extrahepatic abscesses, such as hepatic amebomas, occur infrequently. When gastroenteritis occurs, it is frequently chronic and relapsing.

E. histolytica trophozoites or cysts can be detected microscopically in the stool. The presence of *E. histolytica* in the stool does not confirm its etiologic role in illness, however, and other diarrheal pathogens should be sought as well. Serologic studies may be useful, particularly in patients with extraintestinal infection.[50]

B. Giardiasis

Giardiasis is caused by *Giardia lamblia,* a protozoan parasite which only recently has been recognized as a pathogen.[51] In a recent survey of state and territorial public health laboratories, *G. lamblia* was the most commonly detected intestinal parasite in the U.S., having been found in 3.8% of the stools examined.[48] Because of its low infectious dose[52] *G. lamblia* is frequently transmitted by fecal-oral, person-to-person means, resulting in the demonstrated high prevalence of the parasite in homosexual males[49] and in children,[53] particularly children in day-care centers.[54] It is also an important waterborne pathogen, however. Since the first reported waterborne giardiasis outbreak in Aspen, Colo.[55] in 1964 and 1965, waterborne transmission has been reported more and more frequently, and *G. lamblia* is now the most commonly reported causative pathogen of waterborne outbreaks.[56]

G. lamblia inhabits the proximal small bowel and may cause no morphologic changes in the mucosa or may cause shortening of microvilli with or without inflammation; the exact pathogenic mechanisms are unknown. Asymptomatic infection with *G. lamblia* occurs; symptoms follow an incubation period of 1 to 4 weeks. Patients often have insidious onset of epigastric pain, bloating, fatigue, and intermittent diarrhea with greasy, malodorous stools.

The illness may subside spontaneously within 1 to 4 weeks, but often continues a relapsing pattern over weeks or months with malabsorption and weight loss.

G. lamblia trophozoites or cysts can be detected microscopically in the stool. At least three stools should be examined before a person is considered not to have infection.[57] Patients may require duodenal sampling to confirm the diagnosis.[58] Some patients with typical symptoms respond to antimicrobial therapy despite stool examinations that are repeatedly negative.

V. ACUTE CHEMICAL POISONINGS

Chemicals that contaminate water and cause acute poisoning can be divided into three groups: fluoride, heavy metals (antimony, cadmium, copper, lead, tin, zinc, etc.), and others (pesticides, petroleum products, etc.). The three groups usually contaminate water under different circumstances. Fluoride poisoning occurs most frequently because of failure of fluoridation equipment.[18] Heavy metal poisoning occurs most frequently because of leaching of metal from tubing or containers by corrosive water with a low pH.[18] The circumstances of the other contaminations vary, but many result from accidental spills of chemicals into water supplies.[18,59]

Fortunately, most acute waterborne chemical poisonings cause mild, self-limited illness. Both fluoride and heavy metals cause vomiting, with onset within 1 hr after consumption of the contaminated water. Few, if any, long-term sequelae have been reported. The symptoms of the other poisonings vary with the substance ingested.

While no single type of acute chemical poisoning is very common, chemical poisonings as a group are the most commonly recognized cause of waterborne outbreaks in the U.S.[18]

REFERENCES

1. **Blaser, M. J. and Reller, L. B.**, Campylobacter enteritis, *N. Engl. J. Med.*, 305, 1444, 1981.
2. **Blaser, M. J., Duncan, D. J., Osterholm, M. T., Istre, G. R., and Wang, W.-L.**, Serologic study of two clusters of infection due to *Campylobacter jejuni*, *J. Infect. Dis.*, 147, 820, 1983.
3. **Taylor, D. N., McDermott, K. T., Little, J. R., Wells, J. G., and Blaser, M. J.**, Campylobacter enteritis from untreated water in the Rocky Mountains, *Ann. Intern. Med.*, 99, 38, 1983.
4. **Vogt, R. L., Sours, H. E., Barrett, T., Feldman, R. A., Dickinson, R. J., and Witherell, L.**, Campylobacter enteritis associated with contaminated water, *Ann. Intern. Med.*, 96, 292, 1982.
5. **Snyder, J. D. and Blake, P. A.**, Is cholera a problem for U.S. travelers?, *JAMA*, 247, 2268, 1982.
6. **Blake, P. A., Allegra, D. T., Snyder, J. D., Barrett, T. J., McFarland, L., Caraway, C. T., Feeley, J. C., Craig, J. P., Lee, J. V., Puhr, N. D., and Feldman, R. A.**, Cholera — a possible endemic focus in the United States, *N. Engl. J. Med.*, 302, 305, 1980.
7. **Johnston, J. M., Martin, D. M., Perdue, J., McFarland, L. M., Caraway, C. T., Lippy, E. C., and Blake, P. A.**, Cholera on a Gulf Coast oil rig, *N. Engl. J. Med.*, 309, 523, 1983.
8. **Bart, K. J., Huq, Z., Khan, M., Mosley, W. H., Nuruzzaman, M., and Kibrya, A. K. M. G.**, Seroepidemiologic studies during a simultaneous epidemic of infection with El Tor Ogawa and Classical Inaba *Vibrio cholerae*, *J. Infect. Dis.*, 121, S17, 1970.
9. **Morris, G. K., Merson, M. H., Huq, I., Kibrya, A. K. M. G., and Black, R.**, Comparison of four plating media for isolating *Vibrio cholerae*, *J. Clin. Microbiol.*, 9, 79, 1979.
10. **Clements, M. L., Levine, M. M., Young, C. R., Black, R. E., Lim, Y.-L., Robins-Browne, R. M., and Craig, J. P.**, Magnitude, kinetics, and duration of vibriocidal antibody responses in North Americans after ingestion of *Vibrio cholerae*, *J. Infect. Dis.*, 145, 465, 1982.
11. **Levine, M. M., Young, C. R., Hughes, T. P., O'Donnell, S., Black, R. E., Clements, M. L., Robins-Browne, R., and Lim, Y.-L.**, Duration of serum antitoxin response following *Vibrio cholerae* infection in North Americans: relevance for seroepidemiology, *Am. J. Epidemiol.*, 114, 348, 1981.
12. **Gangarosa, E. J.**, Epidemiology of *Escherichia coli* in the United States, *J. Infect. Dis.*, 137, 634, 1978.

13. **Gorbach, S. L., Kean, B. H., Evans, D. G., Evans, D. E., and Bessudo, D.,** Travelers' diarrhea and toxigenic *Escherichia coli, N. Engl. J. Med.,* 292, 933, 1975.
14. **Rosenberg, M. L., Koplan, J. P., Wachsmuth, I. K., Wells, J. G., Gangarosa, E. J., Guerrant, R. L., and Sack, D. A.,** Epidemic diarrhea at Crater Lake from enterotoxigenic *Escherichia coli, Ann. Intern. Med.,* 86, 714, 1977.
15. **Finegold, S. M. and Martin, W. J.,** *Diagnostic Microbiology,* 6th ed., C. V. Mosby, St. Louis, 1982.
16. **Wachsmuth, I. K.,** *Escherichia coli:* typing and pathogenicity, *Clin. Microbiol. Newslett.,* 2, 4, 1980.
17. **Cohen, M. L. and Gangarosa, E. J.,** Nontyphoid salmonellosis, *South. Med. J.,* 71, 1540, 1978.
18. **Craun, G. F.,** Outbreaks of waterborne disease in the United States: 1971—1978, *J. Am. Water Works Assoc.,* 73, 360, 1981.
19. A Collaborative Report, A waterborne epidemic of salmonellosis in Riverside, California, 1965, *Am. J. Epidemiol.,* 93, 33, 1971.
20. **Blaser, M. J. and Feldman, R. A.,** *Salmonella* bacteremia: reports to the Centers for Disease Control, 1968—1979, *J. Infect. Dis.,* 143, 743, 1981.
21. **Taylor, D. N., Bied, J. M., Munro, J. S., and Feldman, R. A.,** *Salmonella dublin* infections in the United States, 1979—1980, *J. Infect. Dis.,* 146, 322, 1982.
22. **Taylor, D. N., Pollard, R. A., and Blake, P. A.,** Typhoid in the United States and the risk to the international traveler, *J. Infect. Dis.,* 148, 599, 1983.
23. **Feldman, R. E., Baine, W. B., Nitzkin, J. L., Saslaw, M. S., and Pollard, R. A.,** Epidemiology of *Salmonella typhi* infection in a migrant labor camp in Dade County, Florida, *J. Infect. Dis.,* 130, 334, 1974.
24. **Rowland, H. A. K.,** The complications of typhoid fever, *J. Trop. Med. Hyg.,* 64, 143, 1961.
25. **Gilman, R. H., Terminel, M., Levine, M. M., Hernandez-Mendoza, P., and Hornick, R. B.,** Relative efficacy of blood, urine, rectal swab, bone-marrow, and rose-spot cultures for recovery of *Salmonella typhi* in typhoid fever, *Lancet,* 1, 1211, 1975.
26. **Levine, M. M., Grados, O., Gilman, R. H., Woodward, W. E., Solis-Plaza, R., and Waldman, W.,** Diagnostic value of the Widal test in areas endemic for typhoid fever, *Am. J. Trop. Med. Hyg.,* 27, 795, 1978.
27. **DuPont, H. L., Hornick, R. B., Snyder, M. J., Libonati, J. P., Formal, S. B., and Gangarosa, E. J.,** Immunity in shigellosis. II. Protection induced by oral live vaccine or primary infection, *J. Infect. Dis.,* 125, 12, 1972.
28. **Rosenberg, M. L., Weissman, J. B., Gangarosa, E. J., Reller, L. B., and Beasley, R. P.,** Shigellosis in the United States: ten-year review of nationwide surveillance, 1964—1973, *Am. J. Epidemiol.,* 104, 543, 1976.
29. **Dritz, S. K., Ainsworth, T. E., Garrard, W. F., Back, A., Palmer, R. D., Boucher, L. A., and River, E.,** Patterns of sexually transmitted diseases in a city, *Lancet,* 2, 3, 1977.
30. **Black, R. E., Craun, G. F., and Blake, P. A.,** Epidemiology of common-source outbreaks of shigellosis in the United States, 1961—1975, *Am. J. Epidemiol.,* 108, 47, 1978.
31. **Kinnaman, C. H. and Beelman, F. C.,** An epidemic of 3000 cases of bacillary dysentery involving a war industry and members of the armed forces, *Am. J. Public Health,* 34, 948, 1944.
32. **Wells, J. G. and Morris, G. K.,** Evaluation of transport methods for isolating *Shigella* spp., *J. Clin. Microbiol.,* 13, 789, 1981.
33. **Kay, B. A., Wachsmuth, K., Gemski, P., Feeley, J. C., Quan, T. J., and Brenner, D. J.,** Virulence and phenotypic characterization of *Yersinia enterocolitica* isolated from humans in the United States, *J. Clin. Microbiol.,* 17, 128, 1983.
34. **Black, R. E., Jackson, R. J., Tsai, T., Medvesky, M., Shayegani, M., Feeley, J. C., MacLeod, K. I. E., and Wakelee, A. M.,** Epidemic *Yersinia enterocolitica* infection due to contaminated chocolate milk, *N. Engl. J. Med.,* 298, 76, 1978.
35. Centers for Disease Control, Multi-state outbreak of yersiniosis, *Morbid. Mortal. Weekly Rep.,* 31, 505, 1982.
36. Centers for Disease Control, Outbreak of *Yersinia enterocolitica* — Washington state, *Morbid. Mortal. Weekly Rep.,* 31, 562, 1982.
37. **Weissfeld, A. S. and Sonnenwirth, A. C.,** *Yersinia enterocolitica* in adults with gastrointestinal disturbances: need for cold enrichment, *J. Clin. Microbiol.,* 11, 196, 1980.
38. **Alter, M. J.,** National surveillance of viral hepatitis, 1981, *Morbid. Mortal. Weekly Rep.,* 32, 23SS, 1983.
39. Centers for Disease Control, Immune globulins for protection against viral hepatitis, *Morbid. Mortal. Weekly Rep.,* 30, 423, 1981.
40. **Bradley, D. W., Maynard, J. E., Hindman, S. H., Hornbeck, C. L., Fields, H. A., McCaustland, K. A., and Cook, E. H.,** Serodiagnosis of viral hepatitis A: detection of acute-phase immunoglobulin M anti-hepatitis A virus by radioimmunoassay, *J. Clin. Microbiol.,* 5, 521, 1977.
41. **Blacklow, N. R. and Cukor, G.,** Viral gastroenteritis, *N. Engl. J. Med.,* 304, 397, 1981.

42. **Kaplan, J. E., Gary, W., Baron, R. C., Singh, N., Schonberger, L. B., Feldman, R. A., and Greenberg, H. B.**, Epidemiology of Norwalk gastroenteritis and the role of the Norwalk virus in outbreaks of acute nonbacterial gastroenteritis, *Ann. Intern. Med.,* 96, 756, 1982.
43. **Kapikian, A. Z., Kim, H. W., Wyatt, R. G., Cline, W. L., Arrobio, J. O., Brandt, C. D., Rodriguez, W. J., Sack, D., A., Chanock, R. M., and Parrott, R. H.**, Human reovirus-like agent as the major pathogen associated with "winter" gastroenteritis in hospitalized infants and young children, *N. Engl. J. Med.,* 294, 965, 1976.
44. **Harris, J. R., Cohen, M. L., and Lippy, E. C.**, Water-related disease outbreaks in the United States, 1981, *J. Infect. Dis.,* 148, 759, 1983.
45. **Black, R. E., Brown, K. H., Becker, S., Alim, A. R. M. A., and Huq, I.**, Longitudinal studies of infectious diseases and physical growth of children in rural Bangladesh. II. Incidence of diarrhea and association with known pathogens, *Am. J. Epidemiol.,* 115, 315, 1982.
46. **Black, R. E., Greenberg, H. B., Kapikian, A. Z., Brown, K. H., and Becker, S.**, Acquisition of serum antibody to Norwalk virus and rotavirus and relation to diarrhea in a longitudinal study of young children in rural Bangladesh, *J. Infect. Dis.,* 145, 483, 1982.
47. **Plorde, J. J.**, Amebiasis, in *Harrison's Principles of Internal Medicine,* 10th ed., Petersdorf, R. G., Adams, R. D., Braunwald, E., Isselbacher, K. J., Martin, J. B., and Wilson, J. D., Eds., McGraw-Hill, New York, 1983, 1182.
48. Centers for Disease Control, Intestinal parasite surveillance — United States, 1976, *Morbid. Mortal. Weekly Rep.,* 27, 167, 1978.
49. **William, D. C., Shookoff, H. B., Felman, Y. H., and DeRamos, S. W.**, High rates of enteric protozoal infections in selected homosexual men attending a venereal disease clinic, *Sex. Transm. Dis.,* 5, 155, 1978.
50. **Kagan, I. G.**, Diagnostic, epidemiologic, and experimental parasitology: immunologic aspects, *Am. J. Trop. Med. Hyg.,* 28, 429, 1979.
51. **Meyer, E. A. and Jarroll, E. L.**, Giardiasis, *Am. J. Epidemiol.,* 111, 1, 1980.
52. **Rendtorff, R. C. and Holt, C. J.**, The experimental transmission of human intestinal protozoan parasites. IV. Attempts to transmit *Endamoeba coli* and *Giardia lamblia* cysts by water, *Am. J. Hyg.,* 60, 327, 1954.
53. **Healy, G. R.**, The presence and absence of *Giardia lamblia* in studies on parasite prevalence in the U.S.A., in *Waterborne Transmission of Giardiasis,* Jakubowski, W. and Hoff, J. C., Eds., U.S. Environmental Protection Agency, Cincinnati, 1978, 92.
54. **Black, R. E., Dykes, A. C. Sinclair, S. C., and Wells, J. G.**, Giardiasis in day-care centers: evidence of person-to-person transmission, *Pediatrics,* 60, 486, 1977.
55. **Moore, G. T., Cross, W. M., McGuire, D., Mollohan, C. S., Gleason, N. N., Healy, G. R., and Newton, L. H.**, Epidemic giardiasis at a ski resort, *N. Engl. J. Med.,* 281, 402, 1969.
56. **Craun, G. F.**, Waterborne giardiasis in the United States: a review, *Am. J. Public Health,* 69, 817, 1979.
57. Council for the American Society of Parasitologists, Procedure suggested for use in examination of clinical specimens for parasitic infections, *J. Parasitol.,* 63, 959, 1977.
58. **Beal, C. R., Viens, P., Grant, R. G. L., and Hughes, J. M.**, A new technique for sampling duodenal contents, *Am. J. Trop. Med. Hyg.,* 19, 349, 1970.
59. Centers for Disease Control, Community water supply contaminated with caustic soda — Georgia, *Morbid. Mortal. Weekly Rep.,* 30, 67, 1981.

Chapter 3

DISEASES CAUSED BY WATER CONTACT

Alfred P. Dufour

TABLE OF CONTENTS

I. Introduction 24
II. Disease and Infections Caused by Water Contact 24
    A. Typhoid Fever and Salmonellosis 24
    B. Shigellosis 26
    C. Viral Gastroenteritis and Hepatitis A 26
    D. Ear Infections 28
    E. Mycobacteriosis 29
    F. Pneumonia 30
    G. Septicemias 32
    H. Wound Infections 33
    I. Leptospirosis 35
    J. Primary Amebic Meningocephalitis 36
    K. Schistosome Dermatitis 37

III. Conclusions 37

References 38

I. INTRODUCTION

Water contact recreation is one of the most popular leisure time activities in the U.S. These activities may take the form of swimming, surfing, scuba diving and water-skiing, wherein there is prolonged contact with the water, or boating and fishing, wherein there is minimal water contact. Water-associated health risks from indigenous or introduced microbes will vary according to the type of activity, but it can be reasonably assumed that swimming, surfing, scuba diving, and water-skiing present a greater risk than either boating or fishing because of the potential for prolonged intimate contact with the water and the greater likelihood of ingesting the water. The risks also will vary with the quality of the water, as there is a much greater risk from contact with water that has been contaminated with domestic waste waters discharged from sewage treatment facilities.

The risks of illness from water contact activities can be segregated into two groups depending on the type of water exposure: (1) exposure to water which contains microorganisms introduced to the environment via sewage treatment plant outfalls and nonpoint sources of human and animal wastes and (2) exposure to water which contains microorganisms indigenous to the aquatic environment. Pathogens associated with enteric infections are usually transmitted by exposure to water contaminated by human or animal wastes. The epitome of this type of infection is typhoid fever, which is caused by *Salmonella typhosa* from human wastes. Leptospirosis is not an enteric infection, but is also generally transmitted through contact with waters contaminated by animal and human wastes. Water which has not been contaminated by human or animal fecal wastes contains microorganisms indigenous to the aquatic environment, and although these bacteria are normally harmless, under certain conditions they may be hazardous to individuals who come in contact with them. The untoward conditions occur when an individual whose body defense mechanisms have been compromised, either by a break in the integrity of the skin or because of the use of immunosuppressive agents, or because of unnatural tissue stresses caused by accidental events, such as a near drowning. *Vibrio parahaemolyticus* is a good example of a bacterial species capable of causing infection in individuals with weakened body defense mechanisms. *Vibrio* species are usually not pathogenic for healthy individuals who swim in water containing relatively high densities of this organism. The risk of illness or infection from indigenous aquatic organisms is usually increased by the presence of organic industrial and domestic wastes, which cause bacterial growth to occur. Under these conditions the organisms can easily reach densities far above ambient levels, thereby increasing the probability of exposure.

Epidemiological studies of selected populations exposed to water contaminated with sewage treatment plant discharges have provided the primary evidence associating disease with water contact activities, swimming, and water quality. Outbreaks have provided additional evidence associating disease in a group of individuals with a common water contact exposure. Reports of sporadic cases of disease associated with water contact are another source of information which have helped to define relationships between illness and exposure to surface waters.

This chapter examines the various categories of disease associated with water contact and attempts to define the relationship between pathogens, the water environment, and the water-associated health risks of these pathogens.

II. DISEASE AND INFECTIONS CAUSED BY WATER CONTACT

A. Typhoid Fever and Salmonellosis

Typhoid fever was one of the earliest reported diseases to be associated with water contact. Discher[1] reported on an outbreak of typhoid fever in 1888 where 49 cases were found to be associated with swimming in the Elbe River in Germany and another in 1892 which affected

10 soldiers who had bathed in the Danube River. About 2 decades later, Reece[2] reported an outbreak of typhoid fever in Walmer, England among marine recruits who swam in a seawater-filled pool as a part of their training. The pool, which was filled by opening gates on an incoming tide, had been contaminated with effluents discharged from nearby sewage outfalls. Of 34 typhoid cases, 20 were attributed to swimming in the pool.

Ciampolini[3] reported on the probable cause of typhoid fever cases which occurred in the city of New Haven, Conn. in 1923. This frequently quoted report did not establish a direct relationship between these cases of typhoid and bathing in the grossly polluted water of New Haven harbor, but circumstantial evidence was presented which indicated that swimming in the harbor may have been one of the means of transmission. Swimming activity was suspected for a number of reasons. First, the health center district, which included the harbor area, had a much higher incidence of typhoid than the city of New Haven as a whole. Second, food, milk, and drinking water were ruled out as sources of infection. Lastly, 10 of the 32 cases from the health center district had been swimming in the harbor. Although the evidence is not strong, it is reasonable to assume that the harbor water contained *S. typhosa*, since Winslow and Moxon[4] reported at a later date that the coliform density in harbor water samples was extremely high. A report[5] published by the City of New York Department of Health in 1932 presented conclusions similar to those of Ciampolini. As in Ciampolini's report, direct evidence relating water contact to typhoid fever was not presented, but the elimination of other potential sources of disease suggested that swimming activity was the likely source of the infections.

In 1947 Martin[6] reported an interesting series of cases of paratyphoid fever which occurred in individuals who used a bathing pool on the Waverly River in Beccles, Suffolk, England. Bacteriophage typing of pathogens from infected individuals who had swam in the pool had the same phagetype as *S. paratyphi* isolated from water samples collected upstream and downstream from the pool. A survey of homes in the vicinity of the river revealed that a building upstream from the pool was discharging fecal wastes directly into the stream. Two healthy carriers of *S. paratyphi* were found among the users of the wastewater system of the building. The phagetype of *S. paratyphi* isolated from the carriers was the same as that of *Salmonella* strains isolated from the infected swimmers. This study is unique among the early disease outbreak reports because it clearly established the source of the pathogen, the role of water in the transmission of the disease, and the etiologic agent in the infected swimmers.

The most recent outbreak[7] of swimming-associated typhoid fever occurred in Western Australia in 1958. Of 15 cases of typhoid, 10 were attributed to swimming at a city beach, which was occasionally contaminated with high concentrations of fecal coliforms. The evidence associating the disease with swimming was considered marginal because a common source of the pathogens, such as a carrier, was not identified. Multiple phage types were noted among the *S. typhi* strains isolated from patients and this led observers to assume that the "mixed bag" of Salmonellae must have originated in sewage rather than from a common source.

Swimming-associated disease outbreaks caused by *Salmonella* species have not been reported since the Australian outbreak. The infrequent occurrence of typhoid and salmonellosis among bathers in the last 20 to 30 years is similar to the observed decrease in the overall incidence of these diseases. Olivieri[8] has shown the dramatic decrease in typhoid fever between 1880 and 1980, and there has been a similar decrease in water- and foodborne typhoid fever during this same period. Many factors, such as the use of chlorine to disinfect drinking water and waste water, increased attention to personal hygiene, the use of antibiotics, the pasteurization of milk, and careful food processing, probably have contributed to the decreasing incidence of typhoid fever and salmonellosis.

B. Shigellosis

Shigella gastroenteritis has seldom been associated with swimming in polluted water, in spite of the fact that only 10 to 100 bacilli are required for an infectious dose.[9] The paucity of swimming-associated disease outbreaks caused by *Shigella* is probably due to the fact that they do not survive for long periods in aquatic environments.[10] Only one outbreak of swimming-associated shigellosis has been reported in recent years. In 1974 in Dubuque, Iowa, 31 of 45 cases of shigellosis were found to be associated with swimming in the Mississippi River.[11] Of the cases, 22 had been swimming in the river 3 days prior to becoming ill and the other 9 were secondary cases. The relationship between shigellosis and swimming was strengthened because Shigellae were isolated from water where the swimming occurred. At least one strain isolated from the bathing area had the same antibiotic sensitivity pattern as strains isolated from cases in a foodborne outbreak which occurred in Dubuque at about the same time as the swimming-associated outbreak. This finding indicated that the source of the *Shigella* probably was a sewage treatment plant serving Dubuque and located about 8 km upriver from the swimming area. Water samples collected along the river shortly after the outbreak were shown to have fecal coliform densities between 400,000 and 5 million/100 mℓ.

C. Viral Gastroenteritis and Hepatitis A

The association between viral illnesses and water contact in recreational waters has been tenuous because of the difficulty of isolating the etiologic agent from aquatic environments and because many viruses are unculturable. Some of the early reported outbreaks of viral illnesses described relationships between disease and exposure where the linkage to viruses was somewhat ambiguous. Such was the case in a 1969 outbreak of hepatitis A attributed to swimming exposure.[12] The outbreak occurred among members of a Boy Scout troop camping at an island recreation area in South Carolina; 14 scouts developed hepatitis A. The clustering of cases suggested that lake water was the common source of the exposure. There was some question, however, about whether lake water was swallowed while swimming or inadvertently consumed from containers used to hold lake water for dousing fires. The accidental drinking of lake water from these containers appeared to be the most plausible source, since only lake water from outside the swimming area exhibited evidence of fecal contamination. Another outbreak,[13] possibly associated with swimming activity occurred at a boys' summer camp in Vermont in 1972. Coxsackie B virus was identified as the etiologic agent of the outbreak. The virus was also isolated from a sample of lake water, which implied that swimming was a possible means of transmission of the pathogen. However, there was a strong clustering of infections among groups of boys living in the same cabins, and this suggested that person-to-person spread of the virus may have been the more important principal means of transmission. The study presented insufficient data to clearly establish swimming as the means of transmission of the enteroviruses.

There was less ambiguity about the cause of an outbreak of GI illnesses at a recreational park in Michigan[14] in 1979. An epidemiological investigation indicated that illnesses were positively associated with swimming activity, and serologic studies identified Norwalk virus as the etiologic agent. Although the means of transmission of the virus was clearly identified, the source of the contamination of the lake could not be determined. Water samples collected before and after the outbreak indicated that the microbial quality of the water was within the currently accepted limit established by the Environmental Protection Agency (EPA) as acceptable for recreational contact water. Although this finding did not contradict the association between illness and swimming activity, it did leave open the question of the source of the virus. This question was answered, however, in one reported swimming-associated outbreak of viral gastroenteritis in Niort, France, where coxsackie A virus was isolated not only from stool specimens of cases but also from a lake water sample.[15] The lake water samples contained *Escherichia coli* at densities between 50 and 1000/100 mℓ, indicating fecal contamination by humans or other warm-blooded animals. This outbreak in France appears to establish to a greater extent than other reported outbreaks the linkage between viruses in water and enteritis associated with water contact.

Water-associated disease outbreaks are usually identified because illnesses, which occur among individuals who have something in common, are clustered spatially or temporally. Since they occur sporadically and without warning, it is often impossible to obtain information that would be useful for associating illness with water contact. Epidemiological studies such as conducted in Madison, Wisc.[16] in 1977 are able to overcome some of these shortcomings. This study was designed to test the hypothesis that swimmers contaminate the water they use and that viruses are transmitted by recreational water. A sample population of children under 16 years of age who visited a pediatric clinic submitted clinical specimens for enterovirus analysis and were questioned about the frequency and location of swimming in the 2 weeks prior to the clinic visit. Of the 975 children who participated in the study, 296 presented enteroviral-like syndromes and 679 children had no symptoms. Nonpolio enteroviruses were isolated from about half of the ill children. Enteroviruses that were isolated included coxsackie A and B viruses, echoviruses, adenoviruses, and viruses listed as untypable or unidentifiable. Children who reported swimming exclusively at beaches had a much higher risk of enteroviral illness than those children who reported no swimming activity. Children who reported swimming exclusively in pools had no increased risk of enteroviral illness. There was no difference in the reported swimming exposure during the 2 weeks prior to the clinic visit between the children who had symptoms and no virus isolated, and the children who were not ill or had no symptoms. The results of the study appeared to confirm that enteric viruses are transmitted from swimmer to swimmer by bathing water; however, it was impossible to identify the exact route of exposure because swimmers were defined only as individuals who visited the beach and no distinction was made between those who immersed their bodies in the water and those who did not. It is possible that person-to-person contact could have accounted for some of the illness in the group of children classified as swimmers, especially in those who did not immerse their bodies in the water, and this would tend to reduce any observed relationship between illness and swimming. Person-to-person transmission was ruled out since no illnesses were observed in pool swimmers, where water transmission was effectively eliminated because the pool water was sufficiently chlorinated to keep enteroviral concentrations very low.

Prospective epidemiologic studies have also been used to identify the relationship between viral gastroenteritis and water contact. The EPA conducted a series of ten studies at marine and fresh water bathing beaches to obtain information on the possible relationship between swimming-associated health effects and the quality of bathing water.[17] Symptomatic illness in swimmers and beach-going nonswimmers was determined by telephone interview 9 to 10 days after a swimming exposure. The symptoms were placed in four categories: GI, respiratory, eye and ear, and "other". Swimmers used waters which ranged from barely acceptable by local standards to relatively unpolluted bathing waters, and this provided possible exposures to determine if there was a relationship between sewage-contaminated water and swimming-associated illness. GI illness was found to be associated with swimming and was directly related to water quality whereas respiratory illness, ear and eye infections, and "other" symptoms were not related to water quality. These findings were not unexpected, since it is reasonable to assume that water contaminated with domestic sewage effluents play a significant role in the transmission of GI disease. Swimmers generally described their illnesses as a mild, acute gastroenteritis of short duration characterized by diarrhea, vomiting, nausea, stomach ache, and fever. Some of the study participants had severe symptoms and either stayed home, stayed in bed, or consulted a physician. The incubation period ranged between 24 and 48 hr, and the disease manifestations lasted for 24 to 72 hr. The illness usually cleared spontaneously, without sequelae. The symptoms of illness were nearly identical with those usually described for Norwalk agent, rotavirus, and other similar viruses.[18] The rapid onset and the short duration of symptoms are unlike those of bacterial gastroenteritis, which usually occur after a longer incubation period and usually are of longer duration. Although the etiological agent of swimming-associated gastroenteritis was not determined, similar results were obtained in several studies at different times and places. This consistency leaves little doubt as to the association between GI illness and swimming, and swimming and the observed increase in swimming-associated gastroenteritis in waters with higher levels of pollution.

D. Ear Infections

Otitis externa is an inflammatory condition of the skin of the external auditory canal. This type of infection is so frequent among swimmers that it is commonly called "swimmer's ear". The infection is characterized by swelling, redness, and pain. Otitis externa is frequently associated with hot and humid weather, maceration of tissues in the ear canal, and trauma to ear tissues.[19] Other factors which influence the occurrence of ear infections are age and the time spent in the water. Individuals less than 18 years old and persons who spend long periods of time in the water are more likely to contract "swimmer's ear".[20] Swimming-associated ear infections are most frequently caused by *Pseudomonas aeruginosa*;[21] however, many other Gram-negative bacteria, as well as a few Gram-positive organisms, such as staphylococci and streptococci, have also been identified as the etiologic agent of otitis externa in swimmers.

The association of ear infections with swimming activity has been confirmed through the use of epidemiological studies. Hoadley and Knight[22] in 1975 conducted a survey of 244 individuals in the area around Jackson Lake, Ga., and found through physicians that ear aches were five times more likely in swimmers than in nonswimmers. The isolation of *P. aeruginosa* from the ear of those individuals with ear aches was also associated with swimming activity. Calderon and Mood[23] conducted a similar survey of individuals who swam in lakes and pools and also found that the proportion of swimmers among individuals with ear aches was significantly higher than the proportion of swimmers among individuals without ear aches. *P. aeruginosa* also was related to otitis externa infections in fresh water swimmers in a study conducted by Seyfried and Cook.[24] Pseudomonads isolated from cases were found to have the same phage and serotypes as strains isolated from bathing water samples. These studies[22-24] have clearly established the association between otitis externa and swimming in fresh water and the role of *P. aeruginosa* as the etiologic agent. In recent years another group of bacteria have been reported as causing ear infections in marine swimmers. Table 1 lists the halophilic *Vibrio* species isolated from the ears of individuals who had been swimming in marine waters. Armstrong et al.[25] isolated *V. parahaemolyticus* from the infected ear of a 57-year-old fisherman. An underlying factor in this case was a perforated tympanic membrane, a compromising situation also noted by Olsen[26] in a 7-year-old boy who contracted otitis media after bathing in seawater. The discharge from the boy's ear yielded a pure culture of *V. parahaemolyticus*. Ghosh and Bowen,[27] in Australia, also reported a case of otitis externa which was associated with *V. parahaemolyticus* and which occurred after a sea water contact. Other Vibrios also have been linked to swimming-associated ear infections. *V. alginolyticus* was the causal organism in six cases of otitis externa and four cases of otitis media reported by Ghosh and Bowen.[27] Pien et al.[28] isolated *V. alginolyticus* from the draining outer ears of three swimmers, and Ryan[29] described a case of otitis externa in a 32-year-old man which was attributed to sea water contact. A perforated tympanic membrane, which occurred during sea bathing, was a contributing factor in a case of otitis media reported by Olsen.[26] Two other *Vibrio* species have been isolated from the ears of individuals with otitis externa. Armstrong et al.[25] reported an ear infection

Table 1
EAR INFECTIONS CAUSED BY *VIBRIO* SPECIES IN MARINE WATER SWIMMERS

Etiologic agent	No. of cases	Type of ear infection	Median age (years)	Ref.
V. alginolyticus	14	Otitis externa (10)[a] Otitis media (4)	<20	27—29
V. parahaemolyticus	4	Otitis externa (2) Otitis media (2)	20	25—27
V. vulnificus	1	Otitis externa	8	25
V. mimicus	2	Otitis externa (1) Otitis media (1)	26	30

[a] Number of cases in parentheses.

in an 8-year-old girl who had a perforated tympanic membrane. *V. vulnificus* was the only organism isolated from the ear. *V. mimicus* was reported by Shandera et al.[30] as the causative organism in two cases of ear infections. The age range of cases of ear infections caused by *Vibrio* species was 7 to 57 years old. Of the cases, 77% were less than 20 years old. Calderon and Mood[23] and Seyfried and Cook[24] also found that younger individuals, those under the age of 20, were more at risk for swimming-associated ear infections caused by pseudomonads than other age groups.

Numerous investigators have shown that warm weather and high humidity conditions are significantly associated with the incidence of "swimmers ear" caused by *Pseudomonas* species. Ear infections caused by *Vibrio* species also appear to be related to warm weather conditions. This relationship may be due to the high densities that *Vibrio* species reach in coastal water environments during the warm weather season.[31] In addition, more individuals tend to go to the beach and remain in the water for long periods of time during warm weather periods.

E. Mycobacteriosis

Water contact infections due to mycobacteria other than *Mycobacterium tuberculosis* usually are associated with sea or brackish waters, although fresh water infections do occur. It is characteristic of mycobacterial infections that they are associated with injuries which occur during swimming activity or near the time of water contact. The extremities are the most common sites of the infections, and the hands and knees are most frequently affected. The injury to these sites is usually an abrasion, laceration, or puncture wound. The lesions are of three types: (1) a self-limited verrucous lesion may be observed; (2) an ulcerative skin lesion which forms a granuloma in the subcutaneous tissues; (3) an infection involving more deep-seated lesions, which form sporotricoid nodules or cause tensynovitis, bursitis, arthritis, and osteomyelitis. The causative agent of these infections is invariably *M. marinum*, which also is referred to in the literature under the names M. *balnei* or *M. platy*. *M. marinum* has been so closely identified with the aquatic activities of humans, such as swimming, diving, fishing, and boating, that Feldman and Hershfield[32] have dubbed this species "the leisure-time pathogen". The infections caused by *M. marinum* in pools are so characteristic that they are commonly called "swimming pool" granulomas. The "swimming pool" designation, however, is a misnomer, since the infections are frequently associated with natural bathing waters, both salt and fresh.

Although the vast majority of water contact infections are caused by *M. marinum*, other mycobacterial species, such as *M. kansasii* and *M. szulgai*, occasionally cause skin lesions much like those caused by *M. marinum*. Cott et al.[33] reported one such case, a 36-year-old

Table 2
CHARACTERIZATION OF INFECTIONS CAUSED BY *M. MARINUM*

No. of cases	Method of injury	Site of lesion	Age range (median)	Median duration of infection before diagnosis (range)
15	Laceration	Foot (4)[a] Knee (3) Hand (7) Arm (1)	10—63 (19)	1 year (3 months—27 years)
12	Abrasion	Knee (6) Hand (3) Foot (1) Neck (1) Elbow (1)	14—53 (29)	8 months (3 months—6 years)
6	Puncture Fish fin Shellfish	Hand (5) Ankle (1)	16—58 (52)	3½ months (2 weeks—2 years)
1	Dolphin bite	Finger	19	1 year

[a] Number of cases in parentheses.

man who developed lesions on his hand about 10 days after abrading it while removing barnacles from the propeller of a boat. A new lesion, which drained a large amount of pus, appeared on his hand 5 months after his initial visit. The drainage material was cultured and incubated in the light and the dark. Colonies growing on the Lowenstein-Jensen medium were pigmented in both the light and the dark, indicating that the organisms were scotochromogenic and a member of Runyon's Group II. The organisms, which were typically acid-fast on staining, were not speciated by the authors. Owens and McBride[34] reported isolating a photochromogenic, acid-fast bacterium from lesions on the left hand and arm of a 16-year-old male. The bacterium was identified as *M. kansasii* based on its maximum growth rate at 37°C. Feldman and Hershfield[32] examined the isolate some years later and identified it as *M. marinum*, causing some ambiguity about the ability of *M. kansasii* to cause skin lesions.

Some of the salient features of *M. marinum* infections are shown in Table 2. With few exceptions, *M. marinum* infections are associated with some event which compromises the integrity of the skin. Lacerations and abrasions occur most frequently on the foot, hand, knee, and elbow, and these are the sites where infections are likely to be seen. Animal bites and puncture wounds from fish or shrimp fins occur less frequently, but infections at these sites are well-documented in the literature. The median age of the 34 reported infections was approximately 36 years, which reflects the activities usually associated with aquatic injuries, i.e., boating and fishing. In fact, in some areas, *M. marinum* infections are considered an occupational hazard among commercial fisherman. An interesting feature of *M. marinum* infections is the long duration between the actual infection event and the diagnosis of the causative organism. Although the lesions are usually tender, they are not painful, and therefore, they may be ignored for long periods. Although the duration of illness in the reported cases ranged from 2 weeks to 27 years, most infections, if left untended, will heal spontaneously within a few years.

F. Pneumonia

The density of potential pathogens in natural surface waters is usually not high enough to create a health hazard even if an aerosol is created which might be inhaled by a susceptible

Table 3
PREDISPOSING FACTORS FOR PNEUMONIAS ASSOCIATED WITH WATERBORNE PATHOGENS

Etiological agent	Exposure	Age	No. of cases	Ref.
Pseudomonas putrefaciens	Near-drowning[a]	56	1	35
Aeromonas hydrophila	Near-drowning[b]	35, 13	2	36
Legionella pneumophila	Near-drowning[b]	48	1	37
L. bozemanii	Boating and diving accident[b]	60, 36	2	38

[a] Salt water.
[b] Fresh water.

individual. Similarly, under normal circumstances, seldom is enough water swallowed so that accidental aspiration of water into the bronchial tree would pose a health risk. There is, however, a circumstance wherein the risk for pneumonia has been shown to be high. Near-drowning episodes, especially in individuals over the age of 35 years, clearly are related to cases of pneumonia. Table 3 lists six cases of pneumonia related to water contact. Four of the cases were associated with near-drowning incidents; one was related to a boating accident; the fifth case occurred in a navy diver who died from acute bronchopneumonia following a diving exercise.

Two of these pneumonia cases were observed after sea water contact. A 56-year-old man was admitted to the hospital after a near-drowning accident.[35] On the 3rd day after admittance, the patient appeared to recover, but 1 day later he had a relapse. After antibiotic treatment the patient recovered. *Pseudomonas putrefaciens* and *Staphylococcus aureus* were isolated from sputum specimens, but the latter was recovered only from initial specimens, whereas *P. putrefaciens* continued to be isolated up to day 24. The second case was a 13-year-old boy who was resuscitated after being submerged in a tidal marsh for about 8 min.[36] Following admission to the hospital, the boy developed cardiac and respiratory failure. *A. hydrophila* was isolated from blood and sputum samples. The patient recovered after treatment.

Pneumonia attributed to *A. hydrophila* was observed in another near-drowning victim, a 34-year-old construction worker who fell off a truck into a drainage ditch face down.[36] The worker suffered a respiratory distress 24 hr after admission to the hospital. *A. hydrophila*, which had antibiotic sensitivities similar to *A. hydrophila* isolated from drainage ditch samples, was cultured from sputum samples. Another near-drowning fresh water incident, which resulted in a pneumonia, was reported from Canada.[37] *Legionella pneumophila* was identified as the causative pathogen after greater than fourfold increases in titer was noted between acute and convalescent sera. The 48-year-old woman recovered after treatment.

Two cases of pneumonia caused by *L. bozemanii* transmitted via water have been reported in the literature.[38] One case, a Navy diver who contracted pneumonia after a training exercise, was identified as legionellosis many years after the incident. Retrospective examination of the preserved organism, which had been isolated from lung tissue with techniques used for rickettsia, indicated that it was related to Legionnaires' Disease Bacillus. Shortly afterward it was identified as a new species called *L. bozemanii*. A second case involved a 60-year-old man who had been thrown from a boat into dirty, swampy water where he remained for 15 or 20 min before being rescued. The near-drowning resulted in a severe pneumonia that culminated in death. *L. bozemanii* was identified as the causative organism.

Table 4
FACTORS ASSOCIATED WITH SEPTICEMIC DISEASE THAT OCCURS AS A RESULT OF WATER CONTACT

Etiological agent	No. of cases	Median ages	Range of ages	No. with predisposing factors[a]	Ref.
Aeromonas hydrophila	7	65	16—81	6	41—43
Chromobacterium violaceum	3	49	28—53	1	45
Vibrio vulnificus	7	60	51—79	6	46—48
V. parahaemolyticus	1	65	—	1	46
V. alginolyticus	1	37	—	—	49

[a] Typical predisposing factors are diabetes, leukemia, and aplastic anemia.

The etiologic agents of the cases described above are commonly found in the aquatic environment. *P. putrefacians,* for instance, was isolated from water samples collected at other ocean bathing beaches near where the victim almost drowned. *Aeromonas* and *Legionella* species are ubiquitous in fresh water environments.[39,40] Aeromonads are especially sensitive to the trophic state of lakes and ponds, and therefore, this organism can be found in very high densities in waters that are contaminated by humans. Such high densities greatly increase the risk of infection in individuals whose pulmonary tissues are subjected to the unnatural insult that results from a near-drowning incident. An obvious conclusion is that the risk for respiratory infection in waters used for recreational purposes is related not only to the quality of the water, but also to the type of exposure. As a body of water becomes more polluted, the chances also increase for pneumonia to result from an accidental immersion of the body where the lungs come in contact with the water.

G. Septicemias

The invasion of the bloodstream of man by organisms that produce characteristic symptoms is commonly referred to as septicemia. Table 4 lists 19 cases of septicemia occurring as a result of water contact. The etiologic agents include *A. hydrophila, Chromobacterium violaceum, V. vulnificus, V. parahaemolyticus,* and *V. alginolyticus.*

A. hydrophila usually is not associated with septicemia conditions; it is more likely to be isolated from wounds or even from stool specimens from cases of gastroenteritis. Most of the cases of septicemia caused by *A. hydrophila* have three factors in common. First, the site of entry is usually through a break in the integrity of the skin caused by an insect bite, a cut on the upper or lower extremities, or a laceration. Second, individuals are directly exposed to water or cold-blooded animals that are found in those waters. Third, and perhaps most important, the body defense mechanisms in the exposed individuals are usually compromised in some way. For instance, typical underlying diseases may be leukemia, diabetes, or aplastic anemia. Most of the reported cases involve individuals over the age of 60 years. Only one of the five cases described by Wolff et al.[41] was under the age of 63, and only one case was not compromised. All of the cases were associated with water exposure either through fishing or accidental immersion. Ampel and Peter[42] reported a case of *Aeromonas* septicemia associated with a 70-year-old man who was extensively burned in a boathouse fire and subsequently immersed his body in an adjacent pond. The incident resulted in second degree burns on 40% of his body. Subsequent blood cultures grew *A. hydrophila.* The authors suggested that the septicemia resulted from exposure to the pond water containing *A. hydrophila,* which gained access to the bloodstream through colonization of the burn injury. Tapper et al.[43] also reported a case of *Aeromonas* septicemia. The patient was a 19-year-old youth who had acute myelomonogtic leukemia. The infection leading to septicemia followed contamination of a mosquito bite by stagnant water. Four of the seven cases shown in Table 4 succumbed to the infection.

Chromobacterium violaceum is a saprophyte that is commonly found in soil and water. Most of the cases of septicemic disease in humans have occurred in the southeastern U.S. Only three cases have been associated with water contact. Two of the cases reported by Macher et al.[44] involved a near-drowning and scuba diving. The near-drowning victim had inhaled and swallowed an overwhelming innoculum of *C. violaceum* in an accident which occurred on a Florida river. The second victim was a 28-year-old male who had a chronic granulomatous disease. He complained about malaise and fever 6 days after scuba diving. Multiple dark pustular cutaneous lesions with erythematous margins developed on his buttocks. Blood cultures were positive for *C. violaceum*. Another case of *C. violaceum* septicemia associated with a near-drowning incident was reported by Starr et al.[45] The 53-year-old male victim complained of fever, malaise, myalgia, and abdominal pain. The fever was persistent and a pustular rash developed over his body. Blood cultures showed luxuriant growth of *C. violaceum*. Except in the case of the scuba diver, who had an underlying disease which facilitated the entry of the organism into the bloodstream, the portal of entry for the other two cases was probably the lungs, which had been subjected to large amounts of water. This insult to the lung tissue probably provided a site where *C. violaceum* entered the bloodstream.

Vibrio species, especially *V. vulnificus*, frequently have been associated with septicemia. Two characteristics of *Vibrio* infections is their rapid onset and the high case fatality rate. Individuals with underlying disease appear to be more susceptible than healthy persons. The incubation period is very short and symptoms usually occur in about 16 hr. *V. vulnificus* was first described in 1976, although it had been reported many times before that date as the "lactose positive" *Vibrio*. A majority of the cases of fulminating septicemia caused by *Vibrio* species are due to *V. vulnificus;* however, *V. parahaemolyticus* and *V. alginolyticus* also have been reported as etiologic agents. The five cases of *V. vulnificus* septicemia and the one case of *V. parahaemolyticus* septicemia reported by Bonner et al.[46] were associated with underlying illnesses. Two of the cases had cirrhosis, two had leukemia, one had diabetes, and another had hemochromatosis. The organisms gained entry to the body through puncture wounds in two cases, lacerated extremities in two cases, a cut on the hand in one of the cases, and an ulcer in one case. All of these wounds and the ulcer were exposed to sea water, usually at the time the wound was obtained. Similar cases were reported by Castillo et al.[47] One was a 68-year-old man who punctured his hand handling an uncooked shrimp and the other was a 79-year-old man who punctured his hand with an uncooked crab claw. Both of these individuals exhibited the rapid onset of symptoms characteristic of this organism. Only the former case had an underlying disease. He had an active case of chronic hepatitis for at least 3 years before the *Vibrio* infection. The case of septicemia caused by *V. alginolyticus*, reported by English and Lindberg,[48] was associated with extensive burn injury suffered by a 37-year-old woman in a boating accident. The burn injury was sustained when gasoline fumes ignited in a boat. The individual was immersed in the water to extinguish the flames during the rescue operation. *V. alginolyticus* was cultured from the blood as well as from biopsy material from the burn areas. Five of the nine cases of *Vibrio* septicemia shown in Table 4 succumbed to the infection. This high case fatality rate may be related to age as well as the underlying disease that was present in most of the cases. The median age of all of the *Vibrio* cases was 60 years, a period of life when the bodies immune defense mechanisms are declining.

H. Wound Infections

Skin infections associated with water contact frequently have been reported in the literature. The infections invariably are related to lacerations or abrasions that come about during

Table 5
CHARACTERISTICS ASSOCIATED WITH SKIN INFECTIONS THAT RESULT FROM WATER CONTACT

Etiological agent	No. of cases	Median ages	Range of ages	No. with predisposing factors[a]	Ref.
Aeromonas sobria	1	19	—	—	49
A. hydrophila	8	19	7—62	1	49—55
Vibrio alginolyticus	7	32	22—50	3	58—62
V. vulnificus	20[b]	58	13—75	6	63—65
V. parahaemolyticus	4	25	9—40	—	46, 56, 57

[a] Typical predisposing factors are lacerations, abrasions, and insect or animal bites.
[b] Age information available only for six victims.

recreational or occupational activity, usually as a result of some type of accident. Wounds incurred as a result of insect or animal bites also have been reported as predisposing or related factors for infections that occur after water exposure. The infections usually occur in the soft tissues and are often self-limited, although some cases are much more serious and have resulted in amputation. The wound infections reported in the literature do not appear to be age related. They have been observed in children 7 years old as well as in individuals as old as 75 years. Table 5 is a summary of data reported in the literature.

Aeromonas species usually are associated with fresh water wound infections, although infections caused by aeromonads are occasionally seen in cases of sea water exposure in coastal areas. Most of the reported infections caused by aeromonads are due to *A. hydrophila*; however, one case has been described wherein *A. sobria* was thought to be a causative pathogen. It was isolated, along with *A. hydrophila*, from a leg puncture wound sustained by a scuba diver.[49] Many of the reported wound infections involved accidents that occurred during the water-related activity. A case described by Smith[50] was related to a leg lacerated by a motorboat propeller, and a case reported by Hanson et al.[51] was precipitated by a diving accident in a shallow lake which resulted in an extensive scalp laceration. A case reported by Fulghum et al.[52] involved a 62-year-old woman who fell down in a boat, causing a crab trap to fall on her leg, producing two puncture lacerations. Other accidents reported by Katz and Smith,[53] Phillips et al.[54] and Rosenthal et al.[55] involved stepping on a nail while in the water, an alligator bite, a fall on a rock jetty which resulted in a leg laceration, and a knee laceration due to water skiing. The age range of the cases was from 7 to 62 years, with three fourths of the individuals being 22 years old or less. The preponderance of younger individuals probably accounts for the type of injuries that were observed. Most of them were the result of athletic activities usually associated with young age groups, i.e., water skiing or diving. None of the infections resulted in death, although one of the cases died a few months after the initial infection had been cleared up, probably from a secondary infection.

Many reports of infections due to *Vibrio* species have been reported in the literature. *V. alginolyticus* and *V. vulnificus* are reported most frequently and *V. parahaemolyticus* only occasionally. Only four cases of *V. parahaemlyticus* infection have been reported, one by Roland,[56] one by Porres et al.[57] and two by Bonner et al.[46] The former cases involved injuries that were sustained on Narragansett Bay in Rhode Island. One injury was to a 40-year-old man who was "clamming" and the other was to a 9-year-old boy who scraped his knee on a barnacle-coated raft. The third and fourth cases were associated with a stingray wound and a laceration on a seashell. Water-associated *V. alginolyticus* infections have been reported more frequently. The wounds associated with *V. alginolyticus* infections were incurred in widely dispersed sites such as the Sea of Cortez,[58] the Pacific Ocean near Vancouver,

Table 6
OUTBREAKS OF LEPTOSPIROSIS ASSOCIATED WITH SWIMMING IN FRESH WATER

Outbreak location	Etiologic agent	No. of cases	Possible source	Ref.
Geneva, Ala.	*L. pomona*	50	Hogs	67
Jackson Hole, Wyo.	*L. canicola*	24	Dogs	66
Standing Rock Indian Reservation, S.D.	*L. pomona* *L. antumnalis*	5	Livestock	68
Kennewick, Wash.	*L. pomona*	61	Cattle	69
Tennessee	*L. interrogans*	7	—	70
Cedar Rapids, Iowa	*L. pomona*	40	Cattle	71

Canada,[59] and near Irvine, Calif.,[60] on Long Island Sound,[61] and the Belgian North Sea.[62] In each case the initiation factor was either an accidental breaking of the surface of the skin or a wound which had not healed. Three of the cases had an associated underlying disease which most likely predisposed the affected individual to infection. The wound infections caused by *V. vulnificus* are very much like those caused by *V. alginolyticus*. The infections are usually related to some event or accident which breaks the integrity of the skin. The events or accidents include lacerations,[63] insect bites,[64] puncture wounds,[63] crab bites,[63] and scratches. Some of the victims had underlying diseases such as alcoholism, stasis ulcers, lymphocytic leukemia, and diabetes. The age distribution seemed to be somewhat higher for cases of *V. vulnificus* infections than for that observed with *V. alginolyticus* infections.

V. vulnificus is also notable because wound infections due to this organism have been observed several hundred miles from coastal water in New Mexico and Oklahoma.[65] One case involved a serious scalp laceration in an individual who became disoriented and jumped into a creek that contained brackish water. The other case was associated with an insect bite and subsequent exposure to saline water in the Great Salt Plains reservoir. *V. vulnificus* was not isolated from either of the water sources, although the patient isolates were able to grow in 0.4 and 0.2% sodium chloride (the salinity of the creek and the reservoir, respectively). These isolations point to the importance of considering halophilic vibrios among the potential pathogens wound infections that occur after exposure to brackish waters, even in inland locations.

I. Leptospirosis

Leptospirosis, which also is known as Weil's disease or hemorrhagic jaundice, is caused by *Leptospira* species and is characterized by a course of headache, chills, fever, myalgia, nausea, and in some cases, neck or joint pain or pain on movement of the eyes. Swimming and water contact activities are two of the major means by which humans are exposed to this pathogen. *Leptospira* usually contaminate water environments via the urine of infected animals such as cattle, dogs, rats, and hogs. Racoons, skunks, and muskrats also are known to harbor *Leptospira* species. The organisms usually gain entry to the body through the skin, the integrity of which has been compromised by cuts, abrasions, or maceration. The organism, if accidently swallowed or inhaled, also can penetrate the mucous membranes of the mouth or nasopharynx. *L. pomona*, *L. caneiola*, and *L. icterohemorrhagia* are the most common genera associated with disease in man, but other genera have also been implicated in cases of leptospirosis.

Some examples of typical leptospirosis outbreaks are given in Table 6. Shaeffer[66] reported an outbreak of leptospirosis with an attack rate of 63%. A group of about 80 young adults and adolescents had been swimming at the local swimming hole fed by a stream in which dead hogs had been seen floating; 3 months after the incident, sera from hogs, as well as other animals, were found to be positive for *L. pomona,* the same genus implicated in the outbreak.

A report of an outbreak in Jackson Hole, Wyo. was interesting because the cause of the illnesses was diagnosed 11 years after the event.[67] The late inquiry was made because at the time of the illness no human cases of leptospirosis had been reported. The fact that bovine epizootics of *L. pomona* were common in that part of the country was apparently ignored. Dogs or other wild animals with access to the pool in which the attack victims had been swimming were the probable source of the agent which caused the outbreak. The cases reported by Jellison et al.[68] were part of a larger group of men who had been swimming in the Grand River near Wakpala, S.D. This report was interesting because it noted that 1 year after the outbreak, a water sample from the river was injected into a guinea pig and the animal developed a titer of 1:5000 against *L. pomona* 35 days later. This suggested that the animals contaminating the streams were probably part of a permanent carrier population.

Nelson et al.[69] reported an outbreak associated with the use of an irrigation canal downstream from a pasture where a herd of cattle were kept. The etiologic agent was isolated from standing water in the pasture and 21% of the cattle were found to be shedding *Leptospira*. An outbreak reported from Tennessee was the first one in the U.S. involving *L. interrogans*.[70] Although the source of the etiologic agent was not identified, wild animals were suspected. Tjalma and Galton[71] reported on an outbreak of leptospirosis caused by *L. pomona.* Creeks in which young adults and children had been swimming were accessible to cows and hogs that were shown to be serologically positive for *L. pomona.*

The risks associated with exposure to swimming holes contaminated with urine from infected animals are almost impossible to eliminate. Reduction of the risks can be accomplished only through informing the public of the health hazards associated with swimming in water that is accessible to domestic or wild animals.

J. Primary Amebic Meningoencephalitis

Primary amebic meningoencephalitis (PAM) is an infection of the central nervous system caused by the free-living amebae *Naeglèria flowleri.* PAM usually occurs in young adults or children who have a recent history of swimming in fresh water lakes or pools, or some other water-associated activity, such as water skiing. Amebae in the water are accidently introduced into the nasal passage, where they then penetrate the mucosa and cribiform plate. The amebae gain access to the meninges via the olfactory nerve fibers and after reaching the brain, they initiate an inflammatory response and cause damage to the tissue. The outcome of *Naegleria* infections are invariably fatal. Death usually occurs within 72 hr after the onset of symptoms. Headache, anorexia, fever, nausea, and vomiting are the main symptoms of PAM. Vision also may be affected.

Naegleria species are free-living amoeba which are ordinarily found in water, soil, and decaying vegetation.[72] Warm industrial effluents promote the growth of *Naegleria*.[73] This species has also been isolated from swimming pool water[74] and tap water.[75]

Cases of PAM caused by *Naegleria* species have occurred world-wide. The first case was reported by Fowler and Carter[76] in 1965 from Australia. The disease has also been noted in Czechoslovakia,[77] Belgium,[78] Great Britain,[79] and New Zealand.[80] In the U.S. cases have been reported from South Carolina,[81] Virginia,[82] Florida,[83] Texas,[84] and Georgia.[85]

The age of PAM victims ranges from 8 to 27 years. The young age distribution of those individuals contracting PAM is probably a behavioral component, since more young adults and children than older adults spend their leisure time pursuing swimming and other water activities. Male swimmers have a greater risk of infection than female swimmers, but again this may be a behavioral component, since males have a tendency to do more swimming than females.

It is notable that most of the cases of PAM have been reported from southern states with only a few exceptions. This phenomenon is no doubt a reflection of the warm climate of these states. *Naegleria* species are known to survive and grow in favorable warm waters of the South, as well as in thermal industrial effluents in more northern climates.

Although PAM cases are invariably fatal, the risk of acquiring this disease is very small. Wellings[86] has estimated that only about one case of naeglerial infection will result for every 2.6 million exposures to waters containing this pathogen.

K. Schistosome Dermatitis

Schistosome dermatitis, also called swimmer's itch, clam digger's itch, cercaria dermatitis, or cutaneous dermatitis is an allergic reaction due to the presence of parasitic flatworm cercariae in the skin. Migratory ducks and other waterfowl are the usual host of the adult flatworm, and humans are an incidental host of this parasite. The life cycle of the schistosome begins with the production of eggs in the normal host. The eggs are excreted in the feces, usually into water, where a larvae emerges. The larvae burrow into snails, the intermediate host, where they form sporocysts. After a suitable developmental period, fork-tailed cercariae are released into the water. The cercariae settle on aquatic vegetation, which is eaten by ducks and the cycle begins again. Humans are a substitute host for the cercariae.

Shortly after the initial exposure to the cercariae, a mild itching occurs, which can last up to 1 hr; 10 to 12 hr later, pruritic papules form. The inflammatory response reaches a peak in 3 to 4 days and then recedes in 1 or 2 weeks. The cercariae do not penetrate beyond the dermis of the skin, and they seldom remain alive in the substitute host for more than 1 or 2 days. Symptoms are accelerated and more severe on subsequent exposures to cercariae, but the duration of the inflamation is much shorter.

Schistosome dermatitis is endemic in Canada, Michigan, Wisconsin, Minnesota, and along the Mississippi River and other flyways used by waterfowl commonly infected with the parasitic flatworms. The disease also has been observed in other areas of the U.S. In 1974, Wills et al.[87] described an outbreak of schistosome dermatitis in Pennsylvania which occurred in children who swam in a farm pond suspected of being responsible for the source of itching sensations and red spots that developed shortly thereafter. The pond was the residence for five species of duck and geese and frequently was visited by transient water fowl, such as mallard ducks. Snails collected from the pond were found to shed high densities of schistosome cercariae, which were assumed to be the cause of the dermatitis. In 1981, 16 cases of cercarial dermatitis were reported in Humbolt County, Calif.[88] The individuals had been swimming in a cove of the Mad River, 10 mi from the Pacific coast. A search for evidence of the parasites at the site of exposure revealed snails of the genus *Physa* in shallow pools along the river. The snails were transported to the laboratory where cercariae were observed emerging from the snails held in dishes of water.

The disease is world-wide, with schistosome dermatitis being reported from Europe, Central and South America, Mexico, Australia, New Zealand, Maylasia, India, and Africa.[89] Public health measures to prevent schistosome dermatitis include informing swimmers of the presence of infected snails and, if necessary, controlling infected snail populations through the appropriate use of molluscicides.

III. CONCLUSIONS

The infections associated with water contact fall into two distinct groups. The first group is enteric illnesses which result from the ingestion of water while swimming in fresh and marine waters which have been contaminated with effluents from sewage treatment plants. Illnesses associated with this type of exposure are similar in both marine and fresh waters, and they have a tendency to affect young rather than old individuals. The risk of illness after exposure to water contaminated with human fecal material is governed by the number of diseased individuals in the population discharging the wastes. If the number of individuals shedding pathogens is small, then the risk of illness after contact with contaminated water will be minimal. The risk becomes greater as the size of the contributing population increases, since there is a greater probability for diseased individuals and thus a larger number of individuals shedding pathogens. The source and nature of these pathogens implies that the risk of illness can be decreased or possibly eliminated using sewage treatment technology and disinfection procedures. The second group of infections are not amenable to elimination by man-made technology, since the etiologic agents, with the exception of *Leptospira* species, are mainly bacteria that are indigenous to the aquatic environment. These native bacteria are normally innocuous for healthy people, but they can be pathogenic under certain circumstances. The opportunistic nature of these bacteria is evident in most of the outbreaks and cases that were reviewed. During or before the water contact, an event usually occurred which compromised the host body defenses in some way, such as laceration or abrasions of the skin, puncture wounds, animal or insect bites, or some underlying disease such as diabetes, leukemia, or hepatic dysfunction. The infections caused by native aquatic bacteria appear to be seasonal in nature, usually occurring during the summer months. Most water contact activity takes place during the warmer months of the year, and it is during this period that the native bacteria reach their highest densities. In some cases there appears to be a regional component as seen with the infections caused by *Chromobacterium* species. *Chromobacterium* infections occur mainly in the southeastern region of the U.S. The reason for this regionalization is not clear since this organism is ubiquitous across the country.

Leptospirosis obviously does not fit either of the patterns described above. Commonly used indicators of fecal contamination will not predict risk of illness from *Leptospira* species in water environments, since the urine of domestic and wild animals does not contain bacteria associated with feces. *Leptospira* does, however, have many characteristics in common with opportunistic pathogens that cause infections or illness in individuals who come in contact with water. For instance, leptospirosis is much more likely to occur in swimmers whose skin has been cut, abraded, or macerated. This characteristic is not an absolute requirement for infection, but it does increase the risk of infection. The seasonality of *Leptospira* infections also is similar to that of bacteria native to aquatic environments in that most cases of leptospirosis occur in the warm summer months. Furthermore, the population at risk tends to be the same as the one most commonly infected by bacteria indigenous to water environments. The method of dealing with health risk due to *Leptospira* is also similar, since educating the public as to the potential health hazards associated with swimming in certain environments is the only means of coping with this public health problem.

The etiologic agents that cause disease are not the same for fresh and marine waters. Although there is some overlap, the infections observed in marine waters are usually due to halophilic *Vibrio* species, whereas the fresh water infections are usually caused by aeromonads, pseudomonads, and mycobacterial species. These differences, the ubiquitous character of most of the pathogens, and their opportunistic nature are the factors which preclude using technology to lessen the risks associated with water contact. It is only through educational means, whereby users of water resources are made aware of potential health hazards and their predisposing condition, that the risks for disease can be lowered.

REFERENCES

1. **Discher, D. M.**, The Scientific Basis for Coliform Organisms as an Index of Bathing Water Safety, Coliform Standards for Recreational Waters, Appendix I, Progress Report, Public Health Activities Committee, *J. San. Eng. Div. Am. Soc. Civil Eng.*, 89, 70, 1963.

2. **Reece, R. J.**, 38th Annual Report to Local Government Board, 1908—09, Suppl. with Report to Medical Officer for 1908—09, Appendix A, 1909, No. 6, 90; cited in **Moore, B.**, Sewage contamination of coastal bathing waters, *Bull. Hyg.*, 29, 689, 1954.
3. **Ciampolini, E.**, A Study of the Typhoid Fever Incidence in the Health Center District of New Haven, Report to City of New Haven, 1921.
4. **Winslow, C. E. A. and Moxon, D.**, Bacterial pollution of bathing beach waters in New Haven harbor, *Am. J. Hyg.*, 8, 299, 1928.
5. Typhoid fever from bathing in polluted waters, *N.Y. Dept. Health Weekly Bull.*, 21, 257, 1932.
6. **Martin, P. H.**, Field investigations of paratyphoid fever with typing of *Salmonella paratyphi* by mean of Vi bacteriophage, *Bull. Hyg.*, 22, 754, 1947.
7. Typhoid Traced to Bathing at a Polluted Beach, *Public Works*, 92, 182, 1961.
8. **Olivieri, V. P.**, Measurement of water quality, in Assessment of Microbiology and Turbidity Standards for Drinking Water, Berger, P. S. and Argaman, Y., Eds., EPA 570-9-83-001, U.S. Environmental Protection Agency, Washington, D.C., 1981.
9. **Dupont, H. L. and Hornick, R. B.**, Clinical approach to infectious diarrheas, *Medicine*, 52, 265, 1973.
10. **Dunlop, S. G.**, Survival of pathogenic organisms in sewage, *Public Works*, 88, 80, 1957.
11. **Rosenberg, M. L., Hazlet, K. K., Schaefer, J., Wells, J. G., and Pruneda, R. C.**, Shigellosis from swimming, *JAMA*, 236, 1849, 1976.
12. **Bryan, J. A., Lehmann, J. D., Setiady, I. F., and Hatch, M. H.**, An outbreak of hepatitis-A associated with recreational lake water, *Am. J. Epidemiol.*, 99, 145, 1974.
13. **Hawley, H. B., Morin, D. P., Geraghty, M. E., Tomkow, J., and Phillips, C. A.**, Coxsackievirus B epidemic at a boys' summer camp. Isolation of virus from swimming water, *JAMA*, 226, 33, 1973.
14. **Koopman, J. S., Eckert, E. A., Greenberg, H. B., Strohm, B. C., Isaacson, R. E., and Monto, A. S.**, Norwalk virus enteric illness acquired by swimming exposure, *Am. J. Epidemiol.*, 115, 173, 1982.
15. **Denis, F. A., Blanchovin, E., DeLignieres, A., and Flamen, P.**, Coxsackie A_{16} infection from lake water, *JAMA*, 228, 1370, 1974.
16. **D'Alessio, D. J., Minor, T. E., Allen, C. I., Tsiatis, A. A., and Nelson, D. B.**, A study of the proportions of swimmers among well controls and children with enterovirus-like illness shedding or not shedding an enterovirus, *Am. J. Epidemiol.*, 113, 533, 1981.
17. **Cabelli, V. J.**, Health Effects Criteria for Marine Recreational Waters, EPA-600/1-80-031, U.S. Environmental Protection Agency, Cincinnati, Ohio, 1981.
18. **Cabelli, V. J.**, Epidemiology of enteric viral infections, in *Viruses and Wastewater Treatment*, Goddard, M. and Butler, M., Eds., Pergamon Press, Elmsford, N.Y., 1981, 291.
19. **Cassisi, N., Cohn, A., Davidson, T., and Witten, B. R.**, Diffuse otitis externa: clinical and microbiologic findings in the course of a multicenter study on a new otic solution, *Ann. Otol. Rhinol. Laryngol.*, 86(Suppl. 39), 1, 1977.
20. **Calderon, R. L.**, Epidemiological Studies of Otitis Externa, M. P. H. Essay, Yale University, New Haven, Conn., 1981.
21. **Wright, D. N. and Alexander, J. M.**, Effect of water on the bacterial flora of swimmer's ears, *Arch. Otolaryngol.*, 99, 15, 1974.
22. **Hoadley, A. W. and Knight, D. E.**, External otitis among swimmers and nonswimmers, *Arch. Environ. Health*, 30, 445, 1975.
23. **Calderon, R. and Mood, E.**, An epidemiological assessment of water quality and "swimmer's ear", *Arch. Environ. Health*, 37, 300, 1982.
24. **Seyfried, P. L. and Cook, R. J.**, Otitis externa infections related to *Pseudomonas aeruginosa* levels in five Ontario lakes, *Can. J. Public Health*, 75, 83, 1984.
25. **Armstrong, C. W., Lake, J. L., and Miller, G. B.**, Extraintestinal infections due to halophilic vibrios, *South. Med. J.*, 76, 571, 1983.
26. **Olsen, H.**, *Vibrio parahaemolyticus* isolated from discharge from the ear in two patients exposed to sea water, *Acta Pathol. Microbiol. Scand.*, 86, 247, 1978.
27. **Ghosh, H. K. and Bowen, T. F.**, Halophilic vibrios from human tissue infections on the Pacific coast of Australia, *Pathology*, 12, 397, 1980.
28. **Pien, F., Lee, K., and Higa, H.**, *Vibrio alginolyticus* infections in Hawaii, *J. Clin. Microbiol.*, 5, 670, 1977.
29. **Ryan, W. J.**, Marine vibrios associated with superficial septic lesions, *J. Clin. Pathol.*, 29, 1014, 1976.
30. **Shandera, W. X., Johnston, J. M., Davis, B. R., and Blake, P. A.**, Disease from infection with *Vibrio mimicus*, a newly recognized *Vibrio* species, *Ann. Intern. Med.*, 99, 169, 1983.
31. **Colwell, R. R., Lovelace, T. E., Wan, L., Kaneko, T., Staley, T., Chen, P. K., and Tubiash, H.**, *Vibrio parahaemolyticus* — isolation, identification, classification and ecology, *J. Milk Food Technol.*, 36, 202, 1973.
32. **Feldman, R. A. and Hershfield, E.**, Mycobacterial skin infection by an unidentified species, *Ann. Intern. Med.*, 80, 445, 1974.

33. **Cott, R. E., Carter, D. M., and Salt, T.,** Cutaneous disease caused by atypical mycobacteria, *Arch. Dermatol.,* 95, 259, 1967.
34. **Owens, D. W. and McBride, M. E.,** Sporotrichoid cutaneous infection with *Mycobacterium kansasii, Arch. Dermatol.,* 100, 51, 1969.
35. **Rosenthal, S. L., Zuger, J. H., and Apollo, E.,** Respiratory colonization with *Pseudomonas putrefaciens* after near-drowning in salt water, *J. Clin. Pathol.,* 64, 382, 1975.
36. **Reines, H. D. and Cook, F. V.,** Pneumonia and bacteremia due to *Aeromonas hydrophila, Chest,* 80, 264, 1981.
37. **Sekal, L. H., Stackiw, W., Buchanan, A. G., and Parker, S. E.,** *Legionella pneumophila* pneumonia, *Can. Med. Assoc. J.,* 126, 116, 1982.
38. **Cordes, L. G., Wilkinson, H. W., Gorman, G. W., Fikes, B. J., and Fraser, D. W.,** Atypical *Legionella*-like organisms: fastidious water-associated bacteria pathogenic for man, *Lancet,* I, 927, 1979.
39. **Hazen, T. C., Fliermans, C. B., Hirsch, R. P., and Esch, G. W.,** Prevalence and distribution of *Aeromonas hydrophila* in the United States, *Appl. Environ. Microbiol.,* 36, 731, 1978.
40. **Fliermans, C. B., Cherry, W. B., Orrison, L. H., and Thaker, L.,** Isolation of *Legionella pneumophila* from nonepidemic-related aquatic habitats, *Appl. Environ. Microbiol.,* 37, 1239, 1981.
41. **Wolff, R. L., Wiseman, S. L., and Kitchens, C. S.,** *Aeromonas hydrophila* bacteremia in ambulatory immunocompromised hosts, *Am J. Med.,* 68, 238, 1980.
42. **Ampel, L. and Peter, G.,** *Aeromonas* bacteraemia in a burn patient, *Lancet,* I, 987, 1981.
43. **Tapper, M. L., MacCarthy, L. R., Mayo, J. B., and Armstrong, D.,** Recurrent *Aeromonas* sepsis in a patient with leukemia, *Am. J. Lin. Pathol.,* 64, 525, 1975.
44. **Macher, A. M., Casale, T. B., and Fauci, A. S.,** Chronic granulomatons disease of childhood and *Chromobacterium violaceum* infections in the southeastern United States, *Ann. Intern. Med.,* 97, 51, 1982.
45. **Starr, A. J., Cribbett, L. S., Poklepovic, J., Friedman, H., and Fuffolo, E. H.,** *Chromobacterium violaceum* presenting as a surgical emergency, *South. Med. J.,* 74, 1137, 1981.
46. **Bonner, J. R., Coker, A. S., Berryman, C. R., and Pollack, H. M.,** Spectrum of *Vibrio* infections in a Gulf Coast community, *Ann. Intern. Med.,* 99, 464, 1983.
47. **Castillo, L. E., Winslow, D. L., and Pankey, G. A.,** Wound infection and septicshock due to *Vibrio vulnificus, Am. J. Trop. Med. Hyg.,* 30, 844, 1981.
48. **English, V. L. and Lindberg, R. B.,** Isolation of *Vibrio alginolyticus* from wounds and blood of a burn patient, *Am. J. Med. Technol.,* 43, 989, 1977.
49. **Joseph, S. W., Daily, O. P., Hunt, W. S., Seidler, R. J., Allen, D. A., and Colwell, R. R.,** *Aeromonas* primary wound infection of a diver in polluted waters, *J. Clin. Microbiol.,* 10, 46, 1979.
50. **Smith, J. A.,** *Aeromonas hydrophila*: analysis of 11 cases, *Can. Med. J.,* 122, 1270, 1980.
51. **Hanson, P. G., Standridge, J., Jarrett, F., and Maki, D. G.,** Freshwater wound infection due to *Aeromonas hydrophila, JAMA,* 238, 1053, 1977.
52. **Fulghum, D. D., Linton, W. R., and Taplin, D.,** Fatal *Aeromonas hydrophila* infection of the skin, *South. Med. J.,* 71, 739, 1978.
53. **Katz, D. and Smith, H.,** *Aeromonas hydrophila* infection of a puncture wound, *Ann. Emergency Med.,* 9, 529, 1980.
54. **Phillips, J. A., Bernhardt, H. E., and Rosenthal, S. G.,** *Aeromonas hydrophila* infections, *Pediatrics,* 53, 110, 1974.
55. **Rosenthal, S. G., Bernhardt, H. E., and Phillips, J. A.,** *Aeromonas hydrophila* wound infection, *Plast. Reconstr. Surg.,* 53, 77, 1974.
56. **Roland, F. P.,** Leg gangrene and endotoxin shock due to *Vibrio parahaemolyticus* — an infection acquired in New England coastal waters, *N. Engl. J. Med.,* 282, 1306, 1970.
57. **Porres, J. M., and Fuchs, L. A.,** Isolation of *Vibrio parahaemolyticus* from a knee wound, *Clin. Orthop. Relat. Res.,* 106, 145, 1975.
58. **Spark, R. P., Fried, M. L., Perry, C., and Watkins, C.,** *Vibrio alginolyticus* wound infection: case report and review, *Ann. Clin. Lab. Sci.,* 9, 133, 1979.
59. **Wagner, K. R. and Crichton, E. P.,** Marine *Vibrio* infections acquired in Canada, *Can. Med. Assoc. J.,* 124, 435, 1981.
60. **Pezzlo, M., Valter, P. J., and Burns, J. M.,** Wound infection associated with *Vibrio alginolyticus, Am. Soc. Clin. Pathol.,* 71, 476, 1979.
61. **Rubin, S. J. and Tilton, R. C.,** Isolation of *Vibrio alginolyticus* from wound infections, *J. Clin. Microbiol.,* 2, 556, 1975.
62. **Aelvoet, G., Kets, R., and Pattyn, S. R.,** Cellulitis caused by *Vibrio alginolyticus, Acta Derm. Venereol.,* 63, 559, 1983.
63. **Blake, P. A., Merson, M. H., Weaver, R. E., Hollis, D. G., and Heublein, P. C.,** Disease caused by a marine *Vibrio, N. Engl. J. Med.,* 300, 1, 1979.
64. **Bachman, B., Boyd, W. P., Lieb, S., and Rodrick, G. E.,** Marine noncholera *Vibrio* infections in Florida, *South. Med. J.,* 76, 296, 1983.

65. **Tacket, C. O., Barrett, T. J., Mann, J. M., Roberts, M. A., and Blake, P. A.,** Wound infections caused by *Vibrio Vulnificus*, a marine *Vibrio*, in inland areas of the United States, *J. Clin. Microbiol.*, 19, 197, 1984.
66. **Schaeffer, M.,** Leptospiral meningitis. Investigation of a water-borne epidemic due to *L. pomona*, *J. Clin. Invest.*, 30, 670, 1951.
67. **Cockburn, T. A., Vavra, J. D., Spencer, S. S., Dann, J. R., Peterson, L. J., and Reinhard, K. R.,** Human leptospirosis associated with a swimming pool, diagnosed after eleven years, *Am. J. Hyg.*, 60, 1, 1954.
68. **Jellison, W. L., Stonner, H. G., and Berg, G. M.,** Leptospirosis among Indians in the Dakotas, *Rocky Mount. Med. J.*, 55, 56, 1958.
69. **Nelson, K. E., Ager, E. A., Galton, M. M., Gillespi, R. W. H., and Sulzer, C. R.,** An outbreak of leptospirosis in Washington state, *Am. J. Epidemiol.*, 98, 336, 1973.
70. Center for Disease Control, Leptospirosis — Tennessee, *Morbid. Mortal. Weekly Rep.*, 25, 84, 1976.
71. **Tjalma, R. A. and Galton, M. M.,** Human leptospirosis in Iowa, *Am. J. Trop. Med. Hyg.*, 14, 387, 1965.
72. Center for Disease Control, *Morbid. Mortal. Weekly Rep.*, 27, 343, 1978.
73. **Duma, R. J.,** Study of Pathogenic Free-living Amebas in Fresh-water Lakes in Virginia, EPA-600/10, U.S. Environmental Protection Agency, Cincinnati, Ohio, 1979.
74. **DeJonckheere, J. F.,** Studies on pathogenic free-living amoebae in swimming pools, *Bull. Inst. Pasteur*, 77, 385, 1979.
75. **Anderson, K. and Jamieson, A.,** Primary amoebic meningoencephalitis, *Lancet*, 1, 902, 1972.
76. **Fowler, M. and Carter, R. F.,** Acute pyogenic meningitis probably due to *Acanthamoeba* sp.: a preliminary report, *Br. Med. J.*, 2, 740, 1965.
77. **Červa, L., and Novák, K.,** Amoebic meningoencephalitis: sixteen fatalities, *Science*, 160, 92, 1968.
78. **Jadin, J. B., Hermanne, J., Robijn, G., Willaent, E., Van Maercke, Y., and Stevens, W.,** Trois cas de meningo-encephalite amibienne primitive observes à Anvers (Belgique), *Ann. Soc. Belge Méd. Trop.*, 51, 255, 1971.
79. **Symmers, W. S.,** Primary amoebic meningoencephalitis in Britain, *Br. Med. J.*, 4, 449, 1969.
80. **Mandel, B. N., Gudex, D. J., Fitchett, M. R., Pullon, D. H. H., Malloch, J. A., David, C. M., and Apthorp, J.,** Acute meningo-encephalitis due to amoebae of the order myxomycetale (slime mould), *N.Z. Med. J.*, 71, 16, 1970.
81. **Darby, C. P., Conradi, S. E., Holbrook, T. W., and Chatellier, C.,** Primary amebic meningoencephalitis, *Am. J. Dis. Child.*, 133, 1025, 1979.
82. **Callicott, J. H., Jr.,** Amebic meningoencephalitis due to free-living amebas of the hartmannella (*Acanthamoeba*) — *Naegleria* group, *Am. J. Clin. Pathol.*, 49, 84, 1968.
83. **Butt, C. G.,** Primary amebic meningoencephalitis, *N. Engl. J. Med.*, 274, 1473, 1966.
84. **Patras, D. and Andujar, J. J.,** Meningoencephalitis due to hartmannella (*Acanthamoeba*), *Am. J. Clin. Pathol.*, 46, 226, 1966.
85. **McCroan, J. E. and Patterson, J.,** Primary Amebic Meningoencephalitis — Georgia, *Morbid. Mortal. Weekly Rep.*, 19, 413, 1970.
86. **Wellings, F. M.,** Amoebic meningoencephalitis, *J. Fla. Med. Assoc.*, 64, 327, 1977.
87. **Wills, W., Fried, B., Carroll, D. F., and Jones, G. E.,** Schistosome dermatitis in Pennsylvania, *Public Health Rep.*, 91, 469, 1976.
88. Cercariae dermatitis among bathers in California; katayama syndrome among travelers to Ethiopia, Center for Disease Control, *Morbid. Mortal. Weekly Rep.* 31, 435, 1982.
89. **Rapp, W. F., Johnson, E. W., and Smith, H. D.,** Schistosome dermatitis or swimmer's itch, *Neb. Med. J.*, 57, 210, 1972.

Chapter 4

CHEMICAL DRINKING WATER CONTAMINANTS AND DISEASE

Gunther F. Craun*

TABLE OF CONTENTS

I.	Drinking Water Standards..44	
II.	Inorganic Constituents 45	
	A.	Nitrate..45
	B.	Water Hardness, Corrosivity, and Sodium 50
	C.	Arsenic and Selenium 56
	D.	Fluoride...58
	E.	Asbestos..58
III.	Organic Contaminants 60	
	A.	By-Products of Chlorination...61
	B.	Volatile Synthetic Organic Contaminants.............................63
References..65		

---

* This chapter was written by Gunther F. Craun in his private capacity. No official support or endorsement by the Environmental Protection Agency or any other agency of the Federal Government is intended or should be inferred.

I. DRINKING WATER STANDARDS

The Safe Drinking Water Act (42 U.S.C. 300f, et seq.) requires the U.S. Environmental Protection Agency (EPA) to establish drinking water regulations for public water supplies to prevent adverse effects on human health. National Interim Primary Drinking Water Regulations[1] for bacteria, turbidity, ten inorganic constituents (Table 1), six pesticides (Table 2), and radionuclides (Table 3) were promulgated on December 24, 1975, and became effective on June 24, 1977. Amendments[2,3] to these regulations include (1) a maximum contaminant level (MCL) of 0.10 mg/ℓ and associated monitoring and reporting requirements for total trihalomethanes in community water systems of over 10,000 populations and (2) monitoring and reporting requirements for sodium and corrosivity characteristics. With the exceptions of bacteria, turbidity, and nitrate, the MCLs for these constituents are based upon possible adverse health effects associated with chronic exposures. National Secondary Drinking Water Regulations[4] have also been established for contaminants which may adversely affect the aesthetic quality of drinking water through taste, odor, color, and appearance (Table 4).

At the present time, the EPA is undertaking a comprehensive reassessment of the interim regulations directed toward identifying additional drinking water contaminants which should be regulated and proposing revised primary drinking water regulations.[5,6] In establishing revised primary drinking water regulations, the EPA is directed to specify recommended maximum contaminant levels (RMCLs) which are nonenforceable health goals in addition to MCLs, which are the enforceable standards. RMCLs are to be established at a level at which "no known or anticipated adverse effects on the health of persons occur and which allows an adequate margin of safety".[5] Congressional guidance on RMCLs for carcinogens states that the EPA "must consider the possible impact of synergistic effects, long-term and multi-stage exposures, and the existence of more susceptible groups in the population" and that the RMCL "must include an adequate margin of safety, unless there is no safe threshold for a contaminant" in which case the RMCL "should be set at zero level".[5] The MCLs must be established as close to RMCLs as is feasible, "with the use of the best technology, treatment techniques and other means which (EPA) finds are generally available taking costs into consideration".[5] For each contaminant a treatment technique rather than an MCL can be specified, but a treatment technique requirement can be established only if it is not economically or technologically feasible to ascertain the level of a contaminant in drinking water. The revised primary drinking water regulations are to be reviewed every 3 years and amended whenever changes in technology, treatment techniques, or other factors permit greater health protection.

The National Primary Drinking Water Regulations are being developed in several phases, each using a similar approach. For each phase, an advance notice of proposed rule making will be published to present the issues and allow for public comment. RMCLs will then be proposed for public comment followed by the promulgation of RMCLs and the publication of MCLs and monitoring and reporting requirements. The MCLs, monitoring and reporting, and other requirements, including generally available treatment technologies, will then be promulgated. Advanced notices of proposed rule making have been issued for volatile synthetic organic chemicals, synthetic organic chemicals, inorganic chemicals, microbiological contaminants, and radionuclides, and the EPA is now considering the regulation of a number of drinking water contaminants in addition to those currently included in the National Interim Primary Drinking Water Regulations (Tables 5 and 6). The mere inclusion of a particular contaminant for consideration in an advanced notice of proposed rule making, however, does not necessarily mean that regulations will be developed. The disinfection by-products, including trihalomethanes, will be reviewed after adequate experience has been obtained with the interim regulation for total trihalomethanes.[5,6] The EPA has recently

Table 1
MAXIMUM CONTAMINANT LEVELS FOR INORGANIC CONSTITUENTS IN PUBLIC WATER SYSTEMS IN THE U.S.[1]

Contaminant	Level (mg/ℓ)
Arsenic	0.05
Barium	1.0
Cadmium	0.010
Chromium	0.05
Lead	0.05
Mercury	0.002
Nitrate (as N)	10.0
Selenium	0.01
Silver	0.05
Fluoride	1.4—2.4[a]

[a] Depending upon annual average of maximum daily air temperature.

Table 2
MAXIMUM CONTAMINANT LEVELS FOR ORGANIC CHEMICALS IN PUBLIC WATER SYSTEMS IN THE U.S.[1]

Compound	Level (mg/ℓ)
Chlorinated hydrocarbons	
Endrin	0.0002
Lindane	0.004
Methoxychlor	0.1
Toxaphene	0.005
Chlorophenoxys	
2,4-D	0.1
2,4,5-TP Silvex®	0.01

established RMCLs (Table 7) for volatile synthetic organic chemicals in drinking water, and the publication of MCLs, monitoring and reporting requirements is the next step toward establishing regulations for these chemicals.[6]

II. INORGANIC CONSTITUENTS

It is not possible in this chapter to provide a comprehensive summary of the health effects of all the inorganic constituents found in drinking water supplies, and selected constituents based on their historical significance and occurrence are reviewed. The reader is directed toward other references for additional information on this subject.[7-12]

A. Nitrate

The MCL set by the EPA for nitrate was established solely to prevent infantile methemoglobinemia. Since Comly[13] first associated infantile methemoglobinemia with nitrate in drinking water in 1945, some 2000 cases have been reported in North America and Europe.[14] Approximately 8% of these cases resulted in death. The development of infantile methemoglobinemia from nitrate in drinking water depends on the bacterial conversion of nitrate to nitrite before or after ingestion and occurs almost exclusively in infants less than 3 months old. Most reported cases have been associated with drinking water containing more than 22 mg/ℓ NO_3–N, but cases have occurred at concentrations below 10 mg/ℓ NO_3–N.[7,15] Infants are more susceptible to the development of methemoglobinemia, but not all infants develop methemoglobinemia when high-nitrate water is ingested.[15,16] Methemoglobinemia can also occur in children and adults if sufficient nitrite is ingested directly, but the EPA has not established an MCL for nitrite in drinking water. Nitrite, whether ingested directly or produced by bacterial conversion during ingestion, is able to oxidate ferrous iron in hemoglobin to ferric iron and convert hemoglobin, the blood pigment that carries oxygen from the lungs to the tissues, to methemoglobin. Because the methemoglobin is incapable of

Table 3
MAXIMUM CONTAMINANT LEVELS FOR RADIONUCLIDES IN PUBLIC WATER SYSTEMS IN THE U.S.[1]

Contaminant	Level
Radium 226 and 228	5 pCi/ℓ
Gross α-particle activity	15 pCi/ℓ
β-particle and photon radioactivity	4 mrem/year dose equivalent

Table 4
SECONDARY DRINKING WATER REGULATIONS FOR PUBLIC WATER SYSTEMS IN THE U.S.[4]

Contaminant	Level (mg/ℓ)
Chloride	250
Color	15 Color units
Copper	1
Corrosivity	Noncorrosive
Foaming agents	0.05
Iron	0.3
Manganese	0.5
Odor	3 Threshold odor number
Hydrogen ion	6.5—8.5 pH value
Sulfate	250
Total dissolved solids (TDS)	500
Zinc	5

Table 5
MICROBIOLOGICAL CONTAMINANTS AND TREATMENT TECHNIQUES BEING CONSIDERED BY THE EPA FOR INCLUSION IN THE NATIONAL PRIMARY DRINKING WATER REGULATIONS[5]

Coliforms
Standard plate count
Viruses
Giardia lamblia
Legionella
Filtration for surface waters
Disinfection requirement

binding molecular oxygen, the physiologic effect is oxygen deprivation or suffocation. A methemoglobin reductase system normally present in the erythrocytes keeps methemoglobin at about 1 to 2% of the total hemoglobin in healthy adults.[17] At methemoglobin levels of approximately 10%, the methemoglobin produces a slate grey appearance of cyanosis and the disorder becomes clinically detectable.[18]

It is estimated that about 1% of the U.S. population using public water supplies are consuming nitrate in excess of the EPA limit, but few cases of methemoglobinemia have been reported in recent years.[15,16] Methemoglobinemia is not a reportable disease in the

Table 6
CHEMICAL CONSTITUENTS AND RADIONUCLIDES BEING CONSIDERED BY THE EPA FOR INCLUSION IN THE NATIONAL PRIMARY DRINKING WATER REGULATIONS[5]

Radionuclides	Inorganics	Organics
β-Particle and photon radioactivity	Aluminum	Acrylamide
Gross α-particle activity	Arsenic	Adipates
Radium 226 and 228	Asbestos	Alachlor
Radon	Barium	Aldicarb
Uranium	Beryllium	Atrazine
	Cadmium	Carbofuran
	Chromium	Chlordane
	Copper	Dalapon
	Cyanide	Dibromochloropropane
	Fluoride	Dibromomethane
	Lead	1,2-Dichloropropane
	Mercury	Dinoseb
	Molybdenum	Dioxin
	Nickel	Diquat
	Nitrate	Endothall
	Selenium	Endrin
	Silver	Epichlorohydrin
	Sodium	Ethylene Dibromide
	Sulfate	Glyphosphate
	Thallium	Hexachlorocyclopentadiene
	Vanadium	Lindane
	Zinc	Methoxychlor
		PAHs
		Pentachlorophenol
		PCBs
		Phthalates
		Pichloram
		Simazine
		Toxaphene
		Toluene
		1,1,2-Trichloroethane
		Vydate
		Xylene
		2,4-D
		2,4,5-TP

U.S. and its true incidence is not known; however, it appears that the incidence of methemoglobinemia caused by nitrate in public water supplies is extremely small in the U.S.[15] Almost all of the 335 cases of methemoglobinemia associated with high-nitrate drinking water in the U.S. occurred in the 1940s and 1950s, and only 1 case of methemoglobinemia has been reported from a public water supply in the U.S.[19] since 1960. No cases of methemoglobinemia have been reported from public water supplies in California, even though several supplies have exceeded the EPA limit for many years, and no cases of clinical methemoglobinemia were found in two epidemiologic studies[18,20,21] in California and Illinois, where water supplies contained as much as 20 mg/ℓ NO_3–N. Few investigators have described cases associated with nitrate in individual water supplies in the U.S. in the past 10 years.[15,16] A study of hospital records from 1973 to 1978 in Nebraska, where numerous individual wells exceed the nitrate limit, revealed only one case of methemoglobinemia possibly associated with high nitrate in individual well water supplies. A study of records from 1960

Table 7
RECOMMENDED MAXIMUM CONTAMINANT LEVELS (RMCLs) FOR VOLATILE SYNTHETIC ORGANIC CHEMICALS IN DRINKING WATER[6]

Contaminant	RMCL (mg/ℓ)
Trichloroethylene	0
Tetrachlorethylene	0
Carbon tetrachloride	0
1,2-Dichloroethane	0
Vinyl chloride	0
1,1-Dichloroethylene	0
Benzene	0
1,1,1-Trichloroethane	0.2
1,4-Dichlorobenzene	0.75

to 1971 at Children's Memorial Hospital in Oklahoma found only ten cases of methemoglobinemia associated with high nitrates in individual well water supplies, and a survey of 37 states in 1979 revealed only two previously unreported cases of methemoglobinemia associated with individual wells over the previous 10 years.[15,16] Cases of infantile methemoglobinemia associated with nitrates in water supplies, however, continue to be reported on a more routine basis in Europe, suggesting that factors other than nitrate concentration of water are more important in the development of this disease in the U.S. today. Changes in infant feeding practices and health education in high-nitrate areas have likely reduced exposure to high-nitrate water and may be the reason for the few clinical cases of methemoglobinemia reported in areas of the U.S. with high-nitrate water.[15,16]

Limited experiments suggest that neither nitrate nor nitrite act directly as carcinogens in animals, but nitrite can interact with secondary or tertiary amines or amides to form *N*-nitroso compounds, nitrosamines and nitrosamides, many of which have been shown to be carcinogenic in several species of laboratory animals.[22,23] Besides *N*-nitroso compounds, gastric nitrite could lead to the formation of numerous other agents with potential genotoxic activity.[23] Evidence indicates that nitrosation can occur when amines and/or amides and nitrate and/or nitrite are ingested simultaneously. Foods, drugs, cosmetics, pesticides, tobacco, and some occupational exposures contribute significant sources of exogenous amines and amides. In addition, amines such as dimethylamine and pyrrolidine are synthesized endogenously. The extent of in vivo nitrosation is influenced by a number of factors and conditions, including the stomach pH, concentration of nitrate, nitrite, nitrosatable amines and/or amides, and the presence of catalysts and inhibitors.[22] Thiocynate and iodide have been shown to enhance nitrosation reactions, and vitamins C and E have been shown to inhibit the reactions. In adults with normal gastric activity, nitrate ingested from foods and water is converted to nitrite mainly by microflora in the saliva. Approximately 25% of ingested nitrate is secreted into saliva, and 20% of the salivary nitrate is reduced to nitrite.[22] This accounts for most of the nitrite present in the stomach, which is the most likely site for in vivo nitrosation reactions. The average dietary nitrite intake for adults in the U.S. is estimated to be 0.77 mg per person per day, and the average amount of salivary nitrite derived from dietary nitrate is estimated to be 3.5 mg per person per day.[23]

Differing lifestyles and dietary habits cause wide variations in the amount of nitrate ingested by various population groups in the U.S., but in the average diet, vegetables contribute some 87% of the daily intake of nitrate.[22] For infants, some milk may be an important source of nitrate. Fruits and juices contribute about 6% of the average daily intake of nitrate, and drinking water contributes about 3%. However, in areas with high-nitrate drinking water, the contribution from water can be much greater. Data for the Sangamon River area of central Illinois indicate drinking water to be the major source of nitrate from dietary sources, contributing about 68% of total daily intake compared with 28% from vegetables.[22] In other areas of the U.S. where the nitrate concentration in well waters is higher, the contribution of water to the total daily intake is likely to exceed this estimate. Some public water supplies in California and a number of individual wells in Illinois, Kansas, Missouri, Nebraska, Oklahoma, and Texas have a nitrate concentration greater than the EPA limit, and in one county in Texas some wells are found to contain 110 to 690 mg/ℓ NO_3–N.[16] In some areas of the U.S., nitrate levels of some water supplies have increased over the years, and as the concentration of nitrate/nitrite used to preserve meat products has decreased, the contribution of nitrate from water has increased.[23] Hartman[23] estimates that on average, the gastric nitrate load from drinking water has increased in the U.S. from 2% in 1925 to 5% in 1936 and 8% in 1981.

Although the concentration of nitrite in foods varies widely, estimates[22] of the average daily intake of dietary nitrite show about 39% is contributed by ingestion of cured meats, 34% by baked goods and cereals, 16% by vegetables, and less than 2% by drinking water. It is generally agreed[22] that the nitrite concentration in drinking water is low and not an important source of nitrite, but the formation of nitrite resulting from the in vivo reduction of ingested nitrate can be significant in areas where drinking waters contain high concentrations of nitrate.

N-Nitroso compounds are also formed in the environment and have been found in some occupational settings, tobacco smoke, alcoholic beverages, cured meat products, cosmetics, pharmaceuticals, pesticides, air, and water. Nitrosamines have been identified in deionized water, industrial waste water, and well water.[22] In well water with a high nitrate content, N-nitrosodimethylamine and N-nitrosodiethylamine were found at levels less than 0.01 μg/ℓ. Volatile nitrosamines have been shown to be absent in drinking waters of a number of large eastern and midwestern cities.[22] Cigarette smoke contributes the largest amount to the total nitrosamine intake, and cured meat products are the most important dietary source of exogenous N-nitroso compounds.[22] The National Academy of Sciences[22] has estimated that the amount of preformed nitrosamines in the diet of the average person is roughly equivalent to the amount formed in vivo from the intake of nitrate and nitrite. However, for special population groups, such as those ingesting high-nitrate water, the increased intake of nitrate could lead to a corresponding increase in the amount of nitrosamines formed in vivo.

Epidemiologic studies of populations in several different countries have suggested an association between high dietary nitrates and increased incidence of cancer of the stomach and esophagus; however, the evidence from these studies is insufficient to clearly establish the role of nitrate as a causal agent of these cancers. Many of the studies were statistical correlations where confounding factors were not adequately assessed, and some studies pounds were not always determined for the appropriate latency necessary for development of the cancers, and in some studies, other plausible causative agents were also identified. The primary basis for the association is a correlation between estimated nitrate consumption and stomach cancer mortality in 12 countries.[22,23] The U.S., Denmark, Sweden, and Norway have a low mortality and low estimated nitrate consumption; Japan and Romania have a high mortality and high estimated consumption.

Some studies have considered exposures from nitrate in drinking water. Geleperin et al.

(see Reference 22) compared cancer mortality rates in three Illinois communities whose water supplies contained different levels of nitrate. They found no significant correlations, but few details about the cancer patients or the communities were provided. Hill et al.[24] compared differences in stomach cancer mortality rates with the nitrate content of drinking water in two towns in England and found that the town with the higher level of nitrate (average — 20 mg/ℓ NO_3–N) had the higher mortality from this cancer. However, socioeconomic status, an important possible confounding factor, was not considered. Hartman and Gingo[25] found no correlation between the nitrate content of drinking water supplies and gastric cancer mortality in Maryland, Texas, and Missouri. The high incidence of esophageal cancer in the Caspian Littoral of Iran was studied, and villages with different rates of esophageal cancer were surveyed for dietary, work, and personal habits.[22] The average daily intake of nitrate and nitrite was not significantly different for high- and low-incidence areas, and a comparison of the nitrate and nitrite content of water showed no elevation in high-risk areas. Eisenbrand et al. (see Reference 22), however, noted there were intermittently high levels of nitrite in the saliva of children in the high-incidence area of Iran, especially on hot days when water intake may not have been sufficient. Chile has vast natural deposits of nitrate and has used large amounts of these in fertilizers. A study conducted there found no association between stomach cancer and exposure to nitrate in drinking water, which was reported to be well below 20 mg/ℓ NO_3–N. Correa et al.[27] reported geographic correlations of stomach cancer incidence with the nitrate content of well water in Colombia, which has one of the highest rates of mortality from stomach cancer. Even though the epidemiologic associations are considered preliminary and suggestive, the National Academy of Sciences has indicated a need for caution in concluding there is a lack of adverse health effects of chronic exposures to drinking water nitrate at the EPA limit and has recommended additional research on this subject.[7]

B. Water Hardness, Corrosivity, and Sodium

Water hardness is defined as the sum of polyvalent cations dissolved in water and described by the amount of soap required to form lather. Calcium and magnesium are the two principal cations responsible for hardness. Hardness occurs in two forms: carbonate hardness caused by bicarbonates of calcium and magnesium, and noncarbonate hardness resulting from the solution of calcium and magnesium with sulfates, chlorides, and other minor anions. Two major softening methods used in water supplies are (1) chemical precipitation by lime or lime and soda ash and (2) ion exchange of calcium and magnesium with monovalent cations such as sodium.

In the past 5 years, several excellent reviews have been published on the subject of hardness and cardiovascular disease.[8,28-32] More than 50 publications discussing the relationship between drinking water quality and cardiovascular disease have been published since 1957, when Kobayashi[33] first reported a statistical correlation between cerebrovascular disease mortality and the acidity of water supplies in Japan. Most of the associations of drinking water and cardiovascular disease have been obtained from statistical-correlational studies of mortality rates in areas having different water characteristics. In general, the studies have shown a negative association between cardiovascular disease mortality and water hardness, i.e., lower cardiovascular mortality is found in areas where the hardness of drinking water is high. This association is presently considered statistical rather than causal. Although biologically plausible in some respects, the epidemiologic evidence is not specific; individual exposure and possible confounding by individual risk indicators have not been considered, the data are inconsistent, and the associations are relatively weak. In addition, hardness and total dissolved solids have also been found to be negatively associated with numerous other causes of death (including total mortality, cancer mortality, sudden death from other than arteriosclerotic heart disease, cirrhosis, and peptic ulcer) and positively associated with fatal accidents, congenital malformations, and chronic obstructive pulmonary disease. The negative association of water hardness with cardiovascular mortality was reported in most, but not all, studies involving large areas. However, many attempts to find similar negative associations within small geographic areas and in comparisons of communities have failed. It is possible that the correlation of water quality with cardiovascular mortality is actually measuring the correlation of some other geographically related factors such as demographic, socioeconomic, and cultural characteristics. Some studies have adjusted for factors such as altitude, climate, latitiude, humidity, air temperature, and socioeconomic status and found that they have influenced the correlation of water quality with cardiovascular mortality.

In Great Britain, cardiovascular mortality is about 30% higher in soft-water areas compared with hard-water areas, and in the U.S. this figure is about 15%. It has been suggested that this would be the maximum possible effect of a water factor in these countries. Neri and colleagues[34,35] estimated the maximum effect of a water factor in Canada to be 20%. Comstock[36] calculated increased risks due to soft water for total cardiovascular mortality for white males aged 45 to 64 to be 25% in the U.S. and 19% in England and Wales. Sharett et al.[37] recently reported data from 484 cities in the U.S., showing cardiovascular mortality rates to be approximately 5% higher in cities using soft water compared with cities using hard water in the same region when the results were adjusted for demographic variables.

The role of water quality has also received considerable attention in the British Regional Heart Study,[38] which was initiated several years ago to provide fundamental information about the causes of cardiovascular disease and to explain the regional variations in ischemic heart disease and stroke in Great Britain. Findings thus far from the British Regional Heart Study indicate that the association of cardiovascular disease with water hardness cannot be explained entirely by other environmental, social, or personal risk factors, and it appears that in Britain the very soft-water areas may have a cardiovascular mortality excess of about 10% after allowance for climatic and social factors.[39] A clinical study of 7735 men aged 40 to 59 in 24 British towns identified personal risk factors of cigarette smoking and blood pressure as contributing to the geographic variations in ischemic heart disease and suggested that the strength of association of water hardness with nonfatal ischemic heart disease is similar to that with cardiovascular mortality.[39] The effect of water hardness is evident only up to about 170 mg/ℓ total hardness.[38,39] Of 34 water parameters measured in the clinical study, 6 were statistically associated with ischemic heart disease mortality and prevalence: alkalinity, calcium, lead, potassium, silicon, and total hardness. Since alkalinity and calcium are so highly correlated with total hardness, their potential contribution could not be separately assessed. Using multiple regression techniques to assess the association of hardness and standardized cardiovascular mortality ratios for 234 British towns while controlling for climate and socioeconomic factors, Shaper et al.[38] estimated a 3 to 4% increase in cardiovascular mortality for every 100 mg/ℓ decrease in total hardness. A study of changes in water hardness and cardiovascular mortality between 1961 and 1971 in 76 county boroughs of England and Wales showed similar results. The 7735 men included in the clinical study are currently being followed for both fatal and nonfatal cardiovascular events, and results of a prospective follow-up epidemiology study will be available in the future. In assessing the practical relevance of water hardness in the etiology of cardiovascular disease, Pocock et al.[39] stress that it is important to recognize that certain personal risk factors have a much greater influence on health, and changes in smoking behavior and adoption of a prudent diet may offer far greater rewards in the prevention of cardiovascular disease than changes in water hardness.

Zeighami et al.[40] recently reported results of a case-comparison epidemiologic study of 505 white male Wisconsin farmers aged 35 years or older who had died from coronary artery disease or cerebrovascular disease and 854 living comparison farmers without evidence of cardiovascular disease. All participants obtained drinking water from individual unchlorinated wells. The drinking water of those farmers who had died from either of these diseases was found to be softer than that of the comparison group. A variety of potential confounders of the relationship were examined, and none were found to explain the difference in water hardness between cases and the comparison. Of 19 other water parameters measured, cadmium, lead, and zinc levels were found to be higher and carbonate hardness was found to be lower in the drinking water of cases than the comparison group. Similar to findings in the British Regional Heart Study, the effect of hardness was noted only up to 200 mg/ℓ. The levels of cadmium and lead were extremely low in these drinking waters and the statistical difference between the two groups was limited to the percentage of water samples with values above the analytic detection limits. Neither calcium nor magnesium levels were statistically different between the groups, and no statistical differences were found for any other parameter measured, including copper, manganese, nitrate, sodium, fluoride, and potassium. The separation of calcium and magnesium components of hardness was reported to be difficult in this data set, and for the logistic regressions in which both calcium and magnesium were included neither were statistically significant. The relationship of disease status and calcium in drinking water, however, was more consistent than the relationship with magnesium in drinking water. Evidence exists of an association between blood pressure and sodium intake, and it has been speculated that sodium intake may also be related to cardiovascular disease. This particular study offered no evidence of a relationship between coronary artery disease or cerebrovascular disease and sodium in drinking water.[40] Two recent reports,[41,42] however, have suggested that in experimental animals chloride rather than sodium may be responsible for increased blood pressure.

Although additional evidence has been obtained in recent years to provide further support for an association between cardiovascular disease and water hardness, it is, nevertheless, difficult at present to specify which water parameter(s) may be important as a water factor and its role in the development of the disease. Water hardness itself is unlikely to have a direct effect on cardiovascular disease, and the search for a water factor is concerned with identifying the possible beneficial aspects or protective effect of hard waters and the possible detrimental aspects of soft waters. The deficiency of an essential element or excess of a toxic metal may be important. Soft waters are low in calcium and magnesium, and if hard water is artificially softened by ion exchange, it may contain high levels of sodium. In general, soft waters are more corrosive than hard waters and may contain higher levels of metals which have been leached from water piping and plumbing materials. Calcium and magnesium have been shown to play a role in the development of hypertension and other cardiovascular system effects in animals and may be the likely beneficial constituents of hard water. High calcium levels in hard water may reduce or prevent the absorption of toxic metals from food and water, such as lead and cadmium, which have also been associated with the development of hypertension in animals. Most clinical effects of toxic metals appear only in animals on a calcium deficient diet. Hard water may also be beneficial in that it can provide an essential element(s) which is deficient in the diet. The intake of both calcium and magnesium can be below the recommended values for a substantial portion of the U.S. population, and hard water can be an important nutritional source for both calcium and magnesium.[43] Magnesium is important for myocardium function, and low levels of magnesium in tissues appear to occur more often in soft-water areas. Rats maintained for 12 weeks on diets moderately or severely deficient in magnesium showed significant elevations in blood pressure compared to controls.[44] Calcium deficiency has been suggested as a risk factor in human hypertension,[45] and increased calcium intake has been shown to attenuate a sodium-induced blood pressure increase in rats.[46] Revis et al.[47] studied the effects of calcium, magnesium, cadmium, and lead on cardiovascular disease in white Carneau pigeons and found that the presence of lead or cadmium in drinking water significantly increased both blood pressure and the number and size of aortic atheroscleratic plaques; however, the presence of calcium reduced both the tissue levels and cardiovascular effects of cadmium or lead. The presence of magnesium in water did not influence the cardiovascular effects of cadmium or lead during the 24 months of exposure.

Epidemiologic studies conducted to date suggest that a slight increased risk of cardiovascular mortality is associated with soft waters, but additional case-comparison and follow-up epidemiologic studies are required before recommendations can be formulated regarding the artificial softening of water supplies. Ion exchange softening will increase the sodium concentration of drinking water and could adversely affect those individuals on low-sodium diets. Whether it is advisable to remove calcium or magnesium from drinking water must await further research. The corrosion of water piping and plumbing materials can be a source of metals in drinking water, but factors other than hardness are also important in determining corrosion. Concentrations of cadmium, chromium, copper, lead, or zinc in water are not necessarily associated with soft water, and more research is needed to better define relationships between the occurrence of various trace metals and water hardness. There are benefits associated with reducing the corrositivity of drinking water and preventing the leaching of metals from plumbing materials, but it is not clear at this time whether this will also result in decreased cardiovascular mortality.

Additional water parameters which have not heretofore been considered may also be important. For example, preliminary data[48] from experimental animals suggest that chlorination of drinking water may play some role. Carneau pigeons fed a diet containing 80% of the recommended daily allowance of calcium showed higher serum cholesterol levels after 3 months of exposure to 10 mg/ℓ chlorine in water compared with pigeons who drank unchlorinated water. Pigeons on a normal diet with adequate calcium showed no difference in serum cholesterol between those exposed to chlorine and those not exposed to chlorine. Autopsies of the high-cholesterol pigeons suggested they had developed more atherosclerotic plaques.

Lead and cadmium are two toxic metals sometimes found in drinking water because of the corrosion of water piping plumbing materials. In the U.S., lead is found more frequently than cadmium in drinking water systems, primarily because of the wide use of lead service pipe in many of the older cities. The EPA has established a maximum contaminant level (MCL) for both lead and cadmium and is currently reviewing the scientific data to determine if these limits provide for adequate margins of safety.[5]

A survey of 969 U.S. community water systems in 1969 showed the cadmium limit was exceeded in the tap water of only three systems and the lead limit exceeded in 14 systems.[49] The maximum concentration reported for cadmium and lead was 3.94 and 0.64 mg/ℓ, respectively. An average lead concentration of 0.013 mg/ℓ was found. Craun et al.[50] measured the inorganic constituents of tap water from 3834 individual households in 35 areas selected to provide a representative population sample of the U.S. and found detectable levels of cadmium and lead in 35 and 90% of the samples, respectively. The mean concentration of cadmium in each of the areas ranged from 0.0002 to 0.0083 mg/ℓ, with an overall mean for all areas of 0.0014 mg/ℓ; the lead concentration ranged from 0.0015 to 0.0318 mg/ℓ with an overall mean of 0.0077 mg/ℓ. When corrosive water is distributed, however, the percentage of samples exceeding the EPA limits for cadmium and lead is much greater and concentrations are higher. For example, corrosive water in Seattle and Boston was found to be responsible for higher concentrations of certain metals in tap water than were measured in the water delivered to the distribution system. A survey[5] in Seattle found that in 61% of the tap water samples, cadmium was present in higher concentrations than was measured in treated water; 7% of the tap water samples exceeded the EPA limit for cadmium. In a survey in Boston,[5] some 13% of tap water samples showed higher concentrations of cadmium than was found in treated water, but none of these samples exceeded the EPA limit. The range of cadmium concentrations measured in tap water was 0.0007 to 0.0050 mg/ℓ in Boston and 0.0025 to 0.0250 mg/ℓ in Seattle. For lead, 95% of the tap water samples in Seattle and 30% of the tap water samples in Boston were found to have higher concentrations in tap water; 65% of the samples in Boston and 24% of the samples in Seattle were found to exceed the EPA limit for lead.[51] Lead concentrations ranged from 0.080 to 0.321 mg/ℓ in Boston and 0.039 to 0.170 mg/ℓ in Seattle. In a follow-up survey in Boston, where lead concentrations in tap water ranged from 0.013 to 0.151 mg/ℓ and 15% of the samples exceeded the EPA limit,[52] an association was found between lead in drinking water and the blood lead of children and adults.[53] Lead levels in tap water have also been found to be high in the U.K., and surveys have indicated that a significant number of samples exceed 0.100 mg/ℓ, primarily because of storage of water in lead-lined tanks. Beattie et al.[54] found lead levels to range from 0.570 to 3.136 mg/ℓ in tap water in Scotland.

Reducing the corrosivity of water can, however, result in lower concentrations of lead and other metals.[55,56] After pH adjustment of delivered water by sodium hydroxide in May 1978, the average lead concentrations of tap water in Boston were reduced from 0.031 to 0.073 mg/ℓ to below 0.050 mg/ℓ.[57] Water treatment by pH adjustment and phosphate addition in Glasgow and Ayr, Scotland, reduced lead concentrations in water, which was accompanied by significant decreases in blood lead levels of the population.[58]

Excessive lead intake results primarily in adverse effects on hematopoietic, GI, renal, immunological, and central and peripheral nervous systems.[59,60] Infants and young children are more susceptible than adults to the biochemical effects of lead, and recent reports have indicated that lead may have subtle effects on behavior development and learning abilities of young children.[59-63] In some cases, drinking water can be a significant source of exposure for lead, and the National Academy of Sciences has suggested that consideration be given to reducing the maximum contaminant level (MCL) for lead, as the current limit may not provide a sufficient margin of safety for the fetus and young growing children.[5]

The EPA limit for cadmium was based on the intake of cadmium necessary to produce proteinuria, and the contribution of cadmium from cigarette smoking was not taken into account.[5] Neither was the potential carcinogenicity of cadmium considered.[5] Exposures other than drinking water are the primary sources of cadmium. The total daily intake from air, water, food, and tobacco ranges from 0.040 mg/ℓ/day for the rural nonsmoker on a low-cadmium diet to 0.190 mg/ℓ/day for the urban smoker on a high-cadmium diet.[5] Drinking water contributes less than 5% to the total intake of cadmium. The EPA is currently considering revising the MCL for cadmium and has requested public comment on the effects of cadmium on reproductive, nervous, and cardiovascular systems and its possible carcinogenic effects.[5]

Sodium in drinking waters originates from natural and man-made sources, such as from the use of sodium chloride as a de-icing agent on highways, and is sometimes added during water treatment. The concentration of sodium in drinking water from a survey of 2100 water systems which supplied about 50% of the U.S. population ranged from 0.4 to 1900 mg/ℓ.[7] About 42% of the water systems had sodium concentrations greater than 20mg/ℓ and 5% exceeded 250 mg/ℓ; samples collected from 3834 individual households in 35 areas selected to represent a statistical sample of the U.S. population showed an overall mean of 27.7 mg/ℓ sodium, with the range of means for each of the areas being 4.0 to 79.7 mg/ℓ.[50]

The suggested adequate, safe intake of sodium ranges from 1100 to 3300 mg/day for normal adults and 115 to 750 mg/day for infants.[64] Estimates of the sodium intake of U.S. adults range from 1600 to 9600 mg/day, as sodium is added to many foods during processing, cooking, and prior to consumption.[65] Intake from food is generally the major source of sodium, but in instances where high sodium occurs in drinking water, a significant portion of the total sodium intake can be contributed to water.

Evidence suggests that chronic excessive intake of sodium may be associated with adult hypertension in developed countries.[7,8] The size of the U.S. population predisposed to hypertension when exposed to elevated sodium intake is not known, but about 20% of the U.S. population has hypertension. The American Heart Association advocated a sodium-restricted diet for the long-term management of hypertension.[7,8] Sodium-restricted diets are also required in the treatment of congestive cardiac failure, renal disease, cirrhosis of the liver, toxemia of pregnancy, and Meniere's disease, and may be required for patients on prolonged corticosteroid therapy. Approximately 3% of the U.S. population are currently on prescribed low-sodium diets of either 2000, 1000, or 500 mg/day, and the fraction of sodium allowed from drinking water varies according to diet. Since most foods contain some sodium, it is difficult to provide a nutritionally adequate diet without the daily intake of 440 mg of naturally occurring sodium in food. Where water supplies contain more than 20 mg/ℓ sodium, total dietary sodium restriction to less than 500 mg/day is difficult to achieve and maintain. Since some 40% of the U.S. community water supplies exceed 20 mg/ℓ sodium, it is important that patients on low-sodium diets be informed of the sodium concentration in drinking water.

In Britain, the use of water high in sodium for reconstituting powdered infant formula has been attributed to unexpected infant deaths and infant brain damage from hypernatremia, but the evidence is limited and not felt to be conclusive.[66] In several recent epidemiologic studies[67-70] it has been suggested that a slight increase in blood pressure in children and young adults is associated with sodium concentrations in drinking water; however, questions have been raised regarding several methodologic problems of these studies[71] and similar studies[72-74] in other geographic areas have not associated an increase in blood pressure with sodium concentrations in drinking water. Several studies conducted by Tuthill et al.[69,70] and Calabrese et al.[67] have reported statistically significant differences in the mean systolic and diastolic blood pressures of high school sophomores and third graders in a community Massachusetts whose drinking water contained 108 mg/ℓ sodium compared with a neighboring community of similar socioeconomic and demographic characteristics having 8 mg/ℓ sodium in drinking water. Mean elevations of 3 to 5 mmHg in blood pressure among the normotensive children residing in the community with higher sodium were attributed by these investigators, at least partially, to the differences in drinking water sodium levels which represented approximately 25% of the difference in total sodium intake between the two communities. Preliminary studies by Tuthill et al.[69] of other children, seventh graders in Texas, Oklahoma, and Ohio, where drinking water levels of sodium were 25 and 275 mg/ℓ, only provide partial support for the findings in Massachusetts. In these studies, increased systolic blood pressure in females appeared to be associated with higher levels of sodium in water; diastolic blood pressure did not appear to be affected. In males, neither systolic or diastolic blood pressure was higher in the areas with higher water levels of sodium except when a group of students in Texoma with high water sodium at both home and school was compared with a group having low water sodium at both home and school; in this instance, systolic blood pressure was 5 mmHg higher in the group with high water sodium, but no difference was observed for diastolic blood pressure. Hofman et al.[68] examined the relationship between sodium in drinking water and blood pressure in 348 Dutch schoolchildren, aged 7 to 12 years, born and living in three areas with comparable demographic characteristics but different sodium levels in drinking water: 30 mg/ℓ for at least 15 years, 170 mg/ℓ for at least 15 years, and 161 mg/ℓ for 1 year but 23 mg/ℓ before that time. The mean values of systolical diastolic blood pressure for males and females were higher in the high-sodium areas, ranging from about 2 to 4 mmHg after adjustment for various covariates. No differences in blood pressure were noted for children in the area with 170 mg/ℓ sodium for at least 15 years and the area with 161 mg/ℓ sodium for only 1 year, and this suggests that the association may be one of relatively short-term nature. Hallenbeck et al.[73] found mean female and male systolic blood pressures not to be significantly higher in high school juniors and seniors living in a community with 405 mg/ℓ sodium in drinking water compared with a community of similar socioeconomic and demographic characteristics having only 4 mg/ℓ sodium in drinking water. However, the male and female diastolic blood pressures were higher by approximately 2 mmHg in the community with higher water sodium. Faust[72] found no evidence that the level of sodium in drinking water was related to blood pressure in a cohort of 295 rural residents of Livingston County, Mich., where well water sodium ranged from 33 to 296 mg/ℓ, before softening to 142 to 583 mg/ℓ after softening. Studies in Iowa[74] of 2000 schoolchildren in 8 communities whose drinking water sodium levels ranged from less than 10 to greater than 300 mg/ℓ and of adults in 200 households in 4 communities showed differences of less than 4 mmHg in systolic and dystolic blood pressures, but an association between these differences and water sodium levels was not demonstrated.

In reviewing several of the epidemiologic studies which have reported an association between increased blood pressure and levels of sodium in drinking water, Willett[71] has noted several methodologic problems (with which I agree) and urges that these studies be interpretated with caution: "Overall knowledge strongly suggests that large changes in sodium intake (e.g., several grams per day) do have an effect on blood pressure in at least some individuals, and it is therefore probable that even the relatively small increments in sodium intake (e.g., several grams per day) do have an effect on blood pressure in at least some individuals, and it is therefore probably that even the relatively small increments in sodium intake attributable to certain water supplies have some effect on blood pressure. However, the studies of Tuthill and Calabrese provide little evidence relating to the magnitude of such an effect. Differences in blood pressure attributable to drinking water sodium are likely to be substantially less than the observed differences between the two communities they have studied, and may be undetectable by epidemiologic methods". A more recent experimental study conducted by Calabrese and Tuthill[75] among fourth graders has shown a decrease in blood pressure levels among females, but not males, who received low-sodium bottled water for 3 months. During this period, children matched by gender, school, and baseline blood pressure were randomly allocated to one of three groups receiving (1) high-sodium water bottled from their own community supply, (2) low-sodium water bottled from a comparison community, or (3) high-sodium water manufactured to correspond with sodium levels of the community with higher sodium levels.

Sodium is probably the most difficult and one of the most expensive substances to remove from drinking water.[5] The EPA currently requires only the monitoring and reporting of sodium in public water supplies, but has requested public comment on the need for an MCL for sodium.[5] In attempting to assess the possible adverse health effects of sodium, it is important to evaluate recent studies of experimental animals which suggest that chloride rather than sodium may be responsible for increased blood pressure. The recent epidemiologic studies of the possible association between blood pressure and drinking water sodium must also be critically evaluated. Although the possible effect of small increases of a few mmHg in blood pressure is not likely to be thought of as clinically important for an individual, elevations of this magnitude can be translated into a substantial number of excess cardiovascular deaths and need to be considered as a potential public health problem.[71] The National Academy of Sciences[7,8] feels that sodium should be maintained at the lowest practical levels in water and that practices such as de-icing of highways and water softening, which increase the concenterations of sodium in water supplies, should be discouraged. It has been estimated that 40% of the U.S. population would benefit from a total daily intake of less than 2000 mg of sodium, and the contribution of water to this desired total intake of sodium would be 10% or less for concentrations of 100 mg/ℓ or less in drinking water.[7,8]

C. Arsenic and Selenium

Natural sources account for much of the arsenic and selenium found in drinking water supplies, although the presence of arsenic in soils can be related to its use as a pesticide.[2,7,8]

In the U.S., the arsenic concentration of drinking water ranges from a trace to approximately 0.1 mg/ℓ with higher levels of up to several milligrams per liter found in wells in Indiana, Utah, Alaska, Oregon, and California.[2,7,8,16] Wells in Nova Scotia and the southwest coast of Taiwan have also been found to contain high levels of arsenic.[16] Ground waters may contain significant amounts of selenium in areas where there is an excess of selenium in rocks and soils, and selenium can reach toxic concentrations in wells drilled through seleniferous shales. In seleniferous areas of Wyoming, wells contain enough selenium to be poisonous to man and livestock; high selenium concentrations are also found in wells near Denver, Colo. and 0.210 mg/ℓ selenium was reported in well water on a South Dakota Indian reservation.[7] Water consumed by most populations in the U.S., however, is relatively low in selenium and rarely constitutes a significant source of selenium.[7,8]

Arsenic affects tissues that are rich in oxidative systems, primarily the alimentary tract, kidneys, liver, lungs, and epidermis, and is damaging to capillaries, resulting in hemorrhage into the GI tract, sloughing of mucosal epithelium, renal tubular degeneration, hepatic fatty changes, and necrosis.[7,8] Epidemiologic studies have suggested an association between chronic arsenic exposure and cutaneous lesions, cardiovascular disease, and cancer.[7,8,16] In Taiwan, an increased prevalence of skin cancer, hyperpigmentation, and keratotosis and a higher rate of a peripheral vascular disorder resulting in gangrene of the extremities were seen in a population consuming well water with a 0.6 mg/ℓ or greater concentration of arsenic.[76,77] Residents of Antofagasta, Chile where the drinking water contained 0.8 mg/ℓ arsenic, showed an increased incidence of bronchial and pulmonary disease, cardiovascular pathology (Raynaud's syndrome), hyperpigmentation of the skin, and cutaneous lesions such as leukoderma, melanoderma, hyperkeratosis, and squamous cell carcinoma.[78-80] After water treatment which reduced arsenic levels to 0.08 mg/ℓ, the incidence of cutaneous lesions in Antofagasta decreased from 313 to 19 per 100,000 people per year. High incidences of human skin cancer have also been reported in Teichenstein, Silesia, and Cordoba, Argentina, where populations have been exposed to high levels of arsenic in drinking water.[7] A lifetime probability of 10^{-5} has been estimated as the risk of nonmelanoma skin cancer for a lifetime ingestion of drinking water containing 25×10^{-6} mg/ℓ arsenic.[81] This risk estimate was based primarily upon the epidemiologic studies conducted in Taiwan,[76,77] and Andelman and Barnett[82] have considered whether epidemiologic studies could be conducted in the U.S. to confirm this risk estimate. Populations in the U.S. exposed to arsenic in water were found to have prevalence or incidence rates which were either too small to confirm or predicted to be too small to confirm the Taiwan-based model for risk, and it was felt that it would be unlikely that a water or occupationally exposed population could be found in the U.S. to provide sufficient statistical power for an epidemiologic study to test the model.[82]

Except for limited reports of acute industrial or accidental exposures, effects of selenium exposure for humans must be extrapolated from animal data.[7,8] Inorganic and organic forms of selenium are readily absorbed and are distributed largely to the liver, kidneys, muscle mass, GI tract, and blood. Selenium is essential for domestic animals, but the margin of safety is relatively narrow.[7,8] A low level of selenium is essential to prevent myopathies, liver injury, and congenital abnormalities in domestic and laboratory animals and poultry. There is evidence that selenium may be an essential trace element for humans, and an estimated adequate and safe intake of selenium for adults is felt to be between 0.500 and 0.200 mg, with lower intakes for children and infants.[8] The National Academy of Sciences[8] feels that the current EPA limit of 0.01 mg/ℓ barely provides a minimum nutritional amount of selenium assuming consumption of 2 ℓ/day of water at the limit and suggests re-evaluation of the current limit based on minimum requirements of selenium. There has been concern that selenium is a carcinogen, but the National Academy of Sciences[7,8] does not feel this concern is justified by current scientific evidence. Some epidemiologic studies report a lower incidence of cancer in areas with high selenium and a possible protective effect of selenium has been noted.[83]

D. Fluoride

An extensive survey in 1969 showed that 8.1 million people in 2630 communities in the U.S. were receiving water which contained more than 0.7 mg/ℓ of naturally occurring fluoride; 1 million people in 524 communities received water with more than 2 mg/ℓ fluoride.[84] Fluoride concentrations from 0.7 to 1.2 mg/ℓ in drinking water, depending on the environmental temperature, are generally felt to have a beneficial effect by preventing dental caries in children, and many communities add fluoride to water supplies for this reason.[3,7,8] The scientific data supporting these recommendations are contained in an excellent reference edited by McClure.[85]

Ingestion of excessive fluoride in drinking water can result in mottling of teeth and dental fluorosis in children, and increased density and calcification of bone (osteosclerosis) and skeletal fluorosis in adults.[7,8,16] In deciding upon the maximum acceptable limit of fluoride in drinking water, it is necessary to distinguish between the effects of severe and less severe fluorosis. Severe dental fluorosis generally results in macroscopic defects that disrupt the integrity of the enamel; less severe forms occur as blemishes of intact enamel that impair the appearance of teeth. A recent study in Texas[86] reported that objectionable mottling occurred at 2.3 times the currently accepted optimum water fluoride level, and a recent study in Illinois[87] reported that communities which wished to minimize the occurrence of obvious cosmetic changes should provide less than two times the optimum level. In Illinois[87] only at four times the optimum fluoride level did dental fluorosis increase appreciably; at twice the optimum level fluorosis consisted almost entirely of cosmetic changes, and at three times the optimum level the fluorosis was still mostly cosmetic.

Osteosclerosis has been observed in some populations after the long-term consumption of water containing 4 to 8 mg/ℓ fluoride, but increased bone density has been regarded as a beneficial rather than adverse effect.[3,7,8] Crippling skeletal fluorosis may result from the long-term ingestion of fluoride in amounts greater than 20 to 40 mg/day.[3,7,8] Epidemiologic studies in areas where drinking water is naturally high in fluoride have found no adverse effects except in rare cases, where the concentration is many times more than that recommended, and controlled studies with fluoridation at the 1 mg/ℓ level have reported no adverse effects. Studies reporting effects such as mongolism, cancer mortality, mutagenicity, and birth defects are felt to have been inadequately designed or found lacking substance.[7,8] Two recent studies and reviews which do not support the suggestion of an association between artificially or naturally fluoridated water and cancer mortality are reported by Hoover et al.[88] and Chilvers.[89]

E. Asbestos

Asbestos generally refers to naturally occurring silicate minerals which have a fiberous morphology and commercial utility.[7] Based on crystal structure, the asbestos minerals can be separated into two groups: the serpentine group or chrysotile and the amphibole group, which includes amosite, anthophyllite, actinolyte, tremolite, and crocidolite.[90] Asbestos fibers are found in drinking waters from the erosion of natural mineral deposits, industrial pollution, and the degradation of asbestos cement used to transport aggressive or corrosive water. Over 100 billion asbestos fibers per liter have been found in some northern California water supplies from the natural erosion of serpentine minerals in the watershed.[91] Naturally occurring chrysotile fibers in concentrations of 200 million/ℓ were measured in drinking water samples from Everett, Wash.,[92] and up to 36 million/ℓ from natural sources were found in various water supplies in the San Francisco Bay Area of California.[93] Chrysotile has also been found in water stored in cisterns.[91] Crocidolite fibers were found in ground water in New Jersey, where local bedrock contained crocidolite.[91] The water supply of Duluth, Minn., was found to contain 1 to 65 million amphibole fibers per liter due to the industrial contamination of Lake Superior with taconite tailings.[94] Chrysotile fibers leached from the interior of asbestos cement pipe were found in concentrations of up to 700,000/ℓ in Connecticut[90] and 33 million/ℓ in Pensacola, Fla.[95] Residential tap water samples collected from 538 cities in the U.S. showed that asbestos fibers were detectable in 65% of the samples, with 9% of the samples having a concentration of over 10 million fibers per liter.[91] The length of the vast majority of asbestos fibers found in drinking water was less than 5 μm.[91] Consideration of the population served by the sampled water supplies suggested that less than 5% of the population of the U.S. receives drinking water containing more than 10 million fibers per liter.[91]

Numerous epidemiologic studies have associated the occupational exposure of asbestos with asbestosis, lung cancer, and pleural and peritoneal mesothelioma, and some cases of mesothelioma have been found among wives of asbestos workers and among persons living near mines, factories, and shipyards.[7,91] Excess mortality from cancers of the stomach, colon, and rectum have been noted in asbestos workers,[90] and it has been suggested that cancers occur at remote sites following the inhalation of asbestos because fibers are cleared from the lungs, swallowed, and penetrate the GI wall.[96] Since reports of finding asbestos fibers in drinking waters and beverages in the early 1970's, there has been increasing concern over the possible adverse effects associated with the ingestion of asbestos fibers. A workshop to review the research studies and summarize the current state of knowledge on this subject was held on October 13 and 14, 1982 at the Andrew Breidenbach Environmental Research Center in Cincinnati, Ohio, and the reader is directed to the peer-reviewed proceedings of this workshop which was published in *Environmental Health Perspectives,* Volume 53, November, 1983. A brief discussion of the important conclusions of this workshop are presented here.

Although lung cancer and mesothelioma have been reported in experimental animals caused by inhalation of various types of asbestos,[96] adverse effects in similar animal species have not been demonstrated after the ingestion of asbestos. Two recently completed feeding studies reported at the workshop showed no carcinogenic effects and no adverse effect on survival after the lifetime ingestion of 1% amosite, short-range chrysotile, or intermediate-range chrysotile in the diet of male and female Syrian golden hamsters[97] and of 1% tremolite or amosite in the diet of male and female Fischer 344 rats.[98] A review[96] of studies published prior to the workshop concluded that the long-term, high-level ingestion of various types of asbestos fibers, including amosite, chrysotile, crocidolite, talc, tacontie tailings, and Lake Superior water, failed to produce any definite reproducible, organ-specific carcinogenic effect in experimental animals.

A number of epidemiologic studies have been conducted in the U.S. and Canada, where populations have been exposed to various concentrations and types of asbestos in drinking water, and these were evaluated by Marsh[93] for weaknesses and limitations. Although one or more of these epidemiologic studies found an association between asbestos in drinking water and cancer incidence or mortality due to various cancers, the geographic correlational design of these studies does not permit a definitive conclusion to be made regarding the possible adverse effects of ingested asbestos because data are not obtained for each individual on exposure status and disease outcome and the possible confounding factors cannot be controlled on an individual basis. Several of the studies reported no association between cancer mortality or incidence and asbestos in drinking water, but limitations such as misclassification of exposure, insufficient latency periods, low duration or intensity of exposure, and insensitivity of the summary statistics were identified.[93] Data on individual exposure, disease outcome, and possible confounding factors were obtained by Polissar et al.[99] in a case-comparison epidemiologic study conducted in the Everett, Wash. area, where unusually high concentrations of chrysotile fibers were identified in the drinking water and had likely been present for at least 60 years. Logistic regression analysis was used to estimate cancer risk associated with cumulative asbestos exposure estimated from individual water consumption and residence-workplace history for 382 individuals with cancer of the buccal cavity, pharynx, respiratory system, digestive system, bladder, or kindeys identified from a population-based tumor registry and 462 population-based comparison subjects without identified cancers; covariates included age and smoking among other risk factors. Significantly elevated risks were found only for male stomach and male pharyngeal cancer. It was concluded because these risks were gender specific and based on small numbers that factors other than asbestos in water were likely responsible, and the study provided no convincing evidence for an increased cancer risk from ingested asbestos.[99] It was concluded by Erdreich;[100] that none of the epidemiologic studies of populations exposed to asbestos in drinking water are able to provide quantitative data for estimating levels of asbestos associated with a defined risk nor can they be used to identify clearly safe or unsafe ranges; however, it was noted that the studies suggested the risk associated with the ingestion of asbestos from water is not greater than has been estimated from inhalation studies and may be less. Assuming ingestion of 2 ℓ of water per day and 70 years of exposure, the ambient water level, estimated from three large occupational cohorts exposed for a working lifetime to airborne asbestos necessary to keep excess cancer risk below $1/10^{-5}$, would be 0.3 million fibers per liter.[100] In Canada it was concluded that the risk to health associated with the ingestion of asbestos at levels found in municipal drinking water supplies was so small it could not be detected by currently available epidemiologic techniques.[101]

Regarding future regulatory decisions for the control of asbestos in drinking water, Cotruvo[102] noted that although a GI cancer risk is reported from occupational exposures, the existence of a risk associated with asbestos in drinking water has not been satisfactorily demonstrated, as experimental animal studies have failed to show toxicity via ingestion and the epidemiologic evidence of risk from ingestion of water containing asbestos is not convincing. "Whether or not there is a risk from asbestos in drinking water, however, common sense tells us to deal with an undesirable situation by employing means that are commonly and economically available. Well-known methods can minimize the presence of asbestos fibers in finished drinking water. In the case of natural fiber in raw water, standard or augmented filtration practices are extremely effective. If the source of asbestos fiber is asbetos-cement that is being attacked by corrosive water, then, there is more than sufficient economic reason to correct the corrosivity of the water."[102] Asbestos in drinking water is currently under consideration for regulation[5] by the EPA, as it is estimated that "substantial populations" are exposed to more than the current ambient water level of 0.3 million fibers per liter.[102] The ambient water level for asbestos is based on a scientific assessment of the level of a pollutant estimated to result in a defined risk for a carcinogen as outlined in the Ambient Water Quality Criteria Document for Asbestos and is intended as a guide for regulatory agencies in establishing regulations.[100]

III. ORGANIC CONTAMINANTS

Organic contaminants in drinking water and their possible adverse effects have received increased attention over the past decade, as improvements in analytical techniques and increased sensitivity have allowed more and more organic compounds to be detected at lower levels. In 1974, the EPA reported over 100 organic compounds in drinking water, primarily organic solvents and industrial chemicals in surface waters; by 1979, over 700 organic compounds had been identified in drinking water supplies.[103] In recent years a number of volatile synthetic organic compounds have been identified in ground waters. Because volatilization is restricted in ground waters, concentrations are often higher in ground water supplies than in surface waters receiving industrial effluents.[16]

Approximately 90% by weight of the volatile organic contaminants thought to be present in drinking waters have been identified; however, these contaminants represent only about

10% of the total organic content of water.[104] Of the remaining organic compounds, it is generally felt that only 5 to 10% have been identified.[104] Concentrates of organics from treated drinking waters have been found to be mutagenic in several in vitro systems, but because most volatile compounds are lost during the concentration process, these tests apply primarily to the novolatile compounds. Of the volatile synthetic organic contaminants identified in ground waters, many have been found to be carcinogenic in humans and one or more animal species, but most have not been tested for carcinogenicity.[16]

In addition to the synthetic organic contaminants found in drinking water, volatile chlorinated organic compounds have been found to result from the disinfection of drinking water with chlorine. In 1974, Rook[105] and Bellar et al.[106] reported that chloroform and other trihalomethanes (THMs), such as bromodichloromethane, dibromochloromethane, and bromoform, are formed during the chlorination of drinking water containing bromide and natural organic precursors such as humic acids. Page and Saffioti[107a] subsequently reported an increase in liver and kidney tumors among experimental animals exposed to high doses of chloroform.

A third event which called further attention to the potential problem of organic contaminants in drinking water was a study of cancer rates in Louisiana by Harris (see Reference 8) in 1974. Harris found higher cancer rates among populations using the Mississippi River as a drinking water source compared with populations using other water sources and concluded that chemical pollutants in the Mississippi River were responsible. Although this conclusion was not widely accepted in the scientific community at the time, the study was responsible for stimulating a number of statistical and epidemiologic studies of the possible association between cancer and drinking water quality. Most of the subsequent studies used indirect measures of water quality and compared populations using surface and ground water or chlorinated and unchlorinated water. It was generally assumed that ground waters were less contaminated by industrial discharges than surface waters, but this was later found not to be true in all instances.

This discussion of organic drinking water contaminants must necessarily be brief, as there is considerable research currently being conducted in this area. The two major issues to consider are by-products of chlorination and volatile synthetic organic contaminants.

A. By-Products of Chlorination

Under certain conditions, the reaction of free chlorine with selected precursor substances in drinking water produces a group of halogen-substituted single-carbon compounds which are named as derivatives of methane.[107] Methane, however, does not seem to be involved in the reaction. The important precursors which react with free chlorine to produce these compounds are naturally occurring aquatic humic substances such as humic and fulvic acids and bromide rather than synthetic organic compounds from industrial sources. Under typical circumstances, the predominant compounds produced in drinking water are chloroform and bromodichloromethane, but dibromochloromethane and bromoform are frequently found.[107] The arithmetic sum of the concentration of these four compounds has been defined as total trihalomethanes (TTHMs). The concentration of TTHMs formed is dependent upon the presence and concentration of the precursors, water temperature, pH, chlorine dose, and contact time.[107] Chlorination of both ground and surface waters produced TTHMs, but the concentration of TTHMs is many times higher in surface water supplies.[108] Disinfection with chloramines does not cause the formation of TTHMs.[107]

Surveys of water utilities in the U.S. and results of studies in other countries have shown that the reaction of chlorine to produce TTHMs is widespread and has likely been occurring for as long as chlorine has been in use as a disinfectant. Chloroform is the compound most often formed and has been found in concentrations of several hundred micrograms per liter.[108] Chloroform has been shown to be a carcinogen in mice and rats at high dose levels, and it has been estimated that an incremental risk of 3 to 4 cancers per 10,000 population may be associated with the consumption of 2 ℓ of water containing 0.10 mg/ℓ chloroform daily for 70 years.[107]

Since 1974, a number of statistical and epidemiologic studies have been conducted to assess the relationship between chlorinated drinking water and cancer, and these studies have been extensively reviewed.[8,109-113] The epidemiology subcommittee of the National Academy of Sciences (NAS) Safe Drinking Water Committee reviewed[8] in 1980 all of these studies conducted through 1978. These early studies were descriptive in nature, using indirect measures of exposure and comparing cancer rates among populations receiving surface or ground water with and without chlorination. Several studies determined THM concentrations as a measure of exposure status. Nine of ten descriptive studies found associations between indirectly characterized water quality (e.g., chlorinated surface water) and cancer rates (incidence or mortality) of several sites, and three additional studies found associations between cancer mortality and present-day THM concentrations. The studies which provided information on THM concentrations were felt to be more useful but were still of limited value, as it was recognized that a single measurement of water quality may not provide a reliable estimate of potential exposure over time. Since all of the early studies were descriptive, information was available on exposure and disease for groups of people rather than individuals and doubts remained as to whether the observed associations resulted from drinking water exposures or from confounding factors which had not been taken into account. Although some demographic variables were considered, it is not possible to adequately control for the effects of potential confounding factors in descriptive studies, and these early studies are felt to be most useful in developing hypotheses to be tested in analytic epidemiologic studies where information on exposure, disease, and potential confounding factors are available for individuals. The NAS[8] noted that "the bladder, stomach, large intestine, and rectum, which were cancer sites identified in a number of geographic areas, warrant further study" and recommended that analytical epidemiologic studies be conducted where a quantitative measure of the association can be obtained. The two basic types of analytical studies are case-comparison and cohort. In a case-comparison study, individuals enter the study on the basis of disease and nondisease status, and various exposures are determined. In a cohort study, individuals enter the study on the basis of exposure status. In a prospective cohort study, the disease has not occurred at the time the exposed and nonexposed groups are defined. In a retrospective cohort study, the disease has occurred at the time the exposed and nonexposed groups are defined. In a case-comparison study, the proportion of exposed individuals in a diseased group is compared with the proportion of exposed individuals without the disease, and any number of exposures can be assessed for each disease studied. In a cohort study, disease rates in an exposed group of individuals are compared with disease rates in an exposed group; any number of various diseases can be assessed but only for a single exposure classification.

More recently, a variation of the traditional case-comparison study using data available from death certificates for decedent cases and comparison subjects has been used to obtain information on associations of cancer and water quality.[111,112] Advantages of this approach include considerable cost savings, as no interviews are conducted, but the disadvantages include the loss of important information on residence history, exposure, and confounding factors. Reviews of five studies of this type completed through 1981 concluded that information provided by the studies strengthened the evidence for an association between rectal, colon, and bladder cancer and water quality, but were not sufficient to establish a causal relationship between chlorinated by-products in drinking water and cancer. Because of methodological limitations and the small increased risks observed in these studies, it was not possible to separate possible associations with potential confounding factors which could not be assessed or controlled.

The most recent studies have employed traditional epidemiologic designs and have been reviewed by the author with a discussion of the basic considerations for evaluating and interpreting results of epidemiologic studies. A brief summary is provided here and readers interested in a more detailed discussion are referred to the complete text.[113]

"An epidemiologic study is nothing more than an attempt to describe nature, and it is essential in epidemiologic studies to collect and analyze data in a comparable manner. The epidemiologist seeks to make a quantitative statement about the association between exposure and disease, and rates of similar types can be compared. The basic measures are the rate ratio or relative risk and rate difference or attributable risk, and they are generally expressed as a point estimate (a single value) or as an interval estimate (a range of possible values consistent with the data). The appropriateness or accuracy of this measure is dependent upon components of study design, data collection, and analysis of epidemiologic data. These must be evaluated for each study, and the primary considerations are: 1) precision or lack of random error, and 2) validity or lack of systematic error. Precision is influenced primarily by the size of the study population and the efficiency of information obtained for each individual. An interval estimate is preferable to a point estimate because it accounts for random variability in the data. A point estimate is usually guaranteed to be wrong because it is only one point on an infinite scale of values. A point estimate does serve as an anchor point for a given confidence interval and is useful in that context. Assessing the validity of epidemiologic associations requires a search for potential sources of systematic bias which might have influenced the observed results. Both internal and external validity are important. Internal validity involves the making of a valid inference about the association between exposure and disease in a particular study and must be assured before considering external validity. External validity concerns extending the results of several studies to a target population and is often referred to as scientific generalization or interpretation. The results of epidemiologic studies must be interpreted in the context of other information, and no single study is likely to provide a definitive answer. The interpretation of epidemiologic data requires an awareness by both epidemiologists and non-epidemiologists of the potential shortcomings of such data, and an assessment of the internal validity of each study must be undertaken. Internal validity has three basic components which should be considered:

1. Validity of selection of subjects (selection bias)
2. Validity of information (information or observation bias)
3. Validity of comparison (confounding bias)

Random misclassification must also be considered, as an inaccurate definition of disease or of exposure that applies equally to both groups can only alter the results of a study toward no association between exposure and disease."

In papers prepared for the Fifth Conference on Water Chlorination: Environmental Impact and Health Effects, Cantor[114] reported an association between bladder cancer and use of a chlorinated surface water. Cragle[115] reported an association between the incidence of colon cancer and the home consumption of chlorinated water. In a case-comparison study of 2982 incident cases of bladder cancer and 5782 population-based comparison subjects from ten areas of the U.S., a relative risk of 2.3 (confidence interval = 1.3, 4.2) was observed by Cantor in nonsmokers who had used chlorinated surface water for 60 or more years, compared to nonsmokers who had never used surface or chlorinated water. In a case-comparison study of 200 incident cases of colon cancer and 407 hospital-comparison subjects among North Carolina white residents, Cragle observed odds ratios of 1.38, 2.15, and 3.36 for home consumption of chlorinated water for 16 or more years and colon cancer in 60, 70, and 80 year olds. These studies are suggestive of an association between bladder and colon cancer and consumption of chlorinated water over a long period of time in nonsmokers and for a moderate period of time in an elderly population, but additional case-comparison studies are required before conclusions can be extended to other populations. Retrospective cohort studies should also be conducted where suitable cohorts can be assembled and followed. Both of these study designs, however, require an accurate assessment of past exposure to contaminants of interest for a sufficient length of time (which may be greater than 20 years) to minimize random misclassification.

B. Volatile Synthetic Organic Contaminants

Community and individual ground water supplies in the U.S. have been contaminated by synthetic organic compounds, often in concentrations much higher than are found in the

Table 8
OCCURRENCE IN GROUND WATER OF VOLATILE SYNTHETIC ORGANIC CONTAMINANTS PROPOSED FOR REGULATION BY THE EPA[6,16]

	EPA Survey[6]			State surveys[6,16]	
	Concentration		Percent of systems with detectable concentration	Maximum concentration ($\mu g/\ell$)	Percent of systems with detectable concentration
	Median ($\mu g/\ell$)	Range ($\mu g/\ell$)			
Benzene	3.0	0.5—15	0.6	330	1
Vinyl chloride	1.1	1.1	0.2	380	7
Trichloroethylene	1.0	0.2—78	6.4	510,000	15
1,1,1-Trichloroethane	0.8	0.2—18	5.8	5440	2
p-Dichlorobenzene	0.7	0.5—1.3	1.3	1.3	NA
1,2-Dichloroethane	0.6	0.5—1	0.6	2100	7
Tetrachloroethylene	0.5	0.2—23	7.3	1500	17
Carbon tetrachloride	0.4	0.2—16	3.2	1300	14
1,1-Dichloroethylene	0.3	0.2—6.3	6.3	280	NA

Note: NA = not available.

most contaminated surface waters used for water supplies. In the U.S., 24 states have reported finding one or more synthetic organic compounds in community ground water supplies and large-scale contamination of ground water has been reported in New Jersey and California.[16] The data available from state surveys and monitoring programs on the occurrence of organic contaminants in ground waters has been compiled (Table 8), but may not be representative of the extent to which ground waters across the U.S. may be contaminated. Limitations to interpreting the state data include differences among states of available funds to conduct surveys and monitor wells and differences in analytical technology to detect and measure organic contaminants. It is also likely that wells with suspected or known contamination problems were preferentially sampled in the state surveys. A random survey[6] by the EPA of 466 ground water supplies has provided data on the national occurrence of some 29 organic compounds; tetrachloroethylene, trichloroethylene, and 1,1,1-trichloroethane were the compounds most frequently detected (Table 8).

Based on the frequency and level of occurrence, population exposed, potential health aspects, and analytical ability to detect the contaminant, the EPA has proposed RMCLs for nine volatile synthetic organic compounds (Table 7). Seven of these compounds are considered carcinogens and the RMCL is specified as zero: tetrachloroethylene, trichloroethylene, carbon tetrachloride, 1,2-dichloroethane, vinyl chloride, benzene, and 1,1-dichloroethylene. Alternative approaches to selecting an RMCL for these contaminants included setting the RMCL at a non zero level based on a calculated negligible contribution to lifetime risk or setting the RMCL at the analytical detection limit, and these are discussed in detail in the proposed rule making.[6] It was noted that regardless of which alternative was selected for establishing the RMCL, the resulting MCL for a particular substance would not likely be affected, since all approaches would result in "targets that are likely to be below levels that are technically and economically feasible using available technologies."[6] Preliminary analyses have indicated that the MCLs for most of these volatile organic compounds may range from 5 to 50 $\mu g/\ell$.[6] Trichloroethylene, tetrachloroethylene, and 1,1-dichloroethylene have limited animal evidence of carcinogenicity because of either lack of replication in multiple experiments or multiple species or various study limitations, and this must also be taken into consideration.[6] The proposed RMCL for 1,1,1-trichloroethane is derived from a calculated adjusted acceptable daily intake of 1.0 mg/ℓ assuming 20% contribution from drinking water, but a preliminary report on the carcinogenicity of this compound is under review and may result in a revision of the RMCL.[6] The proposed RMCL for p-dichlorobenzene is derived from a calculated adjusted acceptable daily intake of 3.75 mg/ℓ assuming 20% contribution from drinking water to total exposure.[6] Additional data on the occurrence and health effects of these compounds are available in various references.[6-8,16]

REFERENCES

1. National Interim Primary Drinking Water Regulations, U.S. Environmental Protection Agency, EPA-570/9-76-003, Washington, D.C., 1977.
2. National interim primary drinking water regulations; control of trihalomethanes in drinking water; final rule, 40 CFR part 141, *Fed. Regist.*, 44 (231), 68624, 1979.
3. Interim primary drinking water regulations; amendments, 40 CFR Part 141, *Fed. Regist.*, 45 (168), 57332, 1980.
4. *National Secondary Drinking Water Regulations*, U.S. Environmental Protection Agency, EPA-570/9-76-000, Washington, D.C., 1979.
5. National revised primary drinking water regulations; advance notice of proposed rulemaking, 40 CFR part 141, *Fed. Regist.*, 48 (194), 45502, 1983.
6. National primary drinking water regulations; volatile synthetic organic chemicals; proposed rulemaking, 40 CFR part 141, *Fed. Regist.*, 49 (114), 24330, 1984.
7. Committee on Drinking Water, *Drinking Water and Health*, Vol. 1, National Academy of Sciences, Washington, D.C., 1977.
8. Committee on Drinking Water, *Drinking Water and Health*, Vol. 3, National Academy of Sciences, Washington, D.C., 1980.
9. **Friberg, L., Nordberg, G. F., and Vouk, V. B., Eds.**, *Handbook on the Toxicology of Metals*, Elsevier/North-Holland, Amsterdam, 1979.
10. National Research Council, *Geochemistry and the Environment*, Vol. 1, National Academy of Sciences, Washington, D.C., 1974.
11. National Research Council, *Geochemistry and the Environment*, Vol. 2, National Academy of Sciences, Washington, D.C., 1977.
12. National Research Council, *Geochemistry and the Environment*, Vol. 3, National Academy of Sciences, Washington, D.C., 1978.
13. **Comly, H. H.**, Cyanosis in infants caused by nitrate in well water, *JAMA*, 129, 112, 1945.
14. **Walton, G.**, Survey of literature relating to infant methemoglobinemia due to nitrate-contaminated water, *Am. J. Public Health*, 41, 986, 1951.
15. **Craun, G. F.**, An alternative for meeting the nitrate standard for public water supplies, *J. Environ. Health*, 44, 20, 1981.
16. **Craun, G. F.**, Health aspects of groundwater pollution, in *Groundwater Pollution Microbiology*, Bitton, G. and Gerba, C., P., Eds., John Wiley & Sons, New York, 1984, 135.
17. Infant methemoglobinemia: the role of dietary nitrate, *Pediatrics*, 46, 475, 1970.
18. **Winton, E. F., Tardiff, R. G., and McCabe, L. J.**, Nitrate in drinking water, *J. Am. Water Works Assoc.*, 63, 95, 1971.
19. **Vigil, J., Warburton, S., Haynes, W. S., and Kaiser, L. R.**, Nitrates in municipal water supply causes methemoglobinemia in infant, *Public Health Rep.*, 80, 1119, 1965.
20. **Goldsmith, J. R., Rokaw, S. N., and Shearer, L. A.**, Distributions of percentage methemoglobin in several population groups in California, *Int. J. Epidemiol.*, 4, 207, 1975.
21. **Shearer, L. A., Goldsmith, J. R., Young, C., Kearns, O. A., and Tamplin, B. R.**, Methemoglobin levels in infants in an area with high nitrate water supply, *Am. J. Public Health*, 62, 1174, 1972.
22. National Research Council, *The Health Effects of Nitrate, Nitrite, and N-Nitroso Compounds*, National Academy Press, Washington, D.C., 1981.
23. **Hartman, P. E.**, Nitrate/nitrite ingestion and gastric cancer mortality, *Environ. Mutagen.*, 5, 111, 1983.
24. **Hill, M. J., Hawksworth, G., and Tattersall, G.**, Bacteria, nitrosamines and cancer of the stomach, *Br. J. Cancer*, 28, 562, 1973.

25. **Hartman, P. E. and Gingo, D. J.**, Nitrate Content in Some U.S. Water Supplies: Apparent Independence from U.S. Gastric Cancer Mortality by County, 14th Annu. Sci. Meet. Environ. Mutagen. Society, San Antonio, Tex., 1983.
26. **Armijo, R. and Coulson, A. H.**, Epidemiology of stomach cancer in Chile — the role of nitrogen fertilizers, *Int. J. Epidemiol.*, 4, 301, 1975.
27. **Correa, P., Haenszel, W., Cuello, C., Tannenbaum, S., and Archer, M.**, A model for gastric cancer epidemiology, *Lancet*, 2, 58, 1975.
28. **Folsom, A. R. and Prineas, R. J.**, Drinking water composition and blood pressure: a review of the epidemiology, *Am. J. Epidemiol.*, 115, 818, 1982.
29. **Masironi, R. and Shaper, A. G.**, Epidemiological studies of health effects of water from different sources, *Annu. Rev. Nutr.*, 1, 375, 1981.
30. **Comstock, G. W.**, The epidemiologic perspective: water hardness and cardiovascular disease, *J. Environ. Pathol. Toxicol.*, 4, 9, 1980.
31. **Comstock, G. W.**, Water hardness and cardiovascular diseases, *Am. J. Epidemiol.*, 110, 375, 1979.
32. **Sharrett, A. R.**, The role of chemical constituents of drinking water in cardiovascular diseases, *Am. J. Epidemiol.*, 110, 401, 1979.
33. **Kobayashi, J.**, On geographic relationship between the chemical nature of river water and death rate from apoplexy, *Ber. Ohara Inst., Landwirtsch. Biol.*, 11, 12, 1957.
34. **Neri, C. C., Hewitt, D., and Schrieber, G. B.**, Can epidemiology elucidate the water story?, *Am. J. Epidemiol.*, 99, 75, 1974.
35. **Neri, C. C., Hewitt, D., Schreiber, G. B., Anderson, T. W., Mandel, J. S., and Zdrojewsky, A.**, Health aspects of hard and soft waters, *J. Am. Water Works Assoc.*, 67, 403, 1975.
36. **Comstock, G. W.**, Fatal arteriosclerotic heart disease, water at home, and socio-economic characteristics, *Am. J. Epidemiol.*, 94, 1, 1971.
37. **Sharrett, A. R., Morin, M. M., Fabsitz, R. R., and Bailey, K. R.**, Water hardness and cardiovascular mortality, *Proc. Am. Med. Assoc. Symp. Drinking Water and Health*, American Medical Association, Chicago, Ill., in press.
38. **Shaper, A. G., Packham, R. F., and Pocock, S. J.**, The British regional heart study: cardiovascular mortality and water quality, *J. Environ. Pathol. Toxicol.*, 4, 89, 1980.
39. **Pocock, S. J., Shaper, A. G., and Powell, P.**, The British Regional Heart Study: Cardiovascular Disease and Water Quality, Conference on Inorganics in Drinking Water and Cardiovascular Disease, University of Massachusetts, Amherst, May 1—3, 1984.
40. **Zeighami, E. A., Morris, M. D., and Calle, E. E.**, Drinking Water Inorganics and Cardiovascular Disease: A Case-Control Study Among Wisconsin Farmers, Conference on Inorganics in Drinking Water and Cardiovascular Disease, University of Massachusetts, Amherst, May 1—3, 1984.
41. **Kurtz, T. W. and Morris, R. C.**, Dietary chloride as a determinant of sodium-dependent hypertension, *Science*, 222, 1139, 1983.
42. **Whitescarver, S. A., Ott, C. E., Jackson, B. A., Guthrie, G. P., and Kotchen, T. E.**, Salt-sensitive hypertension: contribution of chloride, *Science*, 223, 1430, 1984.
43. **Zeighami, E. A.**, Personal communication.
44. **Altura, B. M., Altura, B. T., Gebrewold, A., Ising, H., and Gunther, T.**, Magnesium deficiency and hypertension: correlation between magnesium-deficient diets and microcirculatory changes in situ, *Science*, 223, 1315, 1984.
45. **McCarron, D. A., Morris, C. D., and Cole, C.**, Dietary calcium in human hypertension, *Science*, 217, 267, 1982.
46. **Douglas, B. H., McCauley, P. T., Bull, R. J., and Revis, N. W.**, Effect of Multiple Element Variation on Blood Pressure, Conference on Inorganics in Drinking Water and Cardiovascular Disease, University of Massachusetts, Amherst, May 1—3, 1984.
47. **Revis, N. W., Schmoyer, R. L., and Bull, R.**, The relationship of minerals commonly found in drinking water to atherosclerosis and hypertension in pigeons, *J. Am. Water Works Assoc.*, 74, 656, 1982.
48. **Raloff, J.**, Calcium, chlorine, and heart disease linked, *Sci. News*, 124, 103, 1983.
49. **McCabe, L. J., Symons, J. A., Lee, R., and Robeck, G. G.**, Survey of community water supply systems, *J. Am. Water Works Assoc.*, 67, 579, 1975.
50. **Craun, G. F., Greathouse, D. G., Ulmer, N. S., and McCabe, L. J.**, Preliminary report of an epidemiologic investigation of the relationship between tap water constituents and cardiovascular disease, *Proc. Am. Water Works Assoc. 97th Annu. Conf.*, Denver, Colo., 1977, 10-2b.
51. **Craun, G. F. and McCabe, L. J.**, Problems associated with metals in drinking water, *J. Am. Water Works Assoc.*, 67, 579, 1975.
52. **Karalekas, P. C., Jr., Craun, G. F., Hammonds, A. F., Ryan, C. R., and Worth, D. J.**, Lead and other trace metals in drinking water in the Boston metropolitan area, *J. N. Engl. Water Works Assoc.*, 90, 150, 1976.

53. **Worth, D., Matranga, A., Lieberman, M., DeVos, E., Karalekas, P., Ryan, C., and Craun, G., F.**, Lead in drinking water: the contribution of household tap water to blood lead levels, in *Environmental Lead*, Lynam, D. R., Piantanida, L. G., and Cole, J. F., Eds., Academic Press, New York, 1981.
54. **Beattie, A. D., Moore, M. R., Devenay, W. T., Miller, A. R., and Goldberg, A.**, Environmental lead pollution in an urban soft-water area, *Br. Med. J.*, 2, 491, 1972.
55. **Kirmeyer, G. J. and Logsdon, G. S.**, Principles of internal corrosion and corrosion monitoring, *J. Am. Water Works Assoc.*, 75, 78, 1983.
56. **Sheiham, I. and Jackson, P. J.**, The scientific basis for control of lead in the drinking water by water treatment, *J. Inst. Water Eng. Sci.*, 33, 491, 1981.
57. **Karalekas, P. C., Jr., Ryan, C. R., and Taylor, F. B.**, Control of lead, copper, and iron pipe corrosion in Boston, *J. Am. Water Works Assoc.*, 75, 92, 1983.
58. **Richards, W. N. and Moore, M. R.**, Lead hazard controlled in Scottish water systems, *J. Am. Water Works Assoc.*, 76, 60, 1984.
59. *Lead in the Human Environment*, National Academy Press, Washington, D.C., 1980.
60. Ambient Water Quality Criteria for Lead, U.S. Environmental Protection Agency, EPA-440/5-80-057, Washington, D.C., 1980.
61. **Waldron, H. A. and Stofen, D.**, *Sub-Clinical Lead Poisoning*, Academic Press, New York, 1974.
62. **Beattie, A. D. Moore, M. R., Goldberg, A., Finlayson, M. J. W., Graham, J. F., Mackie, E. M., Main, J. C., McLaren, D. A., Murdoch, R. M., and Stewart, G. T.**, Role of chronic low-level lead exposure in the aetiology of mental retardation, *Lancet*, 1, 589, 1975.
63. **Needleman, H. L.**, Deficits in psychological and classroom performance of children with elevated dentine lead levels, *N. Engl. J. Med.*, 300, 689, 1979.
64. **Dahl, L. K.**, Salt intake and salt need, *N. Engl. J. Med.*, 258, 1152, 1958.
65. **White, J. M., Wingo, J. G., Alligood, L. M., Cooper, G. R., Gutridge, J., Hydaker, W., Benack, R. T., Dening, J. W., and Taylor, F. B.**, Sodium ion in drinking water, *J. Am. Dietet. Assoc.*, 50, 32, 1967.
66. *Sodium, Chlorides, and Conductivity in Drinking Water*, World Health Organization, Copenhagen, 1979.
67. **Calabrese, E. J., Tuthill, R. W., Klar, J. M., and Sieger, T. L.**, Elevated levels of sodium in community drinking water, *J. Am. Water Works Assoc.*, 72, 646, 1980.
68. **Hofman, A., Valkenburg, H. A., and Vaandrager, G. J.**, Increased blood pressure in schoolchildren related to high sodium levels in drinking water, *J. Epidemiol. Comm. Health*, 34, 179, 1980.
69. **Tuthill, R. W., Sonich, C., Okun, A., and Greathouse, D.**, The influence of naturally and artificially elevated levels of sodium in drinking water on blood pressure in children, *J. Environ. Pathol. Toxicol.*, 4, 173, 1980.
70. **Tuthill, R. W. and Calabrese, E. J.**, Drinking water sodium and blood pressure in children: a second look, *Am. J. Public Health*, 71, 722, 1981.
71. **Willett, W. C.**, Drinking water sodium and blood pressure: a cautious view of the 'second look', *Am. J. Public Health*, 71, 729, 1981.
72. **Faust, H. S.**, Effects of drinking water and total sodium intake on blood pressure, *Am. J. Clin. Nutr.*, 35, 1459, 1982.
73. **Hallenbeck, W. H., Brenniman, G. R., and Anderson, R. J.**, High sodium in drinking water and its effect on blood pressure, *Am. J. Epidemiol.*, 114, 817, 1981.
74. **Pomrehn, P. R., Clarke, W. R., Wallace, R. B., Lauer, R., and Sowers, M. F.**, The Iowa Community Drinking Water Survey, Conference on Inorganics in Drinking Water and Cardiovascular Disease, University of Massachusetts, Amherst, May 1—3, 1984.
75. **Calabrese, E. J. and Tuthill, R. W.**, The Massachusetts Blood Pressure Studies — Part 1, Conference on Inorganics in Drinking Water and Cardiovascular Disease, University of Massachusetts, Amherst, May 1—3, 1984.
76. **Tseng, W. P.**, Effects and dose-response relationships of skin cancer and blackfoot disease with arsenic, *Environ. Health Perspect.*, 19, 109, 1977.
77. **Tseng, W. P., Chu, H. M., How, S. W., Fong, J. M., Lin, C. S., and Yeh, S.**, Prevalence of skin cancer in an epidemic area of chronic arsenicism in Taiwan, *J. Natl. Cancer Inst.*, 40, 435, 1968.
78. **Borgonto, J. M. and Grieber, R.**, Epidemiologic study of arsenicism in the city of Antofagasta, in *Trace Substances in Environmental Health*, Vol. 5, Hemphill, D. D., Ed., University of Missouri, Columbia, 1972.
79. **Zaldivar, R.**, Arsenic contamination of drinking water and foodstuffs causing endemic chronic poisoning, *Beitr. Pathol. Bd.*, 151, 384, 1974.
80. **Borgonto, J. M., Vincent, P., Venturino, H., and Infante, A.**, Arsenic in the drinking water of the city of Antofagasta, *Environ. Health Perspect.*, 19, 103, 1977.
81. **Ambient Water Quality Criteria for Arsenic**, U.S. Environmental Protection Agency, EPA 440/5-80-021, Washington, D.C., 1980.

82. **Andelman, J. B. and Barnett, M.**, Feasibility Study to Resolve Questions on the Relationship of Arsenic in Drinking Water to Skin Cancer, University of Pittsburgh, 1983.
83. **Shamberger, R. J.**, Relation of selenium to cancer. I. Inhibitory effect of selenium on carcinogenesis, *J. Natl. Cancer Inst.*, 44, 931, 1970.
84. Natural Fluoride Content of Community Water Supplies, U.S. Department of Health, Education and Welfare, Washington, D.C., 1969.
85. Fluoride Drinking Waters, U.S. Department of Health, Education and Welfare, PHS Publ. No. 825, Washington, D.C., 1962.
86. **Segreto, V. A., Collins, E. M., Camann, D., and Smith, C. Y.**, A current study of mottled enamel in Texas, *J. Am. Dental Assoc.*, 107, 42, 1983.
87. **Driscoll, W. S., Horowitz, H. S., Meyers, R. J., Heifetz, S. B., Kingman, A., and Zimmerman, E. R.**, Prevalence of dental caries and dental fluorosis in areas with optimal and above-optimal water fluoride concentrations, *J. Am. Dental Assoc.*, 107, 42, 1983.
88. **Hoover, R. N., McKay, F. W., and Fraumeni, J. F.**, Fluoridated drinking water and the occurrence of cancer, *J. Natl. Cancer Inst.*, 57, 757, 1976.
89. **Chilvers, C.**, Cancer mortality and fluoridation of water supplies in 35 U.S. cities, *Int. J. Epidemiol.*, 12, 397, 1983.
90. **Harrington, J. M., Craun, G. F., Meigs, J. W., Landrigan, P. J., Flannery, J. T., and Woodhull, R. S.**, An investigation of the use of asbestos cement pipe for public water supply and the incidence of gastrointestinal cancer in Connecticut, 1935—1973, *Am. J. Epidemiol.*, 107, 96, 1978.
91. **Millette, J. M., Clark, P. J., Stober, J., and Rosenthal, M.**, Asbestos in water supplies of the United States, *Environ. Health Perspect.*, 53, 45, 1983.
92. **Boatman, E. S., Merrill, T., O'Neill, A., Polissar, L., and Millette, J. R.**, Use of quantitative analysis of urine to assess exposure to asbestos fibers in drinking water in the Puget Sound region, *Environ. Health Perspect.*, 53, 131, 1983.
93. **Marsh, G. M.**, Critical review of epidemiologic studies related to ingested asbestos, *Environ. Health Perspect.*, 53, 49, 1983.
94. **Sigurdson, E. E.**, Observations of cancer incidence surveillance in Duluth, Minnesota, *Environ. Health Perspect.*, 53, 61, 1983.
95. **Millette, J. M., Craun, G. F., Stober, J. A., Kraemer, D. F., Tousignant, H. G., Hildago, E., Duboise, R. L., and Benedict, J.**, Epidemiology study of the use of asbestos-cement pipe for the distribution of drinking water in Escambia County, Florida, *Environ. Health Perspect.*, 42, 91, 1983.
96. **Condie, L. W.**, Review of published studies of orally administered asbestos, *Environ. Health Perspect.*, 53, 3, 1983.
97. **McConnell, E. E., Shefner, A. M., Rust, J. H., and Moore, J. A.**, Chronic effects of dietary exposure to amosite and chrysotile asbestos in Syrian golden hamsters, *Environ. Health Perspect.*, 53, 11, 1983.
98. **McConnell, E. E., Rutter, H. A., Ulland, B. M., and Moore, J. A.**, Chronic effects of dietary exposure to amosite asbestos and tremolite in F344 rats, *Environ. Health Perspect.*, 53, 27, 1983.
99. **Polissar, L., Severson, R. K., and Boatman, E. S.**, Cancer risk from asbestos in drinking water: summary of a case-control study in western Washington, *Environ. Health Perspect.*, 53, 57, 1983.
100. **Erdreich, L. S.**, Comparing epidemiologic studies of ingested asbestos for use in risk assessment, *Environ. Health Perspect.*, 53, 99, 1983.
101. **Toft, P. and Meek, M. E.**, Asbestos in water: a Canadian view, *Environ. Health Perspect.*, 53, 177, 1983.
102. **Cotruvo, J. A.**, Asbestos in drinking water: a status report, *Environ. Health Perspect.*, 53, 181, 1983.
103. **Cantor, K. P., Kopfler, F. C., Hoover, R. N., and Strasser, P. H.**, Cancer Epidemiology as Related to Chemicals in Drinking Water, 10th Annu. Symp. Analytical Chemistry of Pollutants, Dortmund, West Germany, May 29, 1980.
104. **Hoel, D. G. and Crump, K. S.**, Waterborne carcinogens: a scientist's view, in *The Scientific Basis of Health and Safety Regulation*, Crandall, R. W. and Love, L. B., Eds., Brookings Institute, Washington, D.C., 1981, 1973.
105. **Rook, J. J.**, Formation of haloforms during chlorination of natural waters, *J. Soc. Water Treat. Exam.*, 23, 234, 1974.
106. **Bellar, T. A., Liehtenberg, J. J., and Kroner, R. C.**, Determining volatile organics at microgram-per-liter levels by gas chromatography, *J. Am. Water Works Assoc.*, 66, 739, 1974.
107. **Symons, J. M., Stevens, A. A., Clark, R. M., Geldreich, E. E., Love, O. T., Jr., and DeMarco, J.**, Treatment Techniques for Controlling Trihalomethanes in Drinking Water, Environmental Protection Agency, EPA-600/2-81-156, Cincinnati, Ohio, 1981.
107a. **Page, N. P. and Saffiotti, U.**, Report on Carcinogenesis Bioassay of Chloroform, Natl. Cancer Institute, Bethesda, Md., 1976.

108. **Burke, T. A., Amsel, J., and Cantor, K. P.,** Trihalomethane variation in public drinking water supplies, in *Water Chlorination: Environmental Impact and Health Effects,* Vol. 4, Jolley, R. L., Brungs, W. A., Cotruvo, J. A., Cumming, R. B., Mattice, J. S., and Jacobs, V. A., Eds., Ann Arbor Science, Ann Arbor, Mich., 1983, 1381.
109. **Wilkins, J. R., III, Reiches, N. A., and Kruse, C. W.,** Organic chemicals in drinking water and cancer, *Am. J. Epidemiol.,* 110, 420, 1979.
110. **Shy, C. M. and Struba, R. J.,** Epidemiologic evidence for human cancer risk associated with organics in drinking water, in *Water Chlorination: Environmental Impact and Health Effects,* Vol. 3, Ann Arbor Science, Ann Arbor, Mich., 1980, 1029.
111. **Crump, K. S. and Guess, H. A.,** Drinking water and cancer: review of recent epidemiological findings and assessment of risks, *Annu. Rev. Public Health,* 3, 339, 1982.
112. **Cantor, K. P.,** Epidemiologic studies of chlorination by-products in drinking water: an overview, in *Water Chlorination: Environmental Impact and Health Effects,* Vol. 4, Jolley, R. L., Brungs, W. A., Cotruvo, J. A., Cumming, R. B., Mattice, J. S., and Jacobs, V. A., Ann Arbor Science, Ann Arbor, Mich., in press.
113. **Craun, G. F.,** Epidemiologic considerations for evaluating associations between the disinfection of drinking water and cancer in humans, in *Water Chlorination: Chemistry, Environmental Impact and Health Effects,* Vol. 5, Jolley, R. L., Bull, R. J., Davis, W. P., Katz, S., Roberts, M. H., Jr., and Jacobs, V. A., Eds., Lewis Publishers, Chelsea, Mich., 1985, 131.
114. **Cantor, K. P., Hoover, R., Hartge, P., Mason, T. J., Silverman, D. T., and Levin, L. I.,** Drinking water source and risk of bladder cancer: a case-control study, in *Water Chlorination: Chemistry, Environmental Impact and Health Effects,* Vol. 5, Jolley, R. L., Bull, R. J., Davis, W. P., Katz, S., Roberts, M. H., Jr., and Jacobs, V. A., Eds., Lewis Publishers, Chelsea, Mich., 1985, 143.
115. **Cragle, D. L., Shy, C. M., Struba, R. J., and Stiff, E. J.,** A case-control study of colon cancer and water chlorination in North Carolina, in *Water Chlorination: Chemistry, Environmental Impact and Health Effects,* Vol. 5, Jolley, R. L., Bull, R. J., Davis, W. P., Katz, S., Roberts, M. H., Jr., and Jacobs, V. A., Eds., Lewis Publishers, Chelsea, Mich., 1985, 151.

Section II: Waterborne Outbreak Statistics

Chapter 5

STATISTICS OF WATERBORNE OUTBREAKS IN THE U.S. (1920—1980)

Gunther F. Craun

TABLE OF CONTENTS

I. Introduction 74

II. Definitions 74

III. Waterborne Outbreaks 75
    A. Reporting 78
    B. Cases of Illness 85
    C. Deaths 86
    D. Seasonal Occurrence 89
    E. Etiologic Agents and Waterborne Diseases 91
        1. Seasonal Variation 96
        2. Importance of Waterborne Transmission 100
        3. Waterborne Diseases 106
            a. Gastroenteritis 108
            b. Typhoid Fever and Hepatitis A 117
            c. Shigellosis 119
            d. Salmonellosis 120
            e. Giardiasis 121
            f. Acute Chemical Poisonings 124
    F. Causes of Waterborne Outbreaks and Disease 124

IV. Waterborne Outbreaks (1971—1980) 142
    A. Occurrence and Distribution of Outbreaks 142
    B. Causes of Outbreaks 142
    C. Etiologic Agents 148

Acknowledgments 155

References 155

I. INTRODUCTION

Statistical data have been collected on waterborne disease outbreaks in the U.S. since 1920.[1-21] During 1920 through 1936, these data were collected and compiled by Gorman and Wolman.[1,2] In subsequent periods,* data were collected by several federal agencies,[3-21] and summaries were prepared by Eliassen and Cummings,[4] Weibel et al.,[5] Craun and McCabe,[8] and Craun et al.[12] The Centers for Disease Control (CDC) and the Environmental Protection Agency (EPA) have conducted waterborne disease investigation and surveillance activities and have published information on waterborne outbreaks annually since 1971.[13-21] All the available data on waterborne outbreaks were compiled and examined to provide a historical perspective of the epidemiology of waterborne disease over the past 61 years.[11] This summary is intended to supplement the reviews and analyses of waterborne outbreaks contained in previously published articles.[1,4,5,8,12]

The reporting of waterborne outbreaks has been and continues to be voluntary, and all of the data on these outbreaks were obtained from the scientific and medical literature and from official sources through the cooperation and assistance of state and local health officials, epidemiologists, and water supply engineers. It is recognized that these data may reflect differences in the reporting and surveillance interests of the states over the past 61 years, but it is felt that these data are reasonably complete for an analysis of the causes and etiologies of waterborne outbreaks.

These data were compiled after reviewing tabulations[2,3] of waterborne outbreaks from 1920 through 1945 and individual reports[6,7] of waterborne outbreaks from 1946 through 1980. An attempt was made to maintain consistency in classifying etiological agents, disease, and types of water systems. Individual reports of investigations were generally unavailable to provide additional information or resolve questions regarding classifications for outbreaks prior to 1946, but the resulting tabulation[11] represents the best available information based on interpretations of footnotes contained in the tabulations[2,3] and descriptions in previous reviews.[1,4]

II. DEFINITIONS

In compilations prior to 1970, water systems experiencing an outbreak were classified as either community (public) or noncommunity (private) water systems. For this summary, all water systems were reclassified as community, noncommunity, and individual according to definitions contained in the Safe Drinking Water Act (P.L. 93-523). Community systems are public or investor-owned water supplies serving large or small communities, subdivisions, and mobile home parks of at least 25 year-round residents or 15 service connections used by year-round residents. Noncommunity systems are seasonally operated water supplies or supplies that serve travelers and transients. These systems must serve an average of at least 25 individuals for 60 or more days during the year and include institutions, industries, camps, parks, hotels, and service stations which maintain their own water supply for employees and the traveling public. Individual systems are water supplies used by residents in areas without community systems or by persons traveling outside of populated areas (e.g., backpackers, campers).

During 1920 through 1936 the criterion of five cases of illness was used to define an outbreak, and only those waterborne outbreaks of five or more cases were included in tabulations[2,3] for this period. In subsequent years, two cases of illness (except for single cases of chemical poisonings) was the criterion used to define an outbreak. Gorman and Wolman[1] reported that 12 waterborne outbreaks were excluded from their analysis of data

---

* No data were collected in 1937.

for 1920 through 1936 because fewer than 5 individuals were affected. It was not possible to include these 12 outbreaks in this summary, as data on the specific outbreaks were unavailable. Using the criterion of 5 cases of illness to define an outbreak during 1936 through 1980 would have resulted in the exclusion of 101 outbreaks, most of which were outbreaks of typhoid fever and chemical poisonings in individual water systems. Rather than lose the information from this large number of outbreaks, it was decided to include outbreaks of less than five cases during 1936 through 1980 and to note the inconsistency in the definition of an outbreak between the two time periods.

In a few instances the number of cases of illness resulting from an outbreak was reported to be unknown or many, and an estimate of the number of actual illnesses was required. When detailed descriptions, such as population at risk and attack rates of smaller populations from special surveys, were available from the outbreak investigation report, it was possible to provide a reasonable estimate of the number of illnesses which might have occurred. When no additional information was available, an estimate of two cases of illness was used even though it was felt that in almost all instances this was a large underestimate of the number of illnesses which had actually occurred. When the number of illnesses was originally reported as a range, the mean was used for this summary except where the maximum number of illnesses reported appeared to be more appropriate based on information contained in the report of the investigation or information used to prepare previous summaries. In the few instances where a single number of estimated cases was substituted for a range, comparison of outbreaks in this tabulation[11] with those of previous tabulations[2,3,6,7] will show the outbreaks affected and the differences in the estimated number of illnesses used for this summary and the number used in previous summaries.

During the period 1920 through 1929, acute GI illness, diarrhea, or dysentery were not differentiated from specific infections of amebic and bacillary dysentery (except for a single outbreak where four cases of bacillary dysentery were identified), and for this summary these illnesses were classified as acute gastroenteritis of undefined etiology. In subsequent periods, these illnesses were differentiated and are listed as such in the tabulation[11] when a specific etiology was noted. Where no etiologic agent was specified in outbreaks of dysentery, illnesses were classified as acute gastroenteritis, and this resulted in the reclassification of 11 outbreaks which had previously been identified as dysentery outbreaks because of the symptomatology of the illness.

III. WATERBORNE OUTBREAKS

During the 61-year period, 1405 waterborne outbreaks were reported in the U.S., and 386,144 cases of illness and 1083 deaths were associated with these outbreaks. Each of the 50 states, the District of Columbia, and Puerto Rico reported at least one outbreak during this period. Of these outbreaks, 43% occurred in community water systems compared with 35% in noncommunity and 22% in individual water systems (Figure 1). The majority of illnesses (87%) and deaths (71%) occurred as the result of outbreaks in community water systems.

The number of outbreaks reported annually ranged from a minimum of 6 in 1957 to a maximum of 60 in 1941 with a mean of 23.4 (SD 11.96) outbreaks and a median of 22.5 outbreaks for the entire period (Figure 2). The data indicate a uniform distribution for waterborne outbreaks reported from 1920 through 1930 with an average of 23 outbreaks annually. Thereafter, the occurrence of waterborne outbreaks appeared to be cyclical, with more than 40 outbreaks being reported annually during 1938 through 1941 and 1979 through 1980 and less than 20 outbreaks reported annually during 1950 through 1970.

During 1920 through 1936, waterborne outbreaks were reported most frequently in community water systems; afterwards, outbreaks were generally reported most frequently in

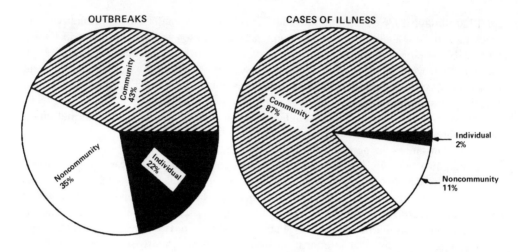

FIGURE 1. Distribution of waterborne outbreaks and illnesses in community, noncommunity, and individual water systems (1920—1980).

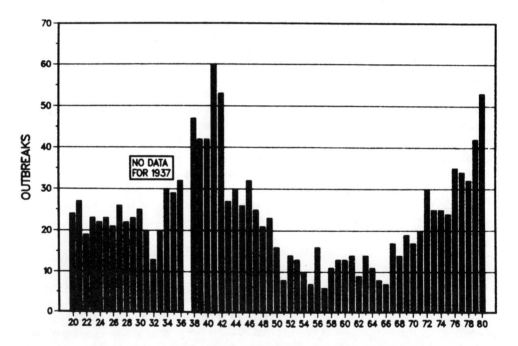

FIGURE 2. Number of waterborne outbreaks occurring each year (1920—1980).

noncommunity water systems (Figures 3 and 4). In most (70%) years after 1938, more outbreaks were reported in noncommunity water systems than were reported in community water systems, and except for the years 1921 through 1929, 1941, and 1955 through 1966, more outbreaks were reported in noncommunity water systems than were reported in individual water systems (Figure 5). More outbreaks were reported in community water systems than were reported in individual water systems for each year except 1930, 1938, 1939, 1941, 1948, 1953, 1955, and 1957 through 1969.

In general, the number of outbreaks reported in community water systems decreased during the late 1920s through the early 1930s and during the mid-1940s through the mid-1950s. Since the mid-1960s, however, the number of outbreaks reported in community water

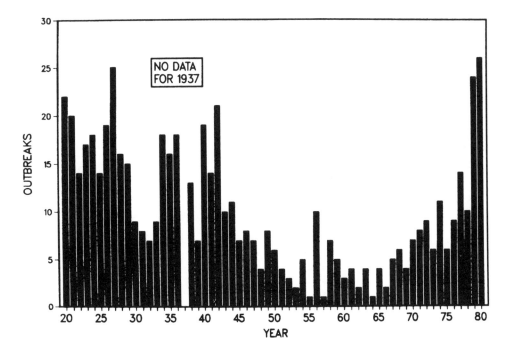

FIGURE 3. Number of waterborne outbreaks occurring in community water systems each year (1920—1980).

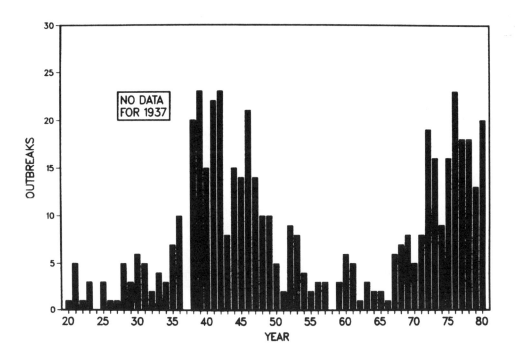

FIGURE 4. Number of waterborne outbreaks occurring in noncommunity water systems each year (1920—1980).

systems has increased dramatically, and the numbers reported in 1979 and 1980 were the most reported in any year since 1942. The number of waterborne outbreaks in noncommunity water systems increased from 1920 through the mid-1940s and subsequently decreased until

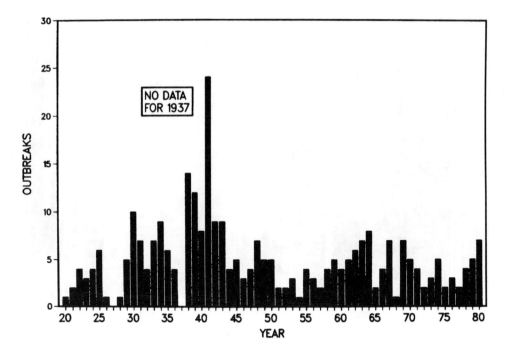

FIGURE 5. Number of waterborne outbreaks occurring in individual water systems each year (1920—1980).

the mid-1960s when they also began to be reported in increased numbers. The number of waterborne outbreaks reported in individual water systems generally increased from 1920 through 1941 and subsequently decreased until the mid-1950s. An increase in reported outbreaks similar to that experienced by community and noncommunity water systems during the late 1960s through 1980 was not seen in individual water systems.

A. Reporting

The decline in the number of waterborne outbreaks which occurred from the 1940s through the late 1960s was not felt to be associated with either major new developments in water treatment technology or a more widespread application of existing treatment technology. Disinfection and filtration, which resulted in a drastic reduction in waterborne disease in the U.S. earlier this century, were developed and applied to most surface water systems prior to this decline in waterborne outbreaks.[22-26] Community water systems were developed for most major cities in the U.S. during the late 1800s and early 1900s, and filtration and disinfection of water to prevent waterborne disease was introduced and applied increasingly in the years prior to 1940 (Tables 1 to 3). Slow sand filtration of municipal drinking water was first applied in the U.S. in 1889 at Albany, N.Y., and the first mechanical or rapid sand filter built along modern lines was installed in 1893 at Lawrence, Mass. By 1904, some 10% of the U.S. urban population (which largely relies on surface water sources) received filtered water, and by 1914 approximately 36% of the urban population received filtered water.

Whipple[22] reported that mortality from typhoid fever was reduced from an average of 104 cases per 100,000 population in Albany and 121 cases per 100,000 population in Lawrence to 26 cases per 100,000 after the filtration of water in both cities. In Binghamton and Watertown, N.Y., the typhoid fever mortality rates decreased from 49 cases per 100,000 to 11 cases per 100,000 and 97 cases per 100,000 to 27 cases per 100,000, respectively. In 1908, Whipple[22] noted that ''filtration affords the safest protection of surface waters against waterborne diseases. This has been proved by long experience abroad, and data

Table 1
GROWTH OF COMMUNITY WATER SYSTEMS IN THE U.S. (1800—1958)[22-26]

Year	No. of water facilities	No. of purification facilities	No. of disinfection facilities
1800	16		
1850	83		
1870	243		
1880	598	1	
1890	1,878	NA[a]	
1896	3,196	NA	
1924	9,850	638	NA
1930	NA	1,531	2,917
1934	10,790	NA	NA
1938	12,760	NA	4,054
1940	NA	1,855	4,590
1948	NA	2,054	6,137
1958	17,808	2,517	8,845

[a] Not available.

Table 2
U.S. POPULATION SERVED BY SAND FILTRATION (1800—1958)[22-26]

Year	Slow sand filtration (population)	Rapid sand filtration (population)	Percent of urban population served by filtered water
1870	0	0	0
1880	30,000	0	<1
1890	35,000	275,000	2
1900	360,000	1,500,000	6
1904	560,000	2,600,000	10
1910	3,883,000	6,922,000	28
1914	5,398,000	11,893,000	36
1924	5,054,000	18,610,000	38
1940	3,579,200	26,751,700	41
1948	2,014,635	39,731,460	47
1958	3,177,380	55,606,245	48

illustrating this truth are constantly being accumulated in this country." Whipple witnessed the period of rapid expansion of water filtration in the U.S. as evidenced by this passage also written in 1908:

"It is gratifying to see that so many of our American cities have awakened to the need of properly safeguarding the quality of the drinking water supplied to the public for drinking purposes. During the last five years (1903—1908), filter plants have been built and put into operation in Charleston, South Carolina, Youngstown, Ohio, Washington, D.C., Providence, Rhode Island, New Haven, Connecticut, Watertown, New York, Indianapolis, Indiana, Philadelphia, Pennsylvania, New Milford, New Jersey, Brooklyn, New York, and Binghamton, New York, Little Falls and Hackensack, New Jersey, Chester, Pennsylvania, and elsewhere. Large filters are under construction in Philadelphia, Pittsburgh, Cincinnati, Louisville, Columbus, Ohio, and New Orleans, and filter plants are being enlarged or improved in Bangor, Maine, Albany, New York, Yonkers, New York. In other cities such as Troy, New York City, Toledo, Ohio, Oakland, California, Grand Rapids, Michigan, Wilmington, Delaware, Minneapolis, Minnesota, Trenton, New Jersey, Lynn, Massachusetts, and Lancaster, Pennsylvania, they have been recommended, and the projects are under consideration."

Table 3
WATER TREATMENT IN THE U.S. (1930—1958)[23-26]

Year	Communities with water systems	Surface water facilities[a] (%)	Facilities with rapid and slow sand purification[a] (%)	Facilities with disinfection[a] (%)	Population served by community water systems	Population served by filtered or disinfected water[b] (%)	Population served by treated surface water[b] (%)
1930	11,000[c]	NA[d]	14	27	NA	NA	NA
1938—1940	13,293	26	16	31	131,669,275	67	35
1948	16,747	30	12	37	150,697,361	83	33
1958	20,459	27	12	43	179,323,175	83	42

[a] Number of facilities divided by number of communities with a water system.
[b] Population served by treated water divided by total population served by community water systems.
[c] Estimated from 1924, 1934, and 1938 data.
[d] Not available.

Although the number of purification (filtration) facilities increased rapidly through 1930 and continued to increase through 1958, the proportion of communities with purification facilities increased only slightly — from 14% in 1930 to 16% in 1940 — and decreased to 12% in 1948 and 1958. The urban population receiving filtered water increased rapidly through 1914 but has increased only moderately since then. In 1914, 36% of the urban population received filtered water; in 1940, 41% received filtered water; and in 1958, 48% received filtered water.

The first continuous application of chlorine to disinfect a municipal water supply in the U.S. occurred in 1908 when sodium hypochlorite was used to disinfect the Boonton reservoir of Jersey City, N.J. The early acceptance of chlorination as a public health safeguard and the development in 1912 of commercial equipment to apply gaseous chlorine to water supplies provided for the rapid expansion of chlorine disinfection through the 1930s. Approximately 2900 community water facilities disinfected drinking water prior to 1930, and although this increased to some 4600 facilities in 1940 and 6137 in 1948, only a moderate increase occurred in the percentage of community water facilities with disinfection capabilities. In 1930, 27% of the community water systems had disinfection facilities; in 1940, 31% provided disinfection, and in 1948, this number increased to 37%. This moderate increase in the percentage of water systems with disinfection facilities and the accompanying increase in the population served by filtered or disinfected water (from 67% in 1940 to 83% in 1948) is not felt to offer sufficient evidence that improved treatment of water supplies was a major reason for the dramatic decrease in the number of reported waterborne outbreaks from 1940 to 1950.

The increased number of waterborne outbreaks observed in the 1930s and 1970s was likewise not felt to be generally associated with a widespread increased contamination of water supplies or with deteriorating water systems with one exception. In the late 1970s, an increased number of waterborne outbreaks of giardiasis was reported, especially in the northeastern, northwestern, and Rocky Mountain states, where surface waters are generally not filtered. Although greater recognition of giardiasis as a waterborne disease and interest in giardiasis surveillance likely resulted in an increased reporting of waterborne outbreaks of giardiasis during this period, increased contamination of watersheds and inadequate treatment of surface water supplies in these areas are felt to be primary reasons for the increased occurrence of these outbreaks.

Because the waterborne outbreak reporting system is voluntary, the differences in the occurrence of waterborne outbreaks during this 61-year period must be largely attributed to differences over the years in the reporting of outbreaks by state and local health departments and interests of others in compiling these reports. Reporting depends primarily uon the capabilities of the state health department to conduct disease surveillance, to follow up reports of unusual occurrences of illness and complaints, and to investigate outbreaks and prepare and forward the reports for compilation. Whether an outbreak will be recognized and investigated depends upon many factors, including the severity of disease, the number of individuals affected, and the type of water system. Not only must the outbreak be recognized and investigated, but information about the outbreak must also be made available to those who evaluate the data and compile the statistics.

To support the assertion that the increase and decrease of outbreaks over this time period are primarily caused by reporting differences, evidence can be obtained by examining the reporting of waterborne outbreaks among the various states. For example, 11 states reported 55.4% of all waterborne outbreaks that occurred during this 61-year period (Table 4). The combined number of outbreaks reported by these 11 states varied for each decade from 42% during 1961 through 1970 to 72% during 1941 through 1950. It is hard to believe that most of the waterborne outbreaks during this 61-year period occurred in these particular states solely because of increased contamination of water supplies, deteriorating water systems, or treatment deficiencies.

Table 4
PROPORTION[a] OF WATERBORNE OUTBREAKS IN STATES REPORTING THE MOST WATERBORNE OUTBREAKS DURING THE PERIOD 1920—1980

State	Time period						
	1920—1930	1931—1940	1941—1950	1951—1960	1961—1970	1971—1980	1920—1980
California	4.7	2.9	3.5	12.6	9.9	6.6	5.6
Oregon	1.2	0	2.2	2.7	7.6	5.0	2.8
Washington	1.6	1.1	1.0	1.8	1.5	4.4	2.0
Colorado	0	0.3	0	3.6	4.6	6.9	2.3
Illinois	5.1	3.2	3.2	7.1	3.1	1.3	3.4
Indiana	3.9	10.2	4.2	1.0	1.0	1.0	4.0
Ohio	3.5	2.2	3.8	7.2	4.6	2.8	3.6
Maryland	8.2	2.6	3.5	0	2.3	1.9	3.4
Pennsylvania	13.3	9.5	2.9	1.8	2.3	20.0	9.8
New York	7.8	27.3	46.0	20.7	3.8	3.8	19.9
Massachusetts	0.8	6.5	1.6	0.9	1.5	1.3	2.3
Total	50.2	65.8	71.9	59.5	42.0	54.7	55.4
U.S.	100	100	100	100	100	100	100

[a] (Number of outbreaks reported by state during period/Number of outbreaks reported for U.S. during period) × 100.

Table 5
COMPARISON[a] OF REPORTED AND EXPECTED[b] WATERBORNE OUTBREAKS IN STATES REPORTING THE MOST WATERBORNE OUTBREAKS DURING THE PERIOD 1920—1980

	Time period					
State	1920—1930	1931—1940	1941—1950	1951—1960	1961—1970	1971—1980
California	0.8	0.5	0.6	2.3[c]	1.8[c]	1.2
Oregon	0.4	0	0.8	1.0	2.7[c]	1.8[c]
Washington	1.6[c]	0.9	0.7	1.5[c]	1.8[c]	4.0[c]
Colorado	0	0.2	0	2.6[c]	4.7[c]	5.5[c]
Illinois	1.5[c]	0.9	0.9	2.1[c]	0.9	0.4
Indiana	1.0	2.6[c]	1.0	0.3	0.3	0.3
Ohio	1.0	0.6	1.1	2.0[c]	1.3	0.8
Maryland	2.4[c]	0.8	1.0	0	0.7	0.6
Pennsylvania	1.4[c]	1.0	0.3	0.2	0.2	2.1[c]
New York	0.4	1.4[c]	2.3[c]	1.1	0.2	0.2
Massachusetts	0.7	4.3[c]	1.0	0.7	1.6[c]	1.0

[a] Observed divided by expected.
[b] Expected number of waterborne outbreaks based on total number of waterborne outbreaks observed in each state and the total number of waterborne outbreaks observed in the U.S. for a given time period.
[c] Reporting 40% or greater than the expected number of outbreaks.

Officials in Pennsylvania, Colorado, and Washington increased their waterborne disease surveillance activities during the 1970s and reported some 31% of all outbreaks during 1971 through 1980 (Table 4). Pennsylvania reported 20% of all waterborne outbreaks in the U.S. during 1971 through 1980 compared with only 2% for the previous 20-year period. Colorado reported 7% of all waterborne outbreaks during 1971 through 1980 compared with 4% for the previous 20-year period and Washington reported more than 4% of all waterborne outbreaks during 1971 through 1980 compared with less than 2% for the previous 20-year period.

The large number of waterborne outbreaks which occurred during 1941 to 1950 can be attributed primarily to an active surveillance program conducted in New York, as this state accounted for 46% of all waterborne outbreaks reported during this period (Table 4). During the previous 21-year period, 18% of all waterborne outbreaks had been reported by New York, but during the subsequent 30-year period of 1951 through 1980, New York reported only 7% of all waterborne outbreaks. The active reporting of waterborne outbreaks by New York during 1938 through 1945 has been described by Eliassen and Cummings,[4] and less active surveillance is one possible explanation for the decreased occurrence of waterborne outbreaks in New York during 1951 through 1980.

Among the 11 states which reported the most outbreaks, no single state consistently reported large numbers of waterborne outbreaks over the entire period (Table 4). Some states reported more waterborne outbreaks in the earlier years of the period than in the latter years, while others reported more waterborne outbreaks in recent years. The number of reported waterborne outbreaks was compared with an expected number based on the marginal probabilities of waterborne outbreaks for each state for each time period (Table 5). Except for 1941 through 1950, when only New York reported more waterborne outbreaks than expected, three to five states reported more waterborne outbreaks than expected in each time period. All of the states except Indiana reported more than the expected number of waterborne outbreaks in at least two time periods, but only California, Oregon, Washington, Colorado, and New York reported more than the expected number in at least two consecutive time periods. A trend over time for any state to observe greater than the expected number of waterborne outbreaks would suggest an increased contamination or deterioration of water supplies, and inconsistencies in observing greater or fewer than the expected number of waterborne outbreaks would suggest surveillance and reporting differences. Because water sources, types of treatment, and contamination problems for water systems are generally similar for states in the same region, it is anticipated that the occurrence of waterborne outbreaks would be similar among the states of a region unless there were differences in surveillance and reporting activities. For example, it is difficult to believe that differences in the occurrence of outbreaks in Ohio, Indiana, and Illinois reflect anything but differences in reporting. If, however, several states within a region consistently observed more than the expected number of waterborne outbreaks over similar time periods, it would be difficult to exclude the possibility of widespread increased contamination or general deterioration of water supplies as the cause of a greater occurrence of outbreaks, since waterborne outbreaks would likely be detected with increased frequency regardless of surveillance and reporting differences among states.

Prior to 1951, no trend over time was apparent for any state to consistently report more waterborne outbreaks than expected nor did states within a region consistently report more than the expected number. This suggests that the occurrence of outbreaks in the U.S. prior to 1951 was likely influenced by inconsistencies in the detection, investigation, and reporting of outbreaks. After 1951, the four western states consistently reported more outbreaks than expected compared with the midwestern and eastern states. Although not shown in Table 3, the New England states of Maine, Connecticut, Massachusetts, New Hampshire, Rhode Island, and Vermont all reported more waterborne outbreaks than expected during 1971 through 1980. This consistency among the northwestern and New England states to report more than the expected number of waterborne outbreaks suggests that increased contamination or deterioration of water supplies in these regions may have contributed to the recently observed increase in waterborne outbreaks. The increase in waterborne giardiasis in these areas caused by use of untreated or disinfected only surface water has already been discussed.

Outbreaks in community water systems, which number about 59,000 and serve about 180 million people, are probably the most likely to be reported. Outbreaks in noncommunity systems, which number about 240,000 and serve about 20 million people, primarily transients, are the next most likely to be reported, although it is often difficult to suspect or investigate a common source when travelers become ill. Travelers are likely to become ill after leaving the area where the illness was contracted, and if medical attention is sought, the illness may be reported to different health authorities who may have no knowledge of similar problems outside their area of jurisdiction and may not suspect a common source. Even if a common source is suspected and investigated, it is often difficult to obtain accurate information on the magnitude of the outbreak because information may not be available to determine who had been exposed or how to contact these individuals. The publicity associated with recent giardiasis outbreaks in travelers, however, has caused physicians and health departments to become more aware of the possibility of common-source outbreaks among travelers and more suspicious when illness is reported after vacation or travel periods. Outbreaks in individual water systems, which serve about 30 million people, are least likely to be reported for several reasons. Many health departments lack the legislative authority or lack sufficient resources to conduct water quality surveillance programs for individual water systems and to adequately investigate sporadic reports of disease in rural areas. In many instances, illness occurring in rural areas is not recognized as possibly transmitted by contaminated water because (1) the illness is mild and medical attention is not sought and (2) if medical attention is obtained, few individuals or only family members are affected, and a common source is not considered (person-to-person transmission is generally assumed). Reportable diseases such as typhoid fever are likely, however, to be investigated even when single cases are reported, and in many instances cases are found to be transmitted by contaminated water.

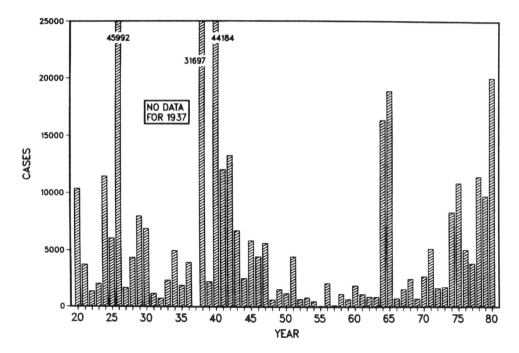

FIGURE 6. Number of cases of waterborne disease occurring each year (1920—1980).

It is difficult to ascertain the number of waterborne outbreaks that go undetected or unreported. One estimate, based on data collected from 1946 to 1970, was that about one half of the waterborne outbreaks in community water systems and about one third of those in noncommunity systems are detected, investigated, and reported.[8] During a 2-year period of improved waterborne disease surveillance in Colorado in 1980 to 1982, 11 waterborne outbreaks were documented compared to 6 waterborne outbreaks during the previous 3-year period.[27] A study of the occurrence of foodborne outbreaks in Washington state indicated that prior to initiation of improved surveillance and investigation only one foodborne outbreak in ten had been recognized and reported.[28] Although this study considered only the reporting of foodborne outbreaks, the results may be applicable to waterborne outbreaks, especially those that occur in noncommunity water systems because of the similarities in recognition and investigation of these kinds of outbreaks. A waterborne outbreak[29] in a residential community of 6500 persons in Florida is a good example of how an active disease surveillance program can help detect outbreaks. Initially, only 10 cases of shigellosis were recognized by health authorities, but further investigation demonstrated an additional 1200 illnesses. If local health authorities had not been conducting shigellosis surveillance, the initial ten cases might never have been recognized as an unusual occurrence, and an outbreak of waterborne disease as large as this might have gone undetected.

B. Cases of Illness

Cases of illness caused by waterborne outbreaks during the 61-year period ranged from a minimum of 76 cases of illness in 1955 to a maximum of 45,992 cases of illness in 1926 with a mean of 6,435.6 (SD 9,297.9) cases of illness and a median of 2634 cases of illness for the entire period (Figure 6). Although the numbers are quite variable from year to year, the number of illnesses generally declined from 1940 through 1955 and increased from 1955 through 1980. Despite the recent increase in reported waterborne outbreaks and the apparent increase in cases of illness caused by these outbreaks, the incidence of waterborne disease has decreased from approximately 8 cases of waterborne disease per 100,000 person-years

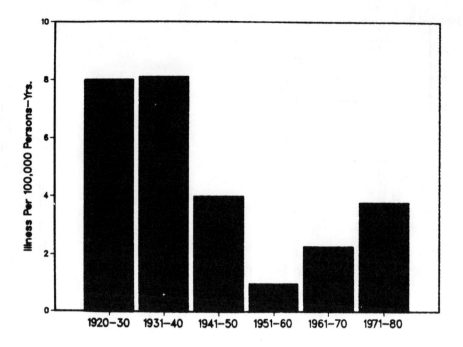

FIGURE 7. Incidence of waterborne disease in the U.S. (1920—1980).

in 1920 to 1940 to approximately 4 cases per 100,000 person-years in 1971 to 1980 (Figure 7).

The magnitude of a waterborne outbreak depends upon the type of water system affected. For the entire 61-year period, community water systems averaged 559 illnesses per outbreak, noncommunity systems averaged 84 illnesses per outbreak, and individual systems averaged 22 illnesses per outbreak. Over one half of the outbreaks in community water systems resulted in fewer than 51 cases of illness, and over 75% of the outbreaks resulted in fewer than 200 cases of illness (Table 6). Only 8% of the outbreaks in community water systems resulted in more than 1000 illnesses. Approximately 92% of the outbreaks reported in noncommunity and individual water systems resulted in fewer than 200 illnesses and 50 illnesses, respectively (Table 6).

No general trend was apparent over time for an increased number of cases of illness in waterborne outbreaks occurring in community, noncommunity, and individual water systems (Figures 8 to 11). For outbreaks in all water systems, the average number of cases of waterborne illness per outbreak varied from 100 to 125 in five time periods, 224 to 256 in three time periods, and 504 to 681 in three time periods (Figure 8). For most time periods, waterborne outbreaks in community water systems averaged 151 to 554 cases of waterborne illness per outbreak (Figure 9). Because of two large outbreaks reported in 1965, 2432 cases of illness per outbreak occurred during 1961 through 1965. The average number of cases of waterborne illness per outbreak in noncommunity water systems varied from 39 during 1956 through 1960 to 166 during 1926 through 1930 (Figure 10) and in individual water systems varied from 9 during 1971 through 1975 to 47 during 1966 through 1970 (Figure 11).

C. Deaths

There were 1083 deaths associated with waterborne outbreaks during this 61-year period. The majority of these deaths occurred because of waterborne outbreaks of typhoid fever (Table 7) and occurred early in the period (Figure 12). Of the deaths, 84% occurred prior to 1936, and less than 1% occurred after 1970. Most deaths were due to typhoid fever

Table 6
MAGNITUDE OF WATERBORNE OUTBREAKS
(1920—1980)

Size of outbreak (cases of illness)	Frequency of occurrence (number of outbreaks)			
	Community systems	Noncommunity systems	Individual systems	All systems
<2	3	0	3	6
2—5	26	35	95	156
6—10	71	50	81	202
11—25	145	119	63	327
26—50	94	124	34	252
51—100	68	82	16	166
101—200	63	50	6	119
201—300	28	14	2	44
301—500	29	14	1	44
501—1000	29	9	1	39
1001—3000	28	3	0	31
3001—5000	9	0	0	9
5001—10,000	5	0	0	5
>10,000	5	0	0	5
Total	603	500	302	1405

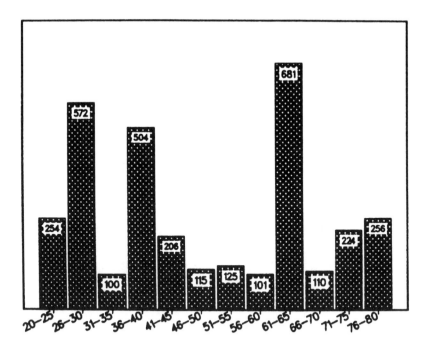

FIGURE 8. Average number of cases of illness per waterborne outbreak (1920—1980).

(87.1%), amebiasis (9.4%), shigellosis (1.3%), and acute chemical poisonings (1.0%). Almost all deaths due to typhoid fever (94%) and amebiasis (96%) occurred prior to 1941. All deaths due to shigellosis and acute chemical poisonings occurred after 1940, and these are the primary causes of deaths in recent waterborne outbreaks (Table 7). Of all deaths, 71% resulted from outbreaks in community water systems, and 21% of all deaths resulted

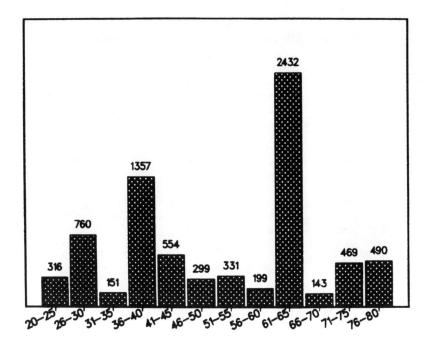

FIGURE 9. Average number of cases of illness per waterborne outbreak in community water systems (1920—1980).

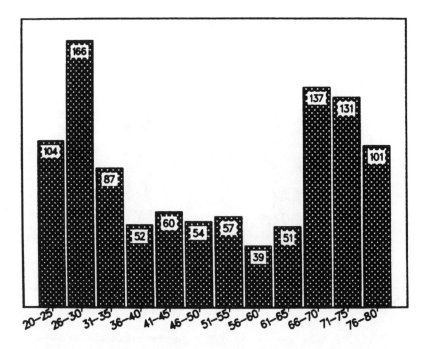

FIGURE 10. Average number of cases of illness per waterborne outbreak in noncommunity water systems (1920—1980).

from outbreaks in individual water systems (Table 8). For waterborne outbreaks which occurred in community systems, typhoid fever and amebiasis were the primary causes of death. Typhoid fever caused the most deaths in noncommunity water system outbreaks while

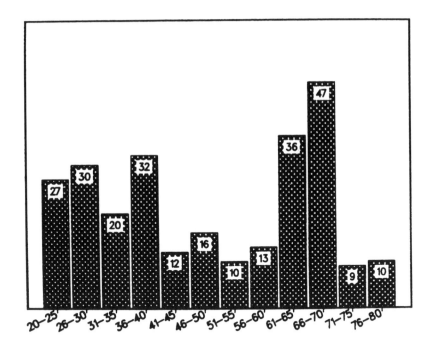

FIGURE 11. Average number of cases of illness per waterborne outbreak in individual water systems (1920—1980).

typhoid fever deaths was about one half that of nontyphoid fever deaths, whereas in noncommunity and individual water systems the mean number of typhoid fever deaths was

Deaths were reported in 231 waterborne outbreaks with a mean of 4.7 deaths occurring in these outbreaks. Deaths due to typhoid fever occurred in 207 waterborne outbreaks. In those outbreaks in community water systems where a death occurred, the mean number of typhoid fever deaths was about one half that of nontyphoid fever deaths, whereas in noncommunity and individual water systems the mean number of typhoid fever deaths was twice that of nontyphoid fever deaths (Table 9).

D. Seasonal Occurrence

A distinct seasonal occurrence of waterborne outbreaks was observed for noncommunity and individual water systems but not for community water systems (Figures 13 to 15). Outbreaks in noncommunity water systems occurred most frequently during June through August, and outbreaks in individual water systems occurred most frequently during June through September. Outbreaks in noncommunity systems affected primarily travelers, campers, restaurant patrons, and visitors to recreational areas, and the seasonal distribution of waterborne outbreaks may be due to either increased contamination of water supplies during the summer or an increased number of susceptible individuals using water supplies which are always contaminated. An analysis of outbreaks in noncommunity water systems shows the seasonal increase in outbreaks to be primarily caused by outbreaks which affect only visitors, and this suggests that the increased use by more susceptible individuals may be important as a possible explanation for the seasonal variation of outbreaks in noncommunity systems. The increased occurrence of outbreaks in individual water supplies during the summer months is caused primarily by outbreaks among rural residents and farm families. It is difficult to make a case for the importance of an increased number of susceptible individuals as an explanation for the seasonal variation of outbreaks in individual water systems, and increased contamination of individual water supplies during the summer may be the primary reason for the seasonal variation in these outbreaks. Increased water demand

Table 7
DEATHS FROM VARIOUS CAUSES IN WATERBORNE OUTBREAKS (1920—1980)

Disease	Time period					
	1920—1930	1931—1940	1941—1950	1951—1960	1961—1970	1971—1980
Typhoid fever	669	220	51	3	—	—
Shigellosis	—	—	7	1	4	2
Amebiasis	—	98	—	2	2	—
Chemical poisoning	—	—	—	4	6	1
Gastroenteritis	—	2	3	—	—	—
Toxigenic *E. coli* diarrhea	—	—	—	—	4	—
Hepatitis A	—	—	—	—	1	—
Salmonellosis	—	—	—	—	3	—

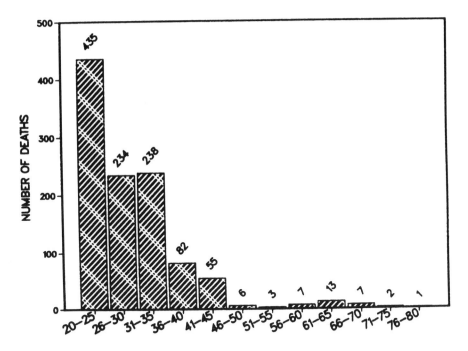

FIGURE 12. Number of deaths associated with waterborne outbreaks (1920—1980).

Table 8
DEATHS ASSOCIATED WITH WATERBORNE OUTBREAKS
IN THE U.S. (1920—1980)

Disease	Type of water supply			
	Community	Noncommunity	Individual	All
Typhoid fever	654	87	202	943
Amebiasis	99	2	1	102
Shigellosis	3	3	8	14
Chemical poisoning	2	0	9	11
Gastroenteritis	3	0	2	5
Toxigenic *E. coli* diarrhea	3	0	1	4
Salmonellosis	3	0	0	3
Hepatitis A	0	0	1	1
Total	767	92	224	1083

placed upon marginal sources of water supply during the summer may also contribute to the summer increase of outbreaks in both noncommunity and individual water systems, but this could not be ascertained from the available data.

E. Etiologic Agents and Waterborne Diseases

An etiologic agent was determined in 56% of the waterborne outbreaks (Table 10). The remaining outbreaks resulted in illness categorized as acute gastroenteritis characterized by symptoms including abdominal cramps, nausea, vomiting, and diarrhea occurring 12 to 48 hr after the consumption of contaminated water. Only during 1920 through 1935 and 1956 through 1970 were more waterborne outbreaks reported with a defined etiology than with an undefined etiology (Figure 16). It is difficult to estimate how many of the gastroenteritis

Table 9
MEAN NUMBER OF DEATHS OCCURRING IN WATERBORNE OUTBREAKS (1920—1980)

Type water supply	Mean number of deaths in outbreaks where a death occurred	
	Typhoid	Other
Community	5.2	10.3
Noncommunity	3.3	1.7
Individual	3.6	1.8
All	4.6	4.6

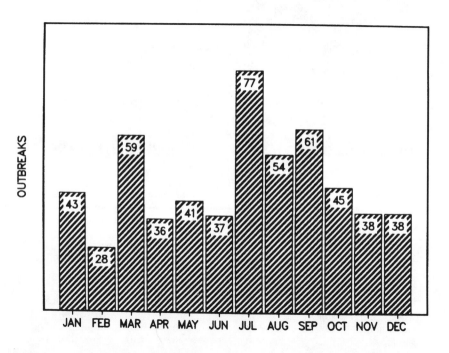

FIGURE 13. Seasonal distribution of waterborne outbreaks in community water systems (1920—1980).

outbreaks were caused by organisms which are commonly transmitted by water, but not identified because clinical specimens were not collected or not collected in a timely manner. In many of the outbreaks, the search for an etiologic agent was limited to those organisms easily cultured, and the etiologic agent was not isolated because the appropriate laboratory analysis was not conducted or not available at the time. These outbreaks could have been caused by unrecognized or less frequently identified etiologic agents. In some of the largest outbreaks occurring in recent years, an etiology could not be established even though there was extensive laboratory analysis of both human specimens and water samples, including appropriate tests for newly recognized bacterial and viral pathogens, parasites, and chemicals. Although several newly recognized etiologic agents have been uncovered in recent years, additional unrecognized agents may be responsible for waterborne gastroenteritis.

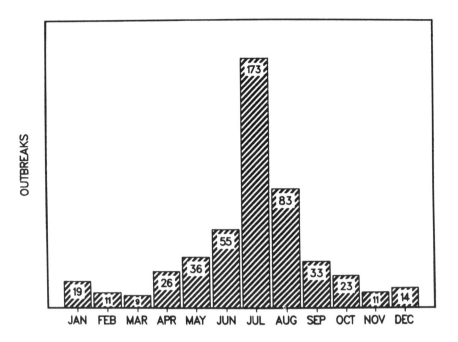

FIGURE 14. Seasonal distribution of waterborne outbreaks in noncommunity water systems (1920—1980).

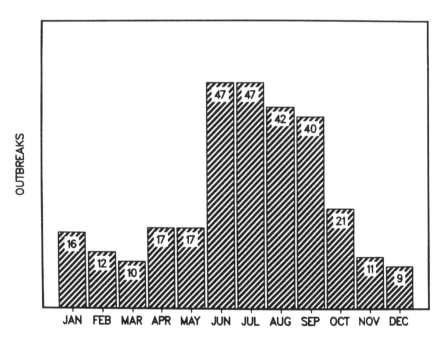

FIGURE 15. Seasonal distribution of waterborne outbreaks in individual water systems (1920—1980).

The diseases most frequently transmitted by contaminated drinking water in the U.S. during 1920 through 1980 were acute gastroenteritis of undetermined etiology, typhoid fever, shigellosis, hepatitis A, giardiasis, and acute chemical poisonings. The pathogen most commonly identified in waterborne outbreaks prior to 1966 was *Salmonella typhi*. During 1966

Table 10
ETIOLOGY OF WATERBORNE OUTBREAKS
(1920—1980)

Time period	Disease	Outbreaks	Cases	Deaths
1920—1925	Typhoid fever	127	7,294	435
	Gastroenteritis	11	27,756	0
1926—1930	Typhoid fever	100	3,072	234
	Gastroenteritis	17	63,902	0
1931—1935	Typhoid fever	85	2,114	140
	Gastroenteritis	25	7,664	0
	Amebiasis	1	1,412	98
	Hepatitis A	1	28	0
1936—1940	Gastroenteritis	91	77,403	2
	Typhoid fever	60	1,281	80
	Shigellosis	10	3,308	0
	Chemical poisoning	1	92	0
	Amebiasis	1	4	0
1941—1945	Gastroenteritis	126	36,118	3
	Typhoid fever	56	1,450	46
	Shigellosis	10	2,817	6
	Salmonellosis	1	12	0
	Paratyphoid fever	2	14	0
	Chemical poisoning	1	30	0
1946—1950	Gastroenteritis	87	10,718	0
	Typhoid fever	18	264	5
	Hepatitis A	5	173	0
	Shigellosis	4	2,321	1
	Paratyphoid fever	1	5	0
	Leptospirosis	1	9	0
	Tularemia	1	4	0
1951—1955	Gastroenteritis	31	5,297	0
	Typhoid fever	7	103	0
	Hepatitis A	7	340	0
	Shigellosis	4	732	1
	Amebiasis	1	31	2
	Salmonellosis	1	2	0
	Poliomyelitis	1	16	0
1956—1960	Gastroenteritis	21	2,306	0
	Typhoid fever	13	128	3
	Hepatitis A	11	417	0
	Shigellosis	7	3,081	0
	Chemical poisoning	3	14	4
	Salmonellosis	2	17	0
	Amebiasis	1	5	0
	Tularemia	1	2	0
1961—1965	Gastroenteritis	18	20,627	0
	Typhoid fever	11	63	0
	Hepatitis A	10	334	0
	Shigellosis	7	520	4
	Chemical poisoning	5	30	6
	Salmonellosis	3	16,425	3
	Giardiasis	1	123	0
	Paratyphoid fever	1	5	0
1966—1970	Gastroenteritis	21	5,922	0
	Hepatitis A	19	562	1
	Shigellosis	14	1,215	0
	Typhoid fever	4	45	0
	Salmonellosis	4	226	0

Table 10 (continued)
ETIOLOGY OF WATERBORNE OUTBREAKS
(1920—1980)

Time period	Disease	Outbreaks	Cases	Deaths
	Toxigenic *E. coli* AGI	4	188	4
	Chemical poisoning	4	15	0
	Amebiasis	3	39	2
	Giardiasis	2	53	0
1971—1975	Gastroenteritis	63	17,752	0
	Shigellosis	14	2,803	0
	Hepatitis A	14	368	0
	Giardiasis	13	5,136	0
	Chemical poisoning	13	513	0
	Typhoid fever	4	222	0
	Salmonellosis	2	37	0
	Toxigenic *E. coli* AGI	1	1,000	0
1976—1980	Gastroenteritis	114	22,093	0
	Giardiasis	26	14,416	0
	Chemical poisoning	25	3,081	1
	Shigellosis	10	2,392	0
	Viral gastroenteritis	10	3,147	0
	Salmonellosis	6	1,113	0
	Campylobacterosis	3	3,821	0
	Hepatitis A	2	95	0

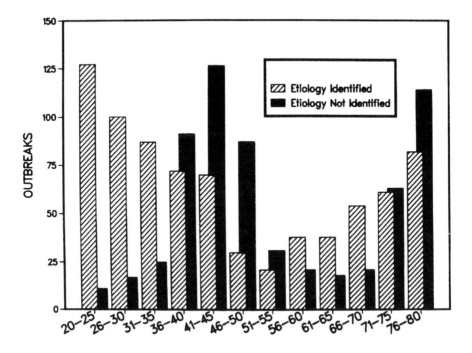

FIGURE 16. Number of waterborne outbreaks of defined and undefined etiology (1920—1980).

through 1975, hepatitis A and *Shigella* were the two most frequently identified causes of waterborne outbreaks, and in the most recent period, *Giardia lamblia,* a flagellated protozoan responsible for giardiasis, was the most commonly identified pathogen.

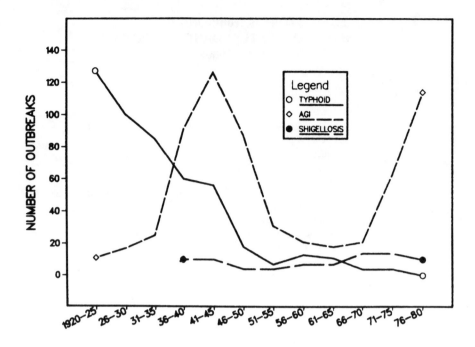

FIGURE 17. Waterborne outbreaks of typhoid fever, shigellosis, and gastroenteritis (1920—1980).

Although waterborne outbreaks of salmonellosis, tularemia, amebiasis, paratyphoid fever, leptospirosis, and poliomyelitis have been reported, the transmission of these diseases in the U.S. by the consumption of contaminated water appears to be of minor importance. In recent years, waterborne viral gastroenteritis has been reported with increasing frequency because of the availability of more sensitive laboratory techniques for identification of viral agents, and it is possible that viral agents, as well as some of the newly recognized bacterial agents (toxigenic *Escherichia coli, Campylobacter fetus, Yersinia enterocolitica*), have been important causes of waterborne outbreaks of gastroenteritis of undetermined etiology.

Waterborne outbreaks of typhoid fever have declined dramatically during this 61-year period (Figure 17). During 1920 through 1925, 127 waterborne outbreaks of typhoid fever were reported, while only 4 were reported during 1971 through 1975 and none were reported during 1976 through 1980. The number of reported waterborne outbreaks of giardiasis and acute chemical poisonings have increased over the past 35 years, especially since 1970 (Figure 18). The number of reported waterborne outbreaks of shigellosis and salmonellosis have been relatively stable over the years, with an average of approximately 9 outbreaks of shigellosis reported during each 5-year period since 1936 (Figure 17) and 2.5 outbreaks of salmonellosis reported during each 5-year period since 1946 (Figure 18). From 1946 through 1975, hepatitis A was an important cause of waterborne outbreaks, but no waterborne outbreaks of hepatitis A were reported during 1976 through 1980 (Figure 18). In general, the occurrence of waterborne outbreaks of acute gastroenteritis of undetermined etiology followed a trend similar to the occurrence of all reported waterborne outbreaks over the 61-year period (Figure 17). After 1935, more waterborne outbreaks of acute gastroenteritis were reported in each 5-year period than any other single illness (Figures 17 and 18).

1. Seasonal Variation

There was a distinct seasonal distribution of waterborne outbreaks of acute gastroenteritis, with the most outbreaks occurring in the months of June, July, and August (Figure 19).

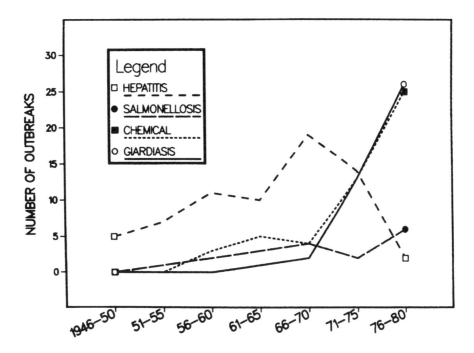

FIGURE 18. Waterborne outbreaks of hepatitis A, salmonellosis, and giardiasis (1920—1980).

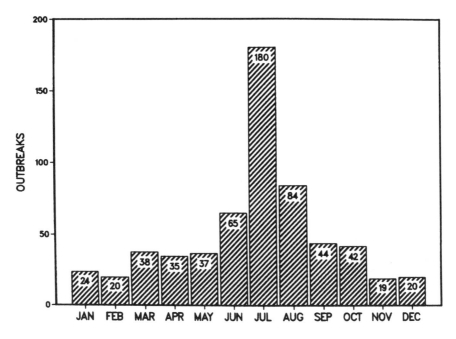

FIGURE 19. Seasonal distribution of waterborne outbreaks of gastroenteritis of undefined etiology (1920—1980).

Most of the waterborne outbreaks of shigellosis occurred during these same months (Figure 20). Seasonal distributions were apparent also for waterborne outbreaks of typhoid fever and hepatitis A, and approximately 40% of these outbreaks occurred during July, August, and September (Figures 21 and 22). Although less data are available, more waterborne

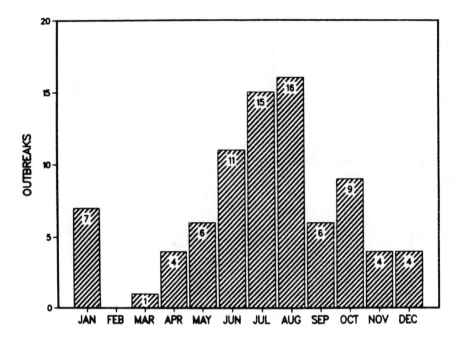

FIGURE 20. Seasonal distribution of waterborne outbreaks of shigellosis (1936—1980).

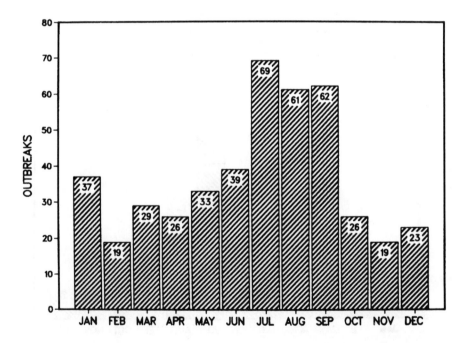

FIGURE 21. Seasonal distribution of waterborne outbreaks of typhoid fever (1920—1980).

outbreaks of salmonellosis and viral gastroenteritis outbreaks occurred in July than any other month (Figures 23 and 24). More waterborne outbreaks of giardiasis occurred in June and September (Figure 25) and a seasonal distribution for these outbreaks is not apparent until outbreaks are categorized according to the population affected (Figure 26). Most of the waterborne outbreaks of giardiasis which affected visitors or travelers were reported during

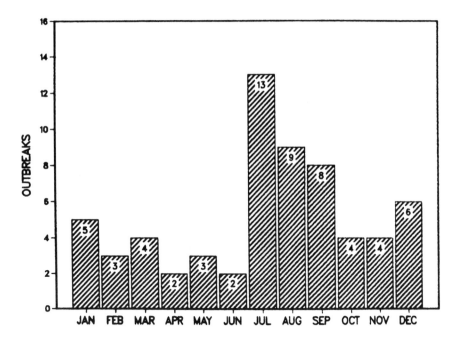

FIGURE 22. Seasonal distribution of waterborne outbreaks of hepatitis A (1931—1980).

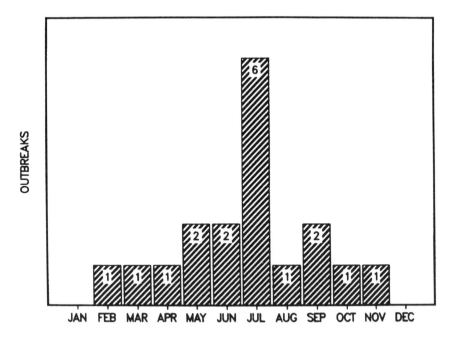

FIGURE 23. Seasonal distribution of waterborne outbreaks of salmonellosis (1941—1980).

May through September, whereas outbreaks affecting usual residents occurred primarily in the spring and fall. Unlike the waterborne outbreaks of infectious disease, acute chemical poisonings exhibited no evidence of a seasonal distribution for occurrence, but more chemical poisonings occurred in November than in any other month (Figure 27).

100 Waterborne Diseases in the United States

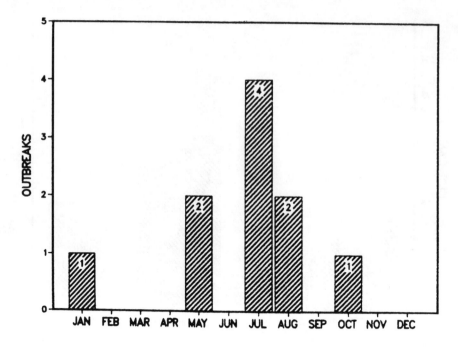

FIGURE 24. Seasonal distribution of waterborne outbreaks of viral gastroenteritis (1978—1980).

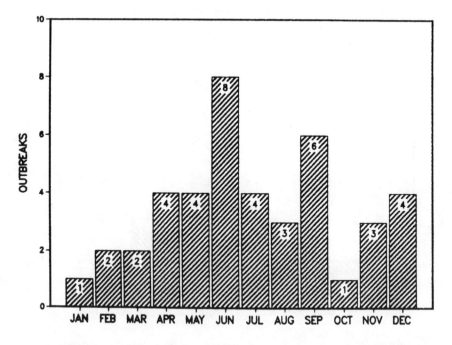

FIGURE 25. Seasonal distribution of waterborne outbreaks of giardiasis (1965—1980).

2. Importance of Waterborne Transmission

Waterborne infectious diseases are generally transmitted by the fecal-oral route from human to human or animal to human, and drinking water is only one of several possible sources of infection. An exception is dracontiasis, or guinea worm disease, which is transmitted exclusively through drinking water, but this disease is unlikely to be contracted in

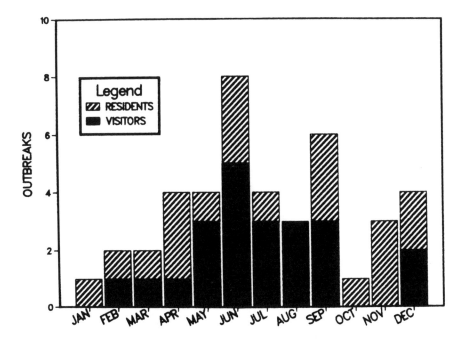

FIGURE 26. Seasonal distribution of waterborne outbreaks of giardiasis in residents and in visitors (1965—1980).

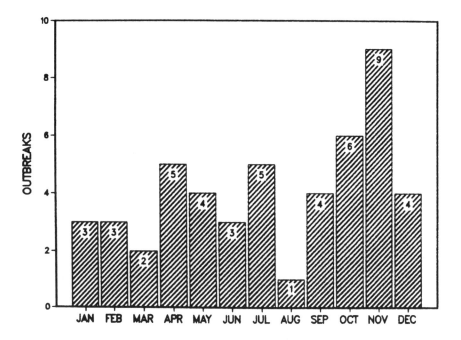

FIGURE 27. Seasonal distribution of waterborne chemical poisonings (1936—1980).

the U.S. Dracontiasis occurs primarily in India, West Africa, northeastern Africa, the Middle East, the West Indies, and northeastern South America, and is transmitted through contaminated step-wells and ponds where infected individuals wade to obtain water. Tularemia, an infectious disease of rodents and rabbits which can also occur in domestic animals, is

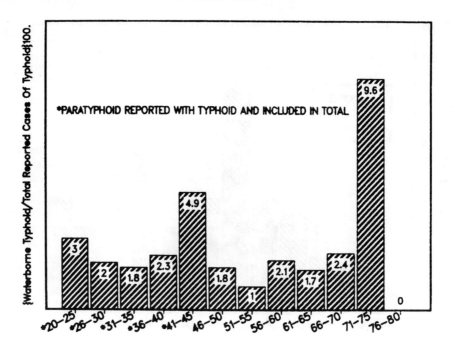

FIGURE 28. Comparison of typhoid fever occurring in waterborne outbreaks and typhoid fever reported from all causes (1920—1980).

occasionally transmitted to man through drinking water, but the more important routes of transmission are contact through wounds, unbroken skin, or conjunctiva, bites of arthropods, inhalation of dust from contaminated areas, and ingestion of insufficiently cooked rabbit. Tularemia is not directly transmitted from one person to another. Only two small outbreaks caused by drinking contaminated water from surface sources have been documented in the U.S. Leptospirosis is frequently an occupational disease of farmers, sewer workers, veterinarians, abattoir and fish workers, and rice and sugar cane field workers who are exposed to urine or tissues of infected animals or to fresh water contaminated by urine of infected domestic or wild animals. An annual average of 75 cases of leptospirosis were reported in the U.S. during 1971 through 1980.[30] The waterborne transmission from nonoccupational sources is generally through the recreational use of contaminated water; however, one waterborne outbreak was reported where drinking water which became contaminated in a storage container caused illness in nine individuals.

Cholera is the classic waterborne disease which is almost always transmitted by ingestion of contaminated water. Fortunately, cholera rarely occurs today in the U.S., but several cases of nonlaboratory-acquired *V. cholerae* infection have recently been reported where the source of infection could not be identified. The largest waterborne outbreak of cholera in the U.S. during this century occurred in 1981 when *V. cholera* 01 was found to be responsible for 17 cases of severe diarrhea caused by sewage contamination of the private water system of an oil rig.[31]

Typhoid fever is an enteric infection often transmitted by contaminated drinking water. Since 1920, no more than 5% of the reported cases of typhoid fever in the U.S. has been transmitted via contaminated drinking water except during 1971 through 1975, when the largest outbreak[32] of typhoid fever since 1936 occurred in a Florida labor camp supplied with contaminated well water (Figure 28). In most periods, cases of typhoid fever associated with waterborne outbreaks represent about 2% of all reported cases of typhoid fever. Contaminated drinking water has also been responsible for the transmission of paratyphoid fever and salmonellosis. Only four small waterborne outbreaks of paratyphoid fever have been

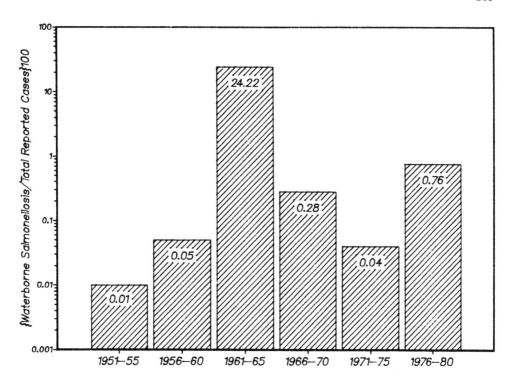

FIGURE 29. Comparison of salmonellosis occurring in waterborne outbreaks and salmonellosis reported from all causes (1951—1980).

reported in the U.S. Some 19 waterborne outbreaks of salmonellosis have been reported, including the large outbreak of 16,000 cases of illness in Riverside, Calif.[33] in 1965. Generally, less than 1% of *Salmonella* gastroenteritis in the U.S. has been transmitted via contaminated drinking water, but during the period which included the Riverside outbreak, 24% of the reported cases of *Salmonella* gastroenteritis occurred as the result of waterborne outbreaks (Figure 29).

Another enteric disease often transmitted by contaminated drinking water in the U.S. is shigellosis. The symptoms are relatively similar to salmonellosis, but *Shigella* infection produces bloody stools and rectal pain more frequently. No more than 6% of the reported cases of shigellosis in the U.S. have occurred as the result of waterborne outbreaks, and generally the cases of shigellosis caused by contaminated drinking water represent about 2% of all reported cases of shigellosis (Figure 30).

Hepatitis A is a viral disease which has long been recognized to be transmitted by contaminated drinking water. The virus produces symptoms of nausea, vomiting, muscle aches, and jaundice occurring approximately 15 to 45 days after exposure. The long incubation period tends to obscure the relationship between illness and consumption of contaminated water, and waterborne outbreaks are often not recognized. Diagnosis is made by history, physical examination, tests of liver function, and serologic studies. In the U.S., generally less than 1% of the reported cases of hepatitis A have resulted from waterborne outbreaks (Figure 31). Batik et al.[34] studied the impact of various water supply characteristics on the endemic rates of hepatitis A in a state identified as having a good surveillance system for hepatitis. Although no statistically significant associations were found between endemic hepatitis A rates and water supply characteristics, the study size was sufficient to estimate that the total variation of hepatitis A rates attributable to water supply characteristics was probably less than 8% of the reported cases of hepatitis A from all causes.

104 Waterborne Diseases in the United States

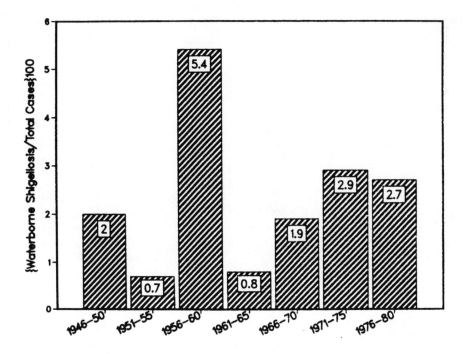

FIGURE 30. Comparison of shigellosis occurring in waterborne outbreaks and shigellosis reported from all causes (1946—1980).

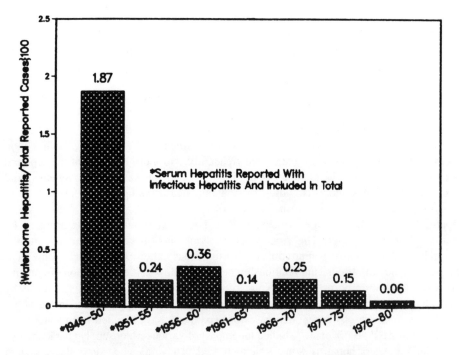

FIGURE 31. Comparison of hepatitis A occurring in waterborne outbreaks and hepatitis A reported from all causes (1946—1980).

In recent years, giardiasis has been a frequent cause of waterborne disease outbreaks in the U.S. The etiologic agent is the protozoan *Giardia lamblia,* which exists in a cyst form that is resistant to the usual disinfection dose and contact time provided for the treatment of drinking water. The organism produces a chronic (more than 1 week) diarrhea syndrome that is accompanied by weight loss, epigastric pain, and flatulence. Diagnosis is established by identifying cysts or trophozoites in fecal material or in fluid from the upper small bowel; however, laboratory confirmation may be difficult. Outbreaks are generally caused by ingestion of untreated stream water in mountainous areas and treated surface waters which have been disinfected but either not filtered or not properly filtered. Beavers have been implicated as the source of infection in some outbreaks where no obvious source of human fecal contamination was found. In the U.S., contaminated drinking water is an important source of transmission of giardiasis, but it is difficult to quantify this because reporting of giardiasis has been sporadic until recently. Cases of giardiasis are optionally reported by certain states, and during 1976 through 1980 an average of 11,700 cases were reported each year.[30] The 14,416 cases of waterborne giardiasis occurring during 1976 through 1980 represent 25% of these optionally reported cases of giardiasis from all causes during this 5-year period. Wild and domestic animals are included in the cycle of transmission, and person-to-person transmission, especially through day-care centers, is also important.

Entamoeba histolytica is the protozoan causing amebic dysentery or amebiasis. Most infections are asymptomatic, and intestinal disease varies from mild abdominal discomfort with diarrhea containing blood to acute dysentery with fever, chills, and bloody or mucoid diarrhea. Some infections result in invasion of the mucosal lining of the intestine, and the organism may be disseminated via the bloodstream, producing abscess of the liver. Humans are the only reservoir and the organism is transmitted by the fecal-oral route. An annual average of 3197 cases of amebiasis were reported[30] in the U.S. during 1971 through 1980. Outbreaks of amebiasis are caused by sewage-contaminated water, but the last reported waterborne outbreak in the U.S. affected only 31 individuals and occurred in 1953. The largest waterborne outbreak of amebiasis occurred in 1933 in Chicago and was caused by sewage entering the water system of a hotel through a cross-connection.[1] This outbreak resulted in at least 1409 cases of amebiasis and 98 deaths among guests and employees of two Chicago hotels.

Poliomyelitis has been suspected of being transmitted via contaminated drinking water. The first reported epidemic of poliomyelitis in the U.S. occurred in New England,[35] and Caverly[36] suggested that the disease might be waterborne. Eight outbreaks of poliomyelitis in Europe and North America have been attributed to contaminated drinking water, but Mosley[37] feels that only the outbreak[38,39] which occurred in Huskerville, Neb. in 1952 had adequate documentation of waterborne transmission. More than 10% of the 347 children in the affected area developed poliomyelitis over a 5-week period and 16 children suffered paralytic disease. During this same period, no cases of paralytic or nonparalytic poliomyelitis occurred among 256 children residing in an adjoining section of the village. The epidemiologic study showed a spatial relationship between the distribution of cases and the location of water closets with flush valves containing no vacuum breakers, and showed a temporal relationship between the outbreak and the occurrence of extreme fluctuations of pressure in the water mains. It was felt that fecal contamination of the water mains could have been caused by backsiphonage, and the evidence pointed to the proximate fecal contamination of the water as the sole factor in explaining the distribution of poliomyelitis within the community. Immunological studies further confirmed that type I poliomyelitis had been transmitted by contaminated water. An outbreak of waterborne gastroenteritis occurred at a restaurant in Michigan in 1970, and poliovirus 2 was reported to have been recovered from the well used as a drinking water source.[40] No cases of poliomyelitis, however, were detected among those who had become ill with nausea, vomiting, and diarrhea.

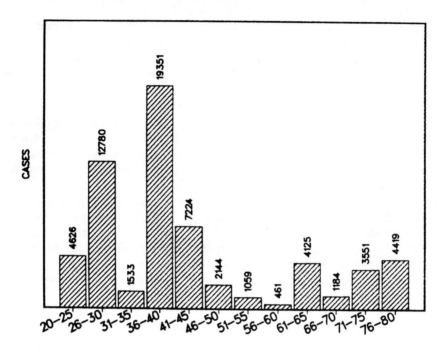

FIGURE 32. Average annual number of cases of waterborne disease of undefined etiology (1920—1980).

Although gastroenteritis caused by *Yersinia enterocolitica* has been suspected to be acquired by drinking contaminated water in the U.S., sufficient documentation of this etiology from waterborne outbreaks was lacking until a small outbreak[31] of 15 cases was reported in Pennsylvania in 1981. Waterborne outbreaks of gastroenteritis in the U.S. have also been caused by enterotoxigenic *Escherichia coli* (ETEC), *Campylobacter fetus* ssp. *jejuni,* and viral agents. These organisms are all transmitted by the human fecal-oral route, and domestic and wild animals are felt to be important reservoirs for *C. jejuni* and *Y. enterocolitica.* Gastroenteritis and diarrhea caused by these organisms are not included in the list of notifiable or optionally reported diseases,[30] and it is difficult to estimate the importance of waterborne transmission for these etiologies.

3. Waterborne Diseases

Some 297,557 cases of waterborne gastroenteritis of undetermined etiology, accounting for about 77% of all waterborne illness, were reported during the 61-year period. The average number of cases of waterborne gastroenteritis varied from 461 cases annually during 1956 through 1960 to 19,351 cases annually during 1936 through 1940 (Figure 32). The 28 waterborne outbreaks of undetermined etiology in 1920 through 1930 resulted in an average of 8332 cases of gastroenteritis annually compared with an annual average of 942 cases of typhoid fever resulting from 227 waterborne outbreaks during the same period (Figure 33). The number of cases of waterborne gastroenteritis exceeded the number of cases of other specifically identified waterborne diseases in each subsequent 5-year period except 1956 through 1960 (Figures 32 to 42).

From 1920 through 1965, more waterborne outbreaks of typhoid fever were reported than were waterborne outbreaks of other defined etiologies, but after 1946 these typhoid outbreaks were relatively small and resulted in less than 100 cases of typhoid annually. From 1936 through 1961, waterborne outbreaks of shigellosis resulted in more cases of illness (Figures 34 to 38) and from 1961 through 1965, waterborne outbreaks of salmonellosis resulted in

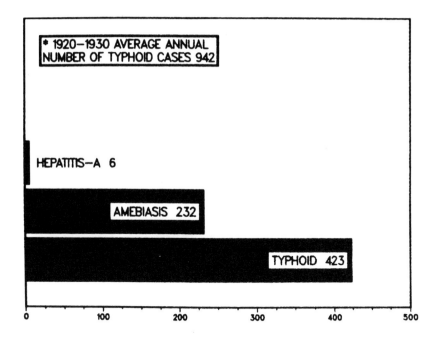

FIGURE 33. Average annual number of cases of waterborne disease of defined etiology (1931—1935).

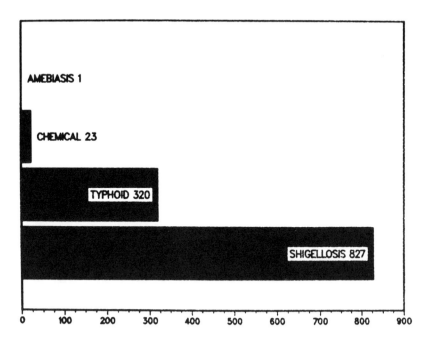

FIGURE 34. Average annual number of cases of waterborne disease of defined etiology (1936—1940).

more cases of illness than occurred in waterborne outbreaks of typhoid fever (Figure 39). During 1966 through 1970, more cases of waterborne shigellosis were reported than any other waterborne disease of defined etiology (Figure 40), and during 1971 through 1980 more cases of waterborne giardiasis were reported than any other waterborne disease of defined etiology (Figures 41 and 42).

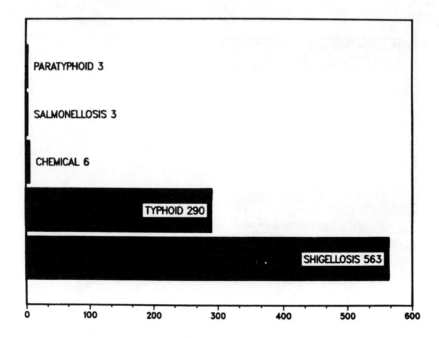

FIGURE 35. Average annual number of cases of waterborne disease of defined etiology (1941—1945).

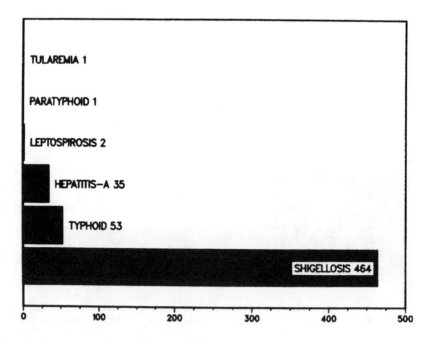

FIGURE 36. Average annual number of cases of waterborne disease of defined etiology (1946—1950).

a. Gastroenteritis

Many causes of acute gastroenteritis are possible, and it has long been suspected that "nonbacterial" causes were important. It is only recently that viral agents have been identified as a cause of acute gastroenteritis, as significant advances in laboratory techniques

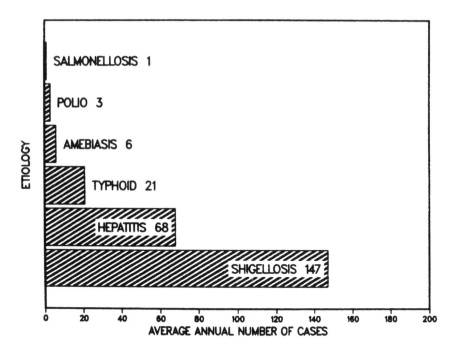

FIGURE 37. Average annual number of cases of waterborne disease of defined etiology (1951—1955).

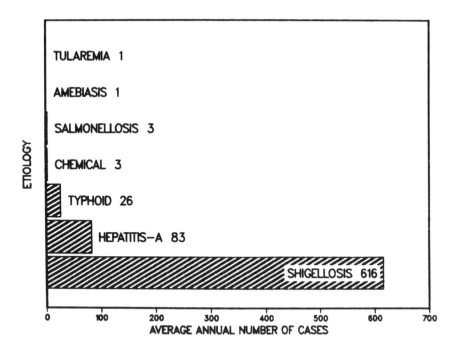

FIGURE 38. Average annual number of cases of waterborne disease of defined etiology (1956—1960).

have occurred, and these agents can now be detected through immunoassays or electron microscopy. Although it was found possible to readily cultivate the major viral respiratory pathogens in cell culture or laboratory animals, efforts to identify and cultivate the viral

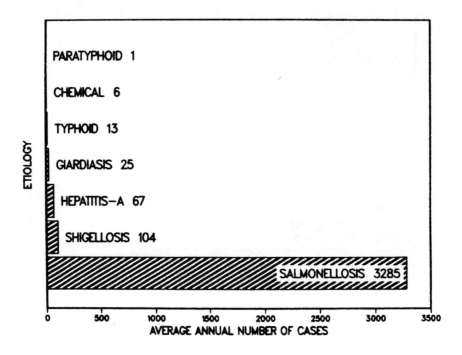

FIGURE 39. Average annual number of cases of waterborne disease of defined etiology (1961—1965).

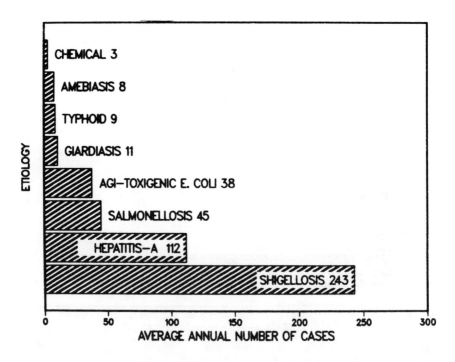

FIGURE 40. Average annual number of cases of waterborne disease of defined etiology (1966—1970).

gastroenteritis agents in cell cultures or laboratory animals have not generally been successful.

Acute infectious nonbacterial gastroenteritis occurs primarily in two epidemiologically distinct forms and is caused in humans by at least two distinct groups of viruses.[41-49] The

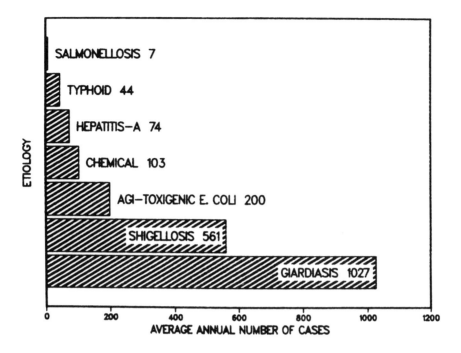

FIGURE 41. Average annual number of cases of waterborne disease of defined etiology (1971—1975).

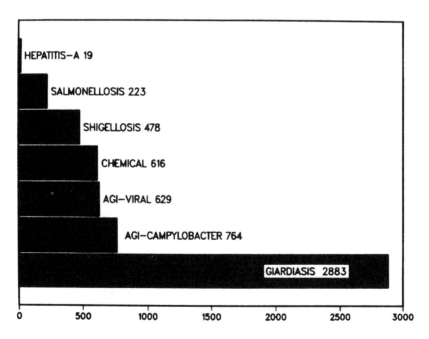

FIGURE 42. Average annual number of cases of waterborne disease of defined etiology (1976—1980).

70-nm rotaviruses have been shown to be a major cause of acute gastroenteritis in infants and young children. This form of illness typically produces severe diarrhea lasting for 5 to 8 days usually accompanied by fever and vomiting. Infants may become severely dehydrated. It is reported that rotavirus infection is responsible for about half the cases of infantile diarrhea that require hospitalization world-wide. In temperate climates the illness is common during the cooler months. Clinical disease may occur among family and adult contacts of ill infants. The illness is usually sporadic but is occasionally epidemic, and although it occurs predominantly in young children, it can affect adults. The other viruses associated with acute nonbacterial gastroenteritis are less well characterized. Virus-like 27-nm particles have been identified in stool filtrates from specimens obtained during gastroenteritis outbreaks in Norwalk, Ohio, Honolulu, Hawaii, and Montgomery County, Maryland. At least three serologically distinct 27-nm particles have been identified. The Norwalk virus is the prototype and most extensively studied of this group, which is referred to as Norwalk-like agents or viruses. Gastroenteritis caused by Norwalk virus is typically an explosive but self-limited illness usually lasting 24 to 48 hr, and symptoms include vomiting, diarrhea, nausea, abdominal cramps, headache, low-grade fever, anorexia, and malaise. It is characteristically epidemic. The Norwalk-like agents have been responsible for family and community outbreaks among older children and adults and are reported to cause one third of the epidemics of viral gastroenteritis that occur in the U.S. Although some rotaviruses have recently been grown by special techniques in cell culture, Norwalk virus has not been cultivated in vitro and produces no disease in inoculated laboratory animals.

Norwalk-like agents are suspected to be the cause of a number of waterborne gastroenteritis outbreaks of undetermined etiology which have occurred in the U.S. Ten waterborne outbreaks of gastroenteritis attributed to Norwalk or Norwalk-like agents on the basis of serologic testing were reported during 1976 through 1980 in the U.S.[50] In addition, two outbreaks of Norwalk GI illness were associated with swimming in recreational lakes in Georgia[51] and in Michigan;[52] two outbreaks were associated with the consumption of oysters,[50,53] and two outbreaks were associated with the consumption of water on cruise ships.[50] Recognizing that most gastroenteritis outbreaks are not specifically investigated for a virus etiology, Kaplan et al.[54] reviewed the records of acute GI outbreaks to determine the proportion of outbreaks that were clinically and epidemiologically consistent with Norwalk-like virus infection using the criteria: negative stool cultures for bacterial pathogens, mean or median duration of illness 12 to 60 hr, vomiting in 50% or more of the cases, and, if known, a mean or median incubation period of 24 to 48 hr. Of the waterborne outbreaks reported during 1976 through 1979, 96 had sufficient data for evaluation, and it was found 22 waterborne outbreaks met the criteria for Norwalk-like virus infection. Only six of these outbreaks had serologic testing for Norwalk virus and had specifically been identified with a viral etiology. The remaining 16 outbreaks had been classified as waterborne gastroenteritis of undetermined etiology. If the classification of waterborne outbreaks were changed to reflect these findings, the statistics would show (1) 15% of the waterborne outbreaks during 1976 through 1979 to be of viral etiology rather than the reported 4% and (2) 48% of the outbreaks during the same period to be of gastroenteritis of undetermined etiology rather than the reported 59%.

The ten reported waterborne outbreaks of viral gastroenteritis caused by Norwalk or Norwalk-like agents occurred primarily in noncommunity water systems and affected 3147 individuals, primarily visitors. Six outbreaks occurred in well water systems where the water received no treatment or the disinfection provided was either inadequate or interrupted. Two outbreaks were caused by cross-connections or backsiphonage. Coliform contamination of the water supply was evident in all outbreaks.

The following outbreaks illustrate that viral agents contribute to waterborne GI disease in the U.S., and substantiate previous reports that outbreaks with a high attack rate, predominance of upper GI symptoms, and relatively short duration of illness are compatible with viral gastroenteritis. Morens et al.[55] investigated an outbreak at a resort camp in Colorado in 1976 where 418 (55%) persons reported gastroenteritis at the camp or within 1 week of leaving it. Symptoms included vomiting (81%), diarrhea (65%), and fever (49%). The median duration of illness was 24 hr. The attack rate increased with increasing consumption of water or iced beverages. The camp was supplied with water by a natural spring in a meadow at the base of a small hill. At the top of the hill was a private cabin with a septic tank approximately 50 ft above the spring. Prior to the outbreak, maintenance personnel noted malfunctioning of the chlorinator at the spring and turned it off for several hours while making repairs. Fluorescein dye flushed into the cabin waste water system rapidly appeared in the spring and in the camp tap water. Routine laboratory tests did not reveal bacterial, viral, or parasitic pathogens, but immune electron microscopy detected 27-nm virus-like particles in two of five diarrheal stool filtrates. Oral administration of one of these bacteria-free filtrates to two volunteers induced a GI illness similar to that observed in the camp visitors. This outbreak was not classified among those with a viral etiology in the waterborne outbreak tabulation[11] primarily because of the lack of serologic evidence, but it is considered to be highly suspect based on the identification of the virus-like particle in the stool specimens.

A survey of 57 groups visiting a camp in northeastern Pennsylvania between May 1, 1978 and June 16, 1978 revealed that 13 groups had experienced illness in over 15% of their members.[56] A total of 350 persons were reported to have been ill. Serum and stool specimens collected from ill members of the last group that visited the camp were negative for bacterial pathogens. However, three of five ill persons had a fourfold titer rise in antibody to a Norwalk-like agent; two controls were negative. The illness in this group was characterized by vomiting (81%), abdominal pain (74%), nausea (67%), and diarrhea (56%). An association was found between the quantity of camp water consumed and illness. A similar questionnaire, administered to two other ill groups, also showed an association between the quantity of camp water consumed and illness. A study of the water system demonstrated the presence of coliforms, inadequate chlorination, and several areas of possible contamination.

On July 27, 1978 an outbreak of gastroenteritis was reported from another summer camp in northeastern Pennsylvania.[56,57] The cases were characterized by abdominal pain (80%), nausea (73%), and vomiting (53%). Headache (47%), diarrhea (38%), and chills (38%) were also prominent findings. The median duration of illness was 2 days. Review of the infirmary records revealed 73 cases of gastroenteritis during the first session of summer camp, which was approximately ten times the rate reported from the previous year. A sharp increase in cases began 48 hr after the arrival of the second-session campers. Consumption of five or more glasses of water a day (or water-containing beverages) was statistically associated with illness. Bacterial samples from the camp water supply revealed fecal coliforms, and although the well water was chlorinated, tests for residual chlorine level revealed no residual until July 28. No new cases were reported after July 29. Laboratory studies of stools from ten patients and ten controls revealed no bacterial pathogens. Three of three paired serum specimens, however, showed fourfold or greater rises to Norwalk-like agent by radioimmunoassay.

An explosive outbreak[58] of 467 cases of Norwalk-related viral gastroenteritis caused by contaminated drinking water occurred in an elementary school in Pierce County, Wash. in May 1978. No bacterial or parasitic pathogens were identified in stool specimens, and immune electron microscopy of three stool specimens did not reveal rotavirus or Norwalk-like agents. Acute and convalescent serum pairs from two of three persons, however, showed fourfold or greater antibody rises to the Norwalk antigen by radioimmunoassay. The water source at the school was contaminated on May 2 when sewage from a plugged septic tank entered a floor drain in the boiler room and likely entered the water system by backsiphonage through a cross-connection at the well-pump pressure tank. Illness compatible with Norwalk-related virus infection began approximately 24 to 48 hr later and affected some 70% of the persons exposed.

The first Norwalk-related outbreak associated with a contamination of a large municipal water system occurred in August 1980 in the Lindale area of Rome, Ga.[59] The outbreak lasted for 1 week and affected over 1500 individuals. Water was likely contaminated through a cross-connection between the community water system and an industrial water system which was found to contain fecal coliforms. The clinical and epidemiologic findings were consistent with the cause being Norwalk-like agent. Rectal swabs from ill persons were negative for bacterial pathogens, and no viral particles were found in four stool specimens examined by immune electron microscopy. Of 19 acute and convalescent serum pairs from patients, 12 demonstrated a fourfold rise in titer of antibody to Norwalk virus antigen.

Rotaviruses have not been shown to be an important cause of waterborne outbreaks, and Kaplan et al.[54] doubt that rotaviruses were a possible etiologic agent for the previously discussed gastroenteritis outbreaks which occurred during 1976 through 1979 and met his criteria for Norwalk-like illness. However, they have caused disease in adults as well as children and do play a role in adult travelers' diarrhea,[42,49] and three reports outside the U.S. have suggested the possibility of rotavirus transmission via drinking water.[60-62] An outbreak of 3172 cases of acute gastroenteritis affecting 30% of the population of a small Swedish town in January 1977 was suspected to be of waterborne origin and caused by rotavirus.[60] Particles with a diameter of 75 nm, similar in appearance to rotavirus, were detected in fecal specimens of acutely ill and convalescent patients. No *Salmonella, Shigella, Yersinia,* or cytopathogenic viruses were found in the ill patients and there was no indication of transmission from children to adults. The disease was characterized by diarrhea, vomiting, headache, muscular pain, and fever. Symptoms disappeared after 1 to 3 days. Observations suggested that the water supply was contaminated with waste water. The contamination of water mains in a village of the U.S.S.R. reportedly caused rotavirus infection in 173 persons.[61] Rotaviruses were reported to have been isolated from stool specimens. All age groups were represented among patients, and children under 7 years of age constituted only 10% of the patients. A contaminated water supply caused an outbreak of gastroenteritis in a private school in Rio de Janeiro in 1980, where rotavirus and *Shigella sonnei* were isolated from stool specimens.[62] The overall attack rate was 75%, and students in all grades were affected. Examination of initial stool specimens from 19 ill children revealed rotavirus in 7, *S. sonnei* in 6, and both pathogens in 4. Second and third collection of stool specimens yielded only *S. sonnei*. Seroconversion for rotavirus was discovered in 18 paired sera in 4 cases and for *S. sonnei* in 7 cases.

Recent waterborne outbreaks have suggested that other viruses not yet fully characterized (or other agents) may be responsible for some waterborne gastroenteritis, as extensive laboratory analyses of both clinical specimens and water samples in several outbreaks could not establish an etiology. As has previously been noted, it is not possible to rule out a bacterial etiology for many of the early waterborne outbreaks of gastroenteritis, as laboratory analyses may not have been available or may not have been conducted as part of the investigation. The recent association of toxigenic *E. coli, Y. enterocolytica,* and *C. fetus* with waterborne gastroenteritis suggests that these may also be important as causes of waterborne outbreaks of undetermined etiology. These bacteria are more difficult to cultivate in the laboratory than are the familiar bacterial pathogens, such as *Salmonella* and *Shigella*. However, sometimes it is possible to use symptoms of the illness to help distinguish bacterial gastroenteritis from viral gastroenteritis. The illness with each of these bacterial pathogens is usually prolonged and vomiting is usually not reported in the majority of cases.

There are at least three types of pathogenic *E. coli*. Invasive strains cause disease manifested by fever and mucoid and occasionally bloody diarrhea. Enterotoxigenic strains cause disease ranging from mild diarrhea to seriously dehydrating cholera-like illness with profuse watery diarrhea, abdominal cramps, and vomiting. The enteropathogenic strains have generally been associated with outbreaks of acute diarrheal disease in newborn nurseries, and although they may produce diarrhea through the elaboration of enterotoxins, the pathogenic mechanisms have not been defined. Enterotoxigenic *E. coli* (ETEC) has been found to be a common cause of diarrhea in travelers to Brazil, Kenya, and Mexico.[63] In a prospective study[63] of travelers to Mexico, median duration of illness was 5 days and median onset of illness was 6 days after arrival.

Four waterborne outbreaks caused by enteropathogenic *E. coli* serotypes 0111:B4 and 0124:B27 were reported in the U.S. in 1967 and 1969, but the strains were not demonstrated to be pathogenic by either toxin production or tissue invasion. The first well-documented waterborne outbreak[64] caused by an ETEC strain occurred in 1975 at Crater Lake National Park and affected more than 1000 individuals. ETEC serotype 06:K15:H16 was isolated from ill persons and from a water sample. The park water supply was contaminated when an obstructed sewer line caused sewage to overflow into a spring. Although the spring was chlorinated, the facilities were improperly designed and did not provide adequate disinfection prior to distribution of the water. Outbreaks of gastroenteritis have been reported in passengers of cruise ships operating from U.S. ports, and several have been associated with contaminated drinking water.[65] A single serotype of *E. coli* that produced heat-labile enterotoxin without producing heat-stable enterotoxin was recovered from most of the passengers who contracted diarrheal illness aboard two successive cruises of a Miami-based cruise ship.[66] Clinical findings were similar to that caused by ETEC that produce only heat-stable and heat-labile enterotoxin. Epidemiologic evidence suggested the illness was associated with drinking water aboard the ship. Waterborne outbreaks of ETEC have also been reported in other countries. In Japan,[67] 956 persons became ill, with diarrhea as the primary symptom, after drinking water at Nagoya airport in September 1973. ETEC serotypes 06:K:H16, 027:K:H7, 0148:K:H28 were isolated from well water and from human stool specimens. Karoly[68] reported a fatal case of illness caused by *E. coli* 0124:K72:B17 in well water in Europe.

The importance of *Y. enterocolitica* as a cause of waterborne gastroenteritis of undetermined etiology in the U.S. is uncertain at this time. The organism has been found in surface waters and unchlorinated well waters in the states of Washington, Wisconsin, New York, California, and Colorado, Australia, Canada, Norway, Denmark, and Belgium and unchlorinated water in Washington state.[69-82] The most common symptoms of *Y. enterocolitica* infection in humans are fever, abdominal pain, and diarrhea, and cases have been reported in association with disease in domestic animals. The number of human cases that are recognized is small and largely dependent on the skill and experience of the clinician and laboratory. *Y. enterocolitica* is frequently isolated from wild and domestic animals. Fecal-oral transmission of the infection has been shown after contact with infected persons or animals and after eating or drinking contaminated food and water. The first sufficiently documented waterborne outbreak[30] in the U.S. was reported in 1981, and prior to that time, evidence for waterborne transmission was only suggestive.[81-82] A waterborne outbreak of gastroenteritis in which *Y. enterocolitica* was isolated from well water occurred[83] at a Montana ski resort in 1974. Illness was epidemiologically associated with drinking water, and chlorination of the wells stopped the outbreak. The significance of finding *Y. enterocolitica* in the well water is unclear because rectal swab cultures from acutely ill persons were not examined for this organism. Christiensen[84] reported isolation of *Y. enterocolitica*, biotype 4, serotype 0:3, from an unchlorinated well water supply in a rural area of Denmark. The well water was suspected as the source of infection for an 8-month-old child hospitalized with yersiniosis. A blood titer against serotype 0:3 was demonstrated. The well water was used in the preparation of baby food, but it could not be established whether enrichment of *Yersinia* in the baby food or direct ingestion of water was responsible for the infection.

C. jejuni has recently been recognized as a leading cause of bacterial gastroenteritis and has been identified as a cause of travelers' diarrhea.[85,86] *Campylobacter* is now isolated as commonly as *Salmonella* and *Shigella* from patients with diarrhea; however, the epidemiology and pathophysiology of campylobacter infection are still not well understood.[87] There are several species of *Campylobacter*, but the major pathogen of humans is *C. jejuni*.[85]

In developed countries with good sanitation, most infections are seen in young adults. In less developed countries, children are most often affected. The incidence of infection increases during the summer months.[85] In North America, Europe, Africa, and Australia, studies have shown *C. jejuni* to be a pathogen in 3 to 11% of patients with diarrhea.[87] Mild infections may produce symptoms lasting for a single day and resembling those seen in viral gastroenteritis, but *C. jejuni* may also cause illness which mimics acute relapses of ulcerative colitis.[85] The predominant symptoms among persons ill enough to seek medical attention are diarrhea, abdominal pain, malaise, fever, nausea, and vomiting; a history of grossly bloody stools is also common.[85] Abdominal pain may be the only symptom and may mimic acute appendicitis.[85] Illness is frequently self-limiting, lasting no more than 1 week, but about 20% of patients have a prolonged illness or relapse. The asymptomatic excretion of *Campylobacter* by humans is uncommon.[85] *Campylobacter* may be transmitted from animal and inanimate reservoirs to humans by direct contact with infected animals or through the ingestion of contaminated food or water.[85] Person-to-person transmission can also occur and transmission has been suggested by infants in day-care centers.[86] *Campylobacter* species are widespread in animals as pathogens and commensals, and the reservoir for *C. jejuni* in nature is large.[86,87] *Campylobacter* organisms found in water are likely due to fecal contamination by wild or domestic animals, as the excretion of these organisms by humans is not widespread.

The first waterborne outbreak of gastroenteritis caused by *C. jejuni* in the U.S. occurred in 1978 in Bennington, Vt. and affected approximately 3000 people (19% of the population).[88] The mean duration of illness was 4.6 days, and reported symptoms included cramps (87%), diarrhea (82%), malaise (73%), headache (47%), nausea (45%), fever (45%), vomiting (21%), and blood diarrhea (9%). *C. jejuni* was cultured from 15 or 42 rectal swabs from ill persons but not from rectal swabs from 23 persons who were not ill. *Campylobacter* organisms were not isolated from water samples or from specimens from wild or domestic animals. *Y. enterocolitica* was also isolated from four rectal swabs, but the data do not suggest that this organism was a cause of the outbreak. The illness was epidemiologically associated with drinking water, and the report of heavy rainfall several days before the outbreak and high turbidity of the water support the hypothesis that increased runoff from the watershed led to contamination of the water which was not filtered. Although the water was disinfected, records showed that throughout the period of the outbreak several areas of the town had no residual chlorine. Waterborne outbreaks have also been reported in Sweden[89] and the U.K.[90] In an outbreak of approximately 2000 people in central Sweden in 1980, *C. jejuni* was isolated from stool specimens in 221 of 263 ill persons.[89] There was strong circumstantial evidence that tap water was the source of infection, and the municipal water system may have been contaminated through cross-connections in the distribution system, as many houses maintained pumps to irrigate gardens with surface water. Reported symptoms included diarrhea (89%), abdominal pain (82%), nausea (57%), headache (56%), fever (53%), vomiting (19%), and blood in stools (6%). In an outbreak of gastroenteritis[90] affecting 234 pupils and 23 staff at a boarding school in the U.K. in 1981, *C. jejuni* was isolated from ill persons and two samples of water from an open storage tank which supplied water to the main school building. The source of the water supply was an unchlorinated well. The most likely source of contamination was fecal material from birds or bats roosting on the storage tower and the nearby clock tower. The transmission of *Campylobacter* enteritis through untreated mountain streams has also been documented in the U.S., and Taylor et al.[91] suggest that backcountry surface water can be an important source of *C. jejuni*. During the summers of 1980 and 1981, *C. jejuni* was isolated from 23% of persons with diarrheal disease acquired in the area of Grand Teton National Park, Wyo.; *G. lamblia* was also isolated from 8% of ill persons. *Campylobacter* enteritis occurred most frequently in young adults who had been hiking in wilderness areas and was statistically associated with drinking

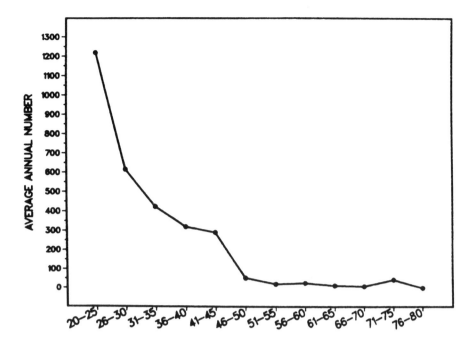

FIGURE 43. Average annual number of cases of typhoid fever occurring in waterborne outbreaks (1920—1980).

untreated surface water. A study[92] of persons with laboratory-confirmed sporadic *C. jejuni* infection and controls matched for age and gender in Colorado in 1981 showed that drinking untreated water, drinking raw milk, and living in a household with a cat were risk factors for acquiring the infection. Camping itself was not associated with infection.

b. Typhoid Fever and Hepatitis A

The annual average number of cases of typhoid fever occurring in waterborne outbreaks decreased from 1216 cases during 1920 through 1925 to 53 cases during 1946 through 1950 and 9 cases during 1966 through 1970 (Figure 43). An average of 44 cases of waterborne typhoid fever occurred during 1971 through 1975 as a result of the large outbreak at a labor camp in Florida[32] in 1973, and no cases of waterborne typhoid fever were reported during 1976 through 1980. The outbreak of 210 suspected and confirmed cases of typhoid fever at the South Dade Migrant Farm Labor Camp in Dade County, Fla. was the largest reported waterborne outbreak of typhoid fever in the U.S. since 1936. No deaths occurred, and the outbreak subsided with few secondary cases and no transmission to individuals not connected with the camp. The water supply of the camp, two shallow wells, were implicated as the vehicle of infection. An engineering evaluation revealed that chlorination of the wells was interrupted prior to the outbreak, and it was felt that the water system became contaminated at that time. The acquifer was composed of solution channels, and the wells had a history of intermittent contamination. A young mentally retarded camp resident was identified as the index case, and the wells could have been contaminated from the adjacent day-care center attended by the index case. The last reported waterborne outbreak of typhoid fever in the U.S. occurred in 1974 in the Dryden area of Chelan County, Wash., and affected five individuals consuming water from a poorly constructed shallow well approximately 15 ft deep. A typhoid carrier was identified in the neighborhood, and it was suspected that he used pit privies, which were confirmed by fluorescein dye studies as a source of contamination of the well. *S. typhi* was also isolated from the well water.

118 *Waterborne Diseases in the United States*

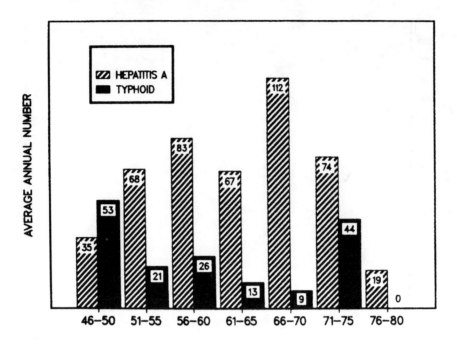

FIGURE 44. Average annual number of cases of typhoid fever and hepatitis A occurring in waterborne outbreaks (1946—1980).

As the number of cases of waterborne typhoid fever declined, cases of waterborne hepatitis A were reported more frequently, and after 1950 the number of cases of hepatitis A resulting from waterborne outbreaks exceeded the cases of typhoid fever transmitted via drinking water (Figure 44). The largest number of cases of hepatitis A resulting from waterborne outbreaks occurred during 1966 through 1970 when an average of 112 cases were reported annually. Since 1946, hepatitis A has been epidemiologically implicated in 69 waterborne outbreaks (Table 11). Half (11) of the outbreaks in community systems occurred as the result of contamination of the distribution system, primarily through cross-connections and back-siphonage. For noncommunity and individual systems, the use of contaminated, untreated ground water was the important factor responsible for waterborne outbreaks of hepatitis A.

Hepatitis A is one of three now-recognized forms of viral hepatitis. Hepatitis A is caused by a picornavirus, is transmitted by the fecal-oral route, does not become chronic, and no chronic virus carriers exist.[93] Hepatitis B is caused by an enveloped virus containing a circular, double-stranded form of DNA, is transmitted parenterally through inoculation of blood or blood products containing virus or through close personal contact with an infected person, and becomes chronic in some cases.[93] Agents of hepatitis non-A, non-B have not been identified, but it is possible to distinguish between a predominantly parenterally transmitted and an orally transmitted form.[93]

A specific laboratory diagnosis of hepatitis A can be obtained by serologic studies or demonstration of hepatitis A particles or specific antigens in feces, and it has been demonstrated that hepatitis A virus may replicate in several cell culture systems.[93] Hepatitis A virus was reported to have been detected in a sample of well water obtained during a waterborne outbreak of 8000 cases of gastroenteritis which occurred among 13,000 residents of Georgetown, Tex. in June 1980.[94] Enteroviruses were also isolated from two of the four wells in the city, which were developed in a limestone aquifer. Approximately 1 month after the gastroenteritis outbreak subsided 36 cases of hepatitis A occurred, compared with a normal incidence of less than 2 cases per month. There is little information on the risk of acquiring hepatitis A from drinking sewage-contaminated water in the U.S., and although

Table 11
CAUSES OF WATERBORNE OUTBREAKS OF HEPATITIS A IN THE U.S. (1946—1980)

	Number of outbreaks	
Cause of outbreaks	Community systems	Noncommunity/ individual systems
Contaminated, untreated surface water	1	11
Contaminated, untreated ground water	4	26
Inadequate or interrupted disinfection	3	6
Contamination through distribution system	11	2
Insufficient data for classification	3	2
Total	22	47

it has been suspected that hepatitis A might be transmitted during large outbreaks of waterborne gastroenteritis, this is the first report[94] which documents a significant occurrence of hepatitis A after a waterborne outbreak of gastroenteritis. Generally, the necessary surveillance has been lacking to confirm this "secondary wave" of hepatitis A cases in waterborne outbreaks of gastroenteritis. Neither of two large outbreaks of gastroenteritis due to sewage contamination at Pico Rivera[95] and Madera[96] was associated with an increase in reported cases of hepatitis A, and only one of 132 cases of gastroenteritis related to sewage-contaminated water in Pennsylvania[97] had any clinical evidence of hepatitis. Generally, outbreaks of waterborne hepatitis A occur without antecedent or sentinel gastroenteritis, and two reviews[98,99] of waterborne hepatitis A have shown that only 9 of 28 waterborne outbreaks of hepatitis A and 7 of 40 waterborne outbreaks of hepatitis A were preceded by gastroenteritis. Rosenberg et al.[100] conducted a survey among those exposed to sewage-contaminated water during the outbreak of toxigenic *E. coli* gastroenteritis at Crater Lake and found only 5 cases of hepatitis A among 2206 persons who drank water but did not receive immune serum globulin within 2 weeks of exposure (an attack rate of 0.23%). No cases of hepatitis A occurred in 320 park staff and family members who had repeatedly been exposed to contaminated water. The incidence of Crater Lake-associated hepatitis A as determined from the three cases reported to state health departments at the time of the outbreak was 12 per 100,000 people per year compared to the reported incidence for the U.S. of 10 per 100,000 per year.

Hepatitis caused by agents not serologically related to hepatitis A or B virus is referred to as hepatitis non-A, non-B. Evidence exists for more than one agent responsible for hepatitis non-A, non-B.[93] Epidemic hepatitis non-A, non-B has been reported from parts of Asia and the Middle East, where the disease has been studied most extensively, and hepatitis with the characteristics of this disease has been seen in Europeans who have traveled to areas where the disease has occurred.[93] Epidemic hepatitis non-A, non-B is thought to be transmitted by the fecal-oral route and contaminated water has been implicated as a cause of outbreaks in India, including the 1955 outbreak of 30,000 cases of hepatitis in Delhi.[101,102]

c. Shigellosis

Shigellosis remains endemic in the U.S., primarily because of direct person-to-person transmission, but common-source outbreaks also occur. During 1961 through 1975, 110 common-source outbreaks and 16,541 cases of illness were attributed to contaminated food and water.[103] The attack rate for both food- and waterborne shigellosis was 47%.[103] The vast majority of waterborne outbreaks of shigellosis occurred because of the contamination

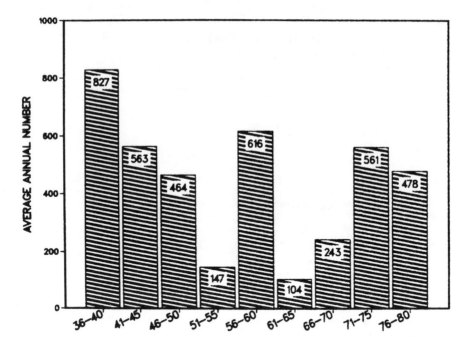

FIGURE 45. Average annual number of cases of shigellosis occurring in waterborne outbreaks (1936—1980).

of untreated ground waters and the lack of adequate disinfection for ground water. *Shigella* has been the most frequently identified pathogen in waterborne outbreaks of ground water systems.

The waterborne transmission of shigellosis has been relatively constant, with approximately 500 to 600 cases reported annually during most periods (Figure 45). The number of cases of waterborne shigellosis reported annually ranged from 104 during 1961 through 1965 to 827 in 1936 through 1940. *S. sonnei* was identified more frequently (56%) in waterborne outbreaks of shigellosis during 1961 through 1980; *S. flexneri* was identified more frequently (43%) during 1936 through 1960 (Table 12). During 1961 through 1980, 5593 (81%) cases of *S. sonnei* gastroenteritis occurred in waterborne outbreaks compared with 2294 (19%) cases 1936 through 1960. During 1961 to 1980, 414 (6%) cases of *S. flexneri* gastroenteritis occurred in waterborne outbreaks compared with 7006 (57%) cases during 1936 through 1960. The change in *Shigella* species causing the majority of waterborne outbreaks and cases of illness from *S. flexneri* to *S. sonnei* cannot be explained by reporting differences, and parallels a similar change in the predominant serotype in nationwide *Shigella* isolations.[103]

d. Salmonellosis

Waterborne outbreaks of salmonellosis have not resulted in many illnesses with the exception of 1961 through 1965, when an average of 3285 cases of salmonellosis occurred in waterborne outbreaks, and 1976 through 1980, when an average of 223 cases occurred (Figure 46). In each of these two periods, a single waterborne outbreak was responsible for the large number of cases: an outbreak of *S. typhimurium* gastroenteritis in Riverside, Calif.[33] in 1965 and an outbreak of 750 cases of *S. typhimurium* gastroenteritis in Suffolk County, N.Y.[103] in 1976. The Riverside, Calif. salmonellosis outbreak in 1965 was one of the largest waterborne outbreaks to occur during the 61-year period.[33] An estimated 16,000 cases of illness occurred, at least 70 individuals were hospitalized, and 3 deaths resulted from this outbreak. The water supply was epidemiologically implicated, and *S. typhimurium* was

Table 12
SHIGELLA ISOLATES FROM WATERBORNE OUTBREAKS (1936—1980)

	Outbreaks		Cases of illness	
Shigella isolate	1936—1960	1961—1980	1936—1960	1961—1980
S. sonnei	12	25	2,294	5,593
S. flexneri	15	10	7,006	414
Unidentified/other	8	10	2,959	923
Total	35	45	12,259	6,930

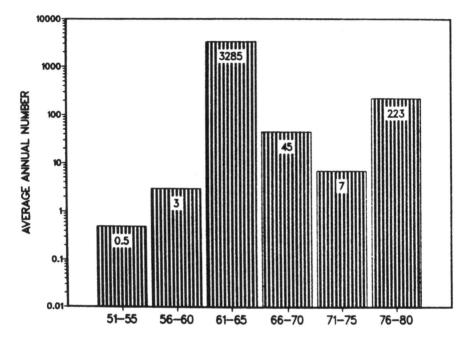

FIGURE 46. Average annual number of cases of salmonellosis occurring in waterborne outbreaks (1951—1980).

isolated from the water. However, the source of contamination was not determined. Chlorination of the ground water source stopped the outbreak. Coliforms were not detected during routine surveillance of the water system prior to the outbreak, and during the investigation it was found that *Salmonella* were more numerous than *E. coli* in the water system. *S. typhimurium* was isolated from stool specimens, and drinking water was epidemiologically implicated in the outbreak of salmonellosis at a large catering facility in Suffolk County, N.Y.[104] A submerged inlet in a clogged slop sink and other cross-connections were found, and it is believed that backsiphonage of contaminated water occurred after activation of booster pumps connected to the system. *S. typhimurium* was isolated from water in the slop sink. No contamination was found in the drinking water, but investigators felt that heavy use of the water system may have flushed the contamination from the system by the time water samples were collected.

e. Giardiasis

The number of cases of waterborne giardiasis occurring in the U.S. has increased dramatically over the past 15 years (Figure 47). The waterborne transmission of *G. lamblia*

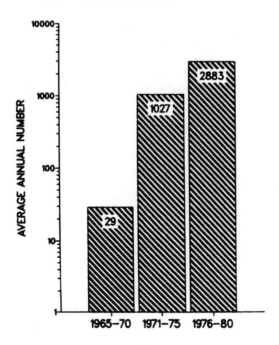

FIGURE 47. Average annual number of cases of giardiasis occurring in waterborne outbreaks (1965—1980).

was suggested as early as 1946 by an epidemiologic investigation of an outbreak of amebiasis attributed to sewage contamination of a water supply in a Tokyo apartment building.[105] *E. histolytica* and *G. lamblia* were recovered from 96 (64%) and 116 (77%) occupants, respectively; *G. lamblia* was isolated from 86% of the occupants who experienced diarrhea with abdominal discomfort and had stools that were negative for *E. histolytica*. Reports of giardiasis in American travelers to the U.S.S.R. in the 1970s generally increased recognition, investigation, and reporting of this disease in the U.S. The first reports involving American travelers appeared in 1970; Leningrad was implicated as the site of acquisition of the infection, and tap water was the probable mode of transmission.[106-108] The first waterborne outbreak of giardiasis documented in the U.S. occurred at Aspen, Colo.[109] during December 1965 through January 1966. The outbreak came to the attention of the Center for Disease Control when a physician there developed characteristic symptoms of giardiasis after returning from a holiday at Aspen. His stools contained no bacterial pathogens, but did contain a large number of *G. lamblia* cysts. A survey of 1094 skiers who had vacationed in Aspen during the 1965 to 1966 ski season showed that at least 123 (11%) had developed similar symptoms. The association of *G. lamblia* with the illness, the absence of other pathogens, and the response to treatment suggested that *G. lamblia* was the agent responsible for illness. Approximately half the water of the city came from a mountain creek and half from three wells. Chlorinators were provided at each water source, but coliform contamination of the water system was noted intermittently during the 1965 to 1966 ski season. Examination of the creek area revealed no obvious sewage contamination. However, fluorescent and detergent tracers placed in the sewage system were detected in two of the three wells, and an engineering evaluation suggested sewage contamination of the wells from leaking sewer mains. *G. lamblia* cysts were isolated from the leaking sewage. A parasitologic survey of Aspen residents showed a difference in prevalence of *G. lamblia* infection between the area of the city served by the wells (6.9%) and the area served by the creek (3.7%). This difference was not statistically significant, and in retrospect it is possible that the creek was also contaminated with *G. lamblia,* although this was not sufficiently evaluated at the time. The largest waterborne outbreak of giardiasis occurred in Rome, N.Y. in November 1974 to June 1975; 350 residents had laboratory-confirmed giardiasis and an epidemiologic study estimated 4800 to 5300 cases of giardiasis.[110] Rome used a surface water source with chlorination as the only treatment. Ammonia and chlorine were added to produce a chloramine for a total combined chlorine residual of 0.8 mg/ℓ. Chloramine requires higher concentrations and longer contact time than free chlorine for effective disinfection. The watershed was sparsley populated, but the presence of human settlements in the watershed suggested that the water supply could have been contaminated by untreated human waste.[111] The bacterial quality of the water source was generally good, but prior to the beginning of the outbreak much higher than normal coliform counts were observed in the untreated water, suggesting that contamination entered the system at this time. Outbreaks in Rome, N.Y. and Bradford, Pa. were responsible for most of the cases of giardiasis in surface water supplies that were disinfected only. An estimated 3500 cases of giardiasis occurred in Bradford[112] during September to December 1979. The first waterborne outbreak of giardiasis involving a filtered water supply occurred in Camas, Wash.[113,114] in the spring of 1976 and affected 600 individuals of a population of 6000. Camas used two water sources: mountain streams and deep wells. *Giardia* cysts were isolated from the raw water in two distribution system storage reservoirs. Well water was not found to be contaminated. The watersheds for the surface water sources were well isolated, had no human habitation, and had extremely limited human activity. Several animals on the watershed were trapped and examined for *Giardia* cysts; three positive beavers were found within foraging distance of the water intakes for Camas. Treatment for the surface water sources consisted of a mixed-media pressure filter and disinfection; no sedimentation was employed prior to filtration. Prior to the outbreak, failure of the chlorination equipment occurred. A number of deficiencies were found in the pressure filters, including ineffective chemical pretreatment. It was reported that the treated water produced by the treatment plant had met both turbidity and coliform standards prior to and during the outbreak. A second outbreak involving a filtered water supply occurred in Berlin, N.H.[115,116] in the spring of 1977 and affected 750 of a population of 15,000.

Waterborne outbreaks of giardiasis have occurred primarily in areas that have traditionally depended upon surface water sources free of gross human sewage contamination where water treatment has been minimal and consisted primarily of disinfection without filtration. Attention to maintaining continuous disinfection may not have been as great as in other areas where surface streams are known to be contaminated with human waste discharges. During 1965 to 1980, 42 waterborne outbreaks of giardiasis were reported. Colorado has reported 19 outbreaks, more than any other state, but this may reflect primarily increased surveillance and investigation. Most of the waterborne outbreaks (67%) and most of the cases of giardiasis (52%) have occurred as the result of consuming untreated surface water or surface water with disinfection as the only treatment (Table 13). Ineffective filtration of surface water was responsible for 5 (12%) outbreaks and 6981 cases (35%) of disease. Community water systems accounted for 24 outbreaks, and 13 occurred in noncommunity systems. Outbreaks have generally occurred in smaller community water systems or noncommunity systems in recreational areas. Studies in Colorado,[117] Minnesota,[118] and Washington[119] have suggested consumption of untreated drinking water to be an important cause of endemic infection in the U.S., and a number of single cases of giardiasis have occurred in travelers, campers, and backpackers where a common source could not be investigated. When investigated, these cases are often reported in an anecdotal manner and have not been included in this tabulation of waterborne outbreaks. An example of under-reporting of waterborne giardiasis due to lack of investigation and imcomplete reporting is provided by Frost and Harter.[119] A follow-up of 883 individual cases of giardiasis routinely reported during 1978 to 1980 in Washington showed that 275 were possibly due to common-source outbreaks: 6 clusters of an least 8 cases involved 70 individuals, and 79 smaller clusters involved a total of 205

Table 13
WATER SYSTEM DEFICIENCIES RESPONSIBLE FOR WATERBORNE GIARDIASIS OUTBREAKS (1965—1980)

	Number of outbreaks			
Cause	Community water system	Noncommunity water system	Individual water system	All systems
Inadequate/interrupted disinfection of surface water (when only treatment)	13	5	—	18
Untreated surface water	1	5	4	10
Inadequate control of filtration and allied treatment	5	—	—	5
Cross-connection	2	1	—	3
Untreated ground water	1	1	—	2
Interrupted disinfection of spring or well	2	—	—	2
Other	—	1	1	2
Total	24	13	5	42

cases. None of these 85 clusters of cases had previously been investigated or reported as outbreaks. On follow-up, several were felt to be of waterborne origin. Ten cases occurred where hikers had consumed water from a stream which was later found to be contaminated with *Giardia* cysts. Common exposure to a work camp accounted for 14 cases where all had consumed water from a stream contaminated with *Giardia* cysts; 11 cases which were exposed to a community water system were later shown to have a much higher rate of giardiasis during the period of the 11 cases than a neighboring town with a different water system; 20 clusters of 43 cases were composed of loggers, game workers, campers, and hikers who drank untreated water while on an outing as a group or during their work.

f. Acute Chemical Poisonings

High concentrations of fluoride, copper, and arsenic in water (Table 14) accounted for 52 outbreaks and 3775 cases of illness. The outbreaks ranged in size from single cases of infantile methemoglobinemia caused by consumption of high nitrates in well water to a large outbreak of several thousand cases of chlordane poisoning in Pittsburgh in 1980. Most of these outbreaks (54%) were caused by contamination of ground water or contamination through cross-connections (Table 15). Of the chemical-caused outbreaks, 13% occurred because of deficiencies in fluoridation practices and 12% occurred as the result of corrosive water causing the leaching of metals from plumbing materials. The seven outbreaks of acute fluoride poisoning were caused by improperly designed or malfunctioning fluoridation equipment and operational problems in the addition of fluoride. These outbreaks illustrate the need for improved design of fluoridation equipment and increased surveillance and operator training. Four of these outbreaks occurred at elementary schools adding fluoride to their own water system.

F. Causes of Waterborne Outbreaks and Disease

Contaminated, untreated ground water was the major cause of waterborne outbreaks (44%) for all types of water systems during this 61-year period (Figure 48). However, most of the cases of waterborne disease (52%) resulted from the inadequate or interrupted treatment of all water sources (Figure 48). In community water systems, most (79%) waterborne outbreaks were caused by contaminated, untreated ground water, deficiencies in water treatment, and

Table 14
CHEMICALS RESPONSIBLE FOR MULTIPLE OUTBREAKS OF WATERBORNE DISEASE (1936—1980)

Chemical	Outbreaks	Illnesses
Fluoride	7	633
Copper	7	106
Arsenic	6	37
Petroleum products	5	111
Chlordane	4	2043
Unidentified pesticides	4	22
Chromates	2	31
Selenium	2	7
Herbicides	2	5
Nitrate	2	2

Table 15
CAUSES OF CHEMICAL POISONINGS (1936—1980)

Cause of outbreak	Outbreaks
Contamination of ground water	14
Cross-connection, backsiphonage	14
Deficiencies in fluoridation practices	7
Leaching of metals from plumbing	6
Deliberate contamination	3
Deficiencies in chemical feed	2
Contamination of storage container	2
Contamination of distribution storage facility	1
Contamination of pipe with cutting oil	1
Contaminated bottled water	1
Insufficient data	1

contamination of the distribution system (Figure 49). Community systems generally provide some water treatment, especially in the larger metropolitan areas, and the interruption or lack of control of this treatment was responsible for most (57%) cases of waterborne disease (Figure 49) and the largest waterborne outbreaks (Table 16). Inadequate filtration and inadequate disinfection in community systems resulted in waterborne outbreaks averaging 2199 illnesses and 1244 illnesses, respectively. Waterborne outbreaks in community water systems using untreated water were smaller: 468 illnesses per outbreak caused by contaminated, untreated well water and 132 illnesses per outbreak caused by contaminated, untreated surface water. In noncommunity and individual water systems, the major cause of both waterborne outbreaks and waterborne disease was the use of contaminated, untreated ground water (Figures 50 and 51). The largest outbreaks in noncommunity systems were caused by cross-connections, backsiphonage, and inadequate disinfection (Table 17). Cross-connections and backsiphonage were responsible for smaller outbreaks in noncommunity systems (203 illnesses per outbreak) than in community systems (538 illnesses per outbreak).

Specific sources of contamination and the water treatment deficiencies responsible for each outbreak were examined and classified (Tables 18 to 25). Insufficient data prevented identification of the source of contamination for most of the outbreaks which occurred as the result of using untreated surface water; however, contamination discovered on the watershed after the occurrence of the outbreak, use of contaminated, untreated surface water as a supplemental or emergency source of supply, and sewage overflows near the water

126 Waterborne Diseases in the United States

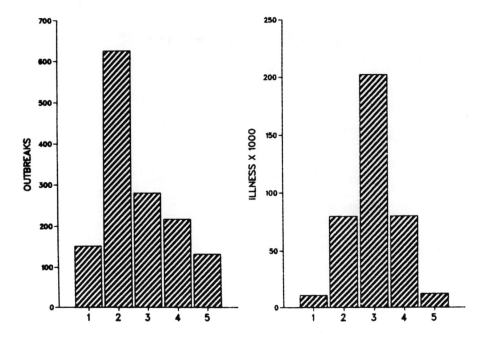

FIGURE 48. Causes of waterborne outbreaks and illnesses (1920—1980). 1 = Contaminated untreated surface water; 2 = contaminated untreated ground water; 3 = inadequate or interrupted treatment; 4 = contamination of distribution system; 5 = miscellaneous deficiencies.

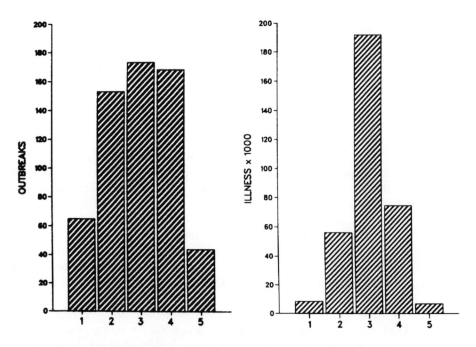

FIGURE 49. Causes of waterborne outbreaks and illnesses in community water systems (1920—1980). 1 = Contaminated untreated surface water; 2 = contaminated untreated ground water; 3 = inadequate or interrupted treatment; 4 = contamination of distribution system; 5 = miscellaneous deficiencies.

intake were the most important causes identified and were responsible for 45% of the outbreaks in all the water systems using untreated surface water. There appeared to be no clear seasonal occurrence of waterborne outbreaks in community water systems caused by

Table 16
MAGNITUDE OF WATERBORNE OUTBREAKS IN COMMUNITY WATER SYSTEMS (1920—1980)

Deficiency	Illness per outbreak
Inadequate filtration/allied treatment	2199
Inadequate disinfection when only treatment	1244
Cross-connection, backsiphonage	538
No treatment of well water	468
Contaminated storage facility	464
Interrupted disinfection when only treatment	256
No treatment of surface water	132
Contaminated mains, service lines, plumbing	104
No treatment of spring water	47

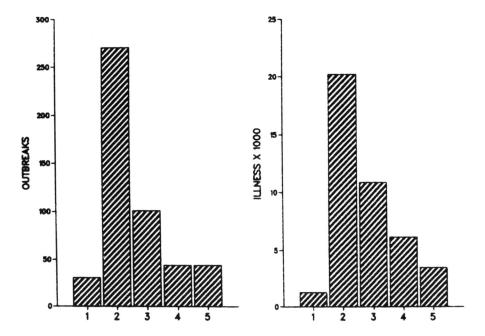

FIGURE 50. Causes of waterborne outbreaks and illnesses in noncommunity water systems (1920—1980). 1 = Contaminated untreated surface water; 2 = contaminated untreated ground water; 3 = inadequate or interrupted treatment; 4 = contamination of distribution system; 5 = miscellaneous deficiencies.

contaminated, untreated surface water, suggesting that contamination is unrelated to season (Figure 52). In noncommunity water systems, the increased occurrence of outbreaks during the summer may be due to the increased use of these systems by more susceptible individuals traveling and camping, as well as increased contamination (Figure 53). The suggestion of a possible increased occurrence of outbreaks in individual surface water systems during the summer may be due to increased use of mountain streams by hikers and campers (Figure 54).

The overflow or seepage of sewage and surface runoff were responsible for 40% of the

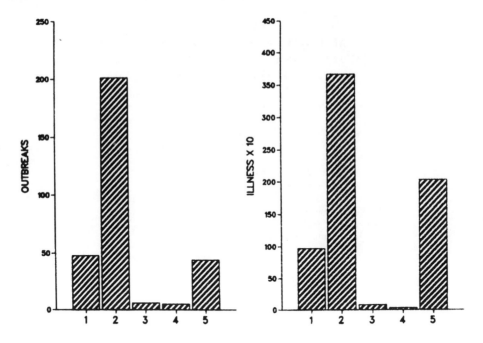

FIGURE 51. Causes of waterborne outbreaks and illnesses in individual water systems (1920—1980). 1 = Contaminated untreated surface water; 2 = contaminated untreated ground water; 3 = inadequate or interrupted treatment; 4 = contamination of distribution system; 5 = miscellaneous deficiencies.

Table 17
MAGNITUDE OF WATERBORNE OUTBREAKS IN NONCOMMUNITY WATER SYSTEMS (1920—1980)

Deficiency	Illness per outbreak
Cross-connection, backsiphonage	203
Inadequate disinfection when only treatment	168
Interrupted disinfection when only treatment	85
No treatment of spring water	77
No treatment of well water	75
Contaminated storage facility	45
No treatment of surface water	31

waterborne outbreaks which occurred as the result of using untreated springs; data were insufficient to determine the source of contamination of springs in approximately half of the outbreaks (Table 19). A distinct seasonal occurrence was noted for outbreaks caused by contamination of untreated springs, and most of these occurred during June through September (Figure 55).

A source of contamination was identified in 76% of the waterborne outbreaks caused by contamination of untreated well water (Table 20). The overflow or seepage of sewage into wells caused 35% of the outbreaks; surface runoff from heavy rains or flooding caused 21% of the outbreaks; contamination of wells through creviced limestone or fissured rock caused 8% of the outbreaks. Other sources of contamination of wells included chemicals, contaminated surface water from streams or rivers, seepage from abandoned wells, and animals entering the well. Outbreaks in community systems using contaminated, untreated well water

Table 18
WATERBORNE OUTBREAKS CAUSED BY USE OF CONTAMINATED, UNTREATED SURFACE WATER (1920—1980)

	Type of water system							
	Community		Non-community		Individual		All	
Deficiency	OB[a]	cases	OB[a]	cases	OB[a]	cases	OB[a]	cases
Contamination on watershed	26	3,498	3	57	12	257	41	3,812
Use of surface water for supplemental source	7	3,613	7	245	2	115	16	3,973
Overflow of sewage or outfall near water intake	3	103	3	39	5	87	11	229
Flooding, heavy rains	2	125	1	93	1	77	4	295
Dead animals in reservoir	—	—	1	100	—	—	1	100
Insufficient data	27	1,228	24	726	28	436	79	2,390
Total	65	8,567	39	1,260	48	972	152	10,799

[a] Number of outbreaks.

Table 19
WATERBORNE OUTBREAKS CAUSED BY USE OF CONTAMINATED UNTREATED GROUND WATER (SPRINGS) (1920—1980)

	Type of water system							
	Community		Non-community		Individual		All	
Deficiency	OB[a]	Cases	OB[a]	Cases	OB[a]	Cases	OB[a]	Cases
Overflow or seepage of sewage	8	238	3	35	5	39	16	312
Surface runoff	11	265	5	162	7	75	23	502
Flooding	2	76	2	123	—	—	4	199
Creviced limestone	1	200	3	213	—	—	4	413
Contamination of raw water transmission line	2	284	1	7	—	—	3	291
Improper construction	—	—	1	26	1	9	2	35
Insufficient data	12	508	18	1961	20	415	50	2884
Total	36	1571	33	2527	33	538	102	4636

[a] Number of outbreaks.

occurred primarily in July, August, and September; outbreaks in noncommunity systems occurred primarily in June, July, and August; outbreaks in individual systems occurred primarily in June, July, August, and September (Figures 56 to 58). During the 61-year period, an etiologic agent was identified in 56% of the outbreaks caused by the contamination of untreated well water (Figure 59). Most of the outbreaks in untreated well water systems where an etiology was identified were caused by *S. typi, Shigella,* and hepatitis A. Examination of the etiologies over the most recent 10-year period shows chemicals, *Shigella,*

Table 20
WATERBORNE OUTBREAKS CAUSED BY USE OF CONTAMINATED, UNTREATED GROUND WATER (WELLS) (1920—1980)

	Type of water system							
	Community		Non-community		Individual		All	
Deficiency	OB[a]	Cases	OB[a]	Cases	OB[a]	Cases	OB[a]	Cases
Overflow or seepage of sewage	28	14,915	104	10,236	52	675	184	25,826
Surface runoff, heavy rains	25	2,492	26	947	34	824	85	4,263
Creviced limestone, fissured rock	9	1,404	19	2,044	12	660	40	4,108
Improper construction, faulty well casing	8	342	10	414	9	141	27	897
Flooding	9	5,883	3	107	5	211	17	6,201
Chemical contamination	3	77	2	16	10	68	15	161
Contamination by stream or river	3	445	6	392	3	48	12	885
Contamination of raw water transmission line	8	10,481	—	—	—	—	8	10,481
Seepage from abandoned well	3	144	1	50	—	—	4	194
Animal in well	1	34	1	238	2	19	4	291
Insufficient data	19	18,480	67	3,309	40	413	126	22,202
Total	116	54,697	239	17,753	167	3,059	522	75,509

[a] Number of outbreaks.

Table 21
WATERBORNE OUTBREAKS CAUSED BY SOURCE CONTAMINATION SUFFICIENT TO OVERWHELM NORMAL DISINFECTION (1920—1980)

	Type of water system							
	Community		Non-community		Individual		All	
Deficiency	OB[a]	Cases	OB[a]	Cases	OB[a]	Cases	OB[a]	Cases
Overflow or seepage of sewage								
Wells	4	16,373	3	137	—	—	7	16,510
Springs	—	—	4	1,287	—	—	4	1,287
Surface water	3	31,990	1	211	—	—	4	32,201
Heavy rains, surface runoff	3	1,215	1	33	—	—	4	1,248
Contamination of raw water transmission line	—	—	1	37	—	—	1	37
Total	10	49,578	10	1,705	—	—	20	51,283

[a] Number of outbreaks.

Table 22
WATERBORNE OUTBREAKS CAUSED BY INADEQUATE TREATMENT (1920—1980)

	Type of water system							
	Community		Non-community		Individual		All	
Deficiency	OB[a]	Cases	OB[a]	Cases	OB[a]	Cases	OB[a]	Cases
Inadequate disinfection when only treatment:								
Surface waters	40	22,087	19	2.684	2	37	61	24,808
Wells	11	13,518	17	1,795	1	13	29	15,326
Springs	5	367	2	60	—	—	7	427
Mixed	4	293	—	—	—	—	4	293
Unknown source	3	4969	—	—	—	—	3	4,969
Interruption of disinfection when only treatment:								
Surface waters	26	2,173	12	1,495	—	—	38	3,668
Wells	16	6,148	29	2,239	—	—	45	8,387
Springs	7	4,672	4	153	1	3	12	4,828
Mixed	—	—	1	40	—	—	1	40
Unknown source	2	42	—	—	—	—	2	42
Inadequate disinfection with other treatment	3	1751	1	7	—	—	4	1,758
Interruption of disinfection with other treatment	7	8,689	1	26	—	—	8	8,715
Inadequate control of filtration and allied treatment	35	76,953	1	100	1	9	37	77,062
Inadequate control of chemical feed	4	92	5	608	—	—	9	700
Total	163	141,654	92	9,207	5	62	260	151,023

[a] Number of outbreaks.

Table 23
WATERBORNE OUTBREAKS CAUSED BY CONTAMINATION OF DISTRIBUTION SYSTEM (1920—1980)

	Type of water system							
	Community		Non-community		Individual		All	
Deficiency	OB[a]	Cases	OB[a]	Cases	OB[a]	Cases	OB[a]	Cases
Cross-connection								
Unspecified	89	57,061	18	4,755	—	—	107	61,816
Backsiphonage	28	5,930	9	714	1	5	38	6,649
Contamination of mains during construction, repair, flushing	9	2,055	2	23	—	—	11	2,078
Contamination of mains when broken or leaking	10	983	1	27	—	—	11	1,010

Table 23 (continued)
WATERBORNE OUTBREAKS CAUSED BY CONTAMINATION OF DISTRIBUTION SYSTEM (1920—1980)

Deficiency	Type of water system							
	Community		Non-community		Individual		All	
	OB[a]	Cases	OB[a]	Cases	OB[a]	Cases	OB[a]	Cases
Contamination or corrosion of household plumbing or service lines	6	69	2	134	1	2	9	205
Corrosion due to CO_2 in soft drink machine or in water coolers	3	84	—	—	—	—	3	84
Water main and sewer in same trench or inadequately separated	3	131	2	71	—	—	5	202
Improper/no disinfection of water main, household plumbing	3	210	—	—	—	—	3	210
Total	151	66,523	34	5,724	2	7	187	72,254

[a] Number of outbreaks

Table 24
ETIOLOGY OF WATERBORNE OUTBREAKS CAUSED BY CROSS-CONNECTIONS, BACKSIPHONAGE (1920—1980)

Disease	Outbreaks	Illnesses
Gastroenteritis	71	55,069
Typhoid fever	37	2,637
Chemical poisoning	14	618
Hepatitis A	8	320
Shigellosis	6	3,367
Giardiasis	3	2,209
Viral gastroenteritis	2	1,967
Salmonellosis	2	1,425
Other	2	853
Total	145	68,465

and hepatitis A to be the most important etiologic agents of outbreaks in untreated well water systems (Figure 60).

After an unexpected heavy contamination completely overwhelmed the normal disinfection required (Table 21), 20 waterborne outbreaks occurred. In most of these outbreaks contamination was caused by the overflow or seepage of sewage. Inadequate disinfection resulted in 108 waterborne outbreaks, and 102 outbreaks were caused by interrupted disinfection (Table 22). In ground water systems, interrupted disinfection was a more important cause of outbreaks than was inadequate disinfection. In surface water systems, inadequate disinfection caused more outbreaks than interrupted disinfection. Inadequate control of filtration

Table 25
WATERBORNE OUTBREAKS CAUSED BY CONTAMINATION OF STORAGE FACILITIES (1920—1980)

	Type of water system							
	Community		Non-community		Individual		All	
Deficiency	OB[a]	Cases	OB[a]	Cases	OB[a]	Cases	OB[a]	Cases
Contamination of distribution storage facility	11	7552	3	45	—	—	14	7597
Contamination of cistern	2	69	3	126	3	24	8	219
Contamination of storage tank	2	196	3	175	—	—	5	371
Improper/no disinfection of new or repaired storage facility	1	58	1	101	—	—	2	159
Insufficient data	1	13	—	—	—	—	1	13
Total	17	7888	10	447	3	24	30	8359

[a] Number of outbreaks.

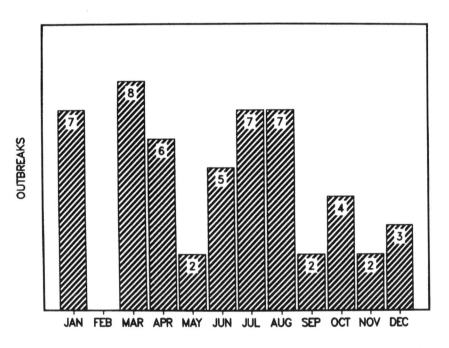

FIGURE 52. Seasonal distribution of waterborne outbreaks in community water systems using untreated surface water (1920—1980).

and allied treatment was responsible for only 14% of the outbreaks caused by inadequate treatment but was responsible for 51% of the cases of illness. A seasonal occurrence of outbreaks caused by treatment deficiencies was apparent in noncommunity, but not in community, water systems (Figures 61 and 62). No seasonal occurrence was noted for outbreaks in filtered water systems (Figure 63). Waterborne outbreaks in surface water systems and well water systems treated only by disinfection occurred primarily in July, but no similar

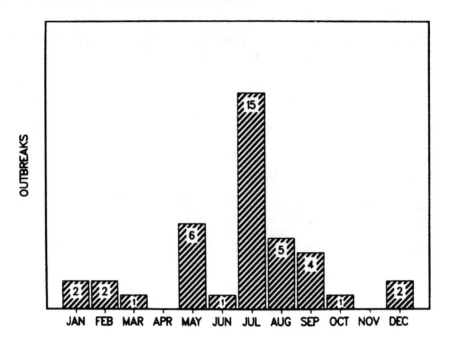

FIGURE 53. Seasonal distribution of waterborne outbreaks in noncommunity water systems using untreated surface water (1920—1980).

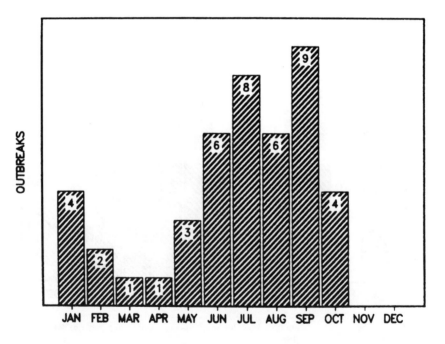

FIGURE 54. Seasonal distribution of waterborne outbreaks in individual water systems using untreated surface water (1920—1980).

occurrence was apparent for outbreaks in systems using springs treated only by disinfection (Figures 64 to 66).

Cross-connections and backsiphonage were responsible for most (78%) outbreaks and disease (95%) caused by distribution system contamination (Table 23). Outbreaks caused

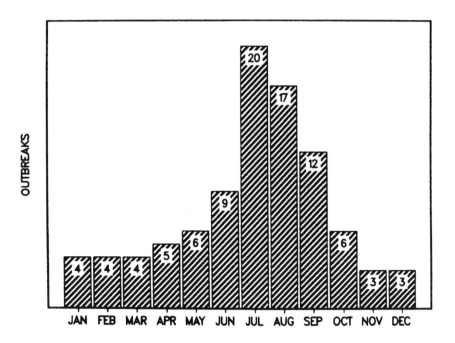

FIGURE 55. Seasonal distribution of waterborne outbreaks in water systems using untreated ground water (springs) (1920—1980).

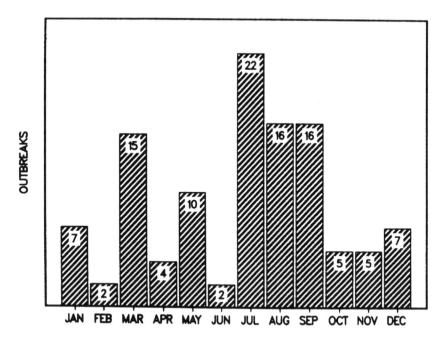

FIGURE 56. Seasonal distribution of waterborne outbreaks in community water systems using untreated ground water (wells) (1920—1980).

by cross-connections and backsiphonage appeared to occur primarily during June through October (Figure 67). Most of the outbreaks (49%) and diseases (81%) caused by cross-connections and backsiphonage were gastroenteritis of unidentified etiology (Table 24); 26% were of typhoid fever and 10% were outbreaks of chemical poisoning.

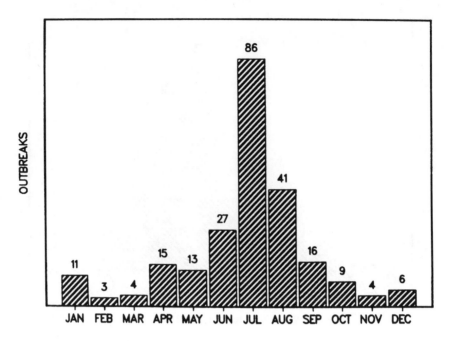

FIGURE 57. Seasonal distribution of waterborne outbreaks in noncommunity water systems using untreated ground water (wells) (1920—1980).

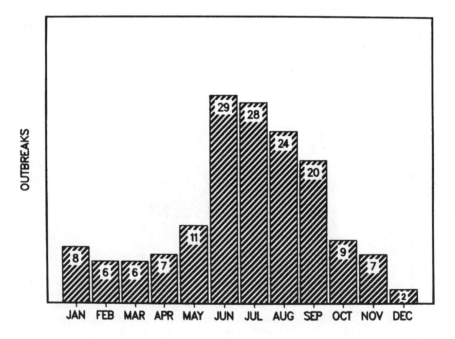

FIGURE 58. Seasonal distribution of waterborne outbreaks in individual water systems using untreated ground water (wells)(1920—1980).

Contamination of distribution reservoirs in community water systems was responsible for most outbreaks and disease caused by contamination of water storage facilities (Table 25). Waterborne outbreaks were caused by several miscellaneous reasons, including insufficient data available for classifying the cause, use of water not intended for drinking, contaminated

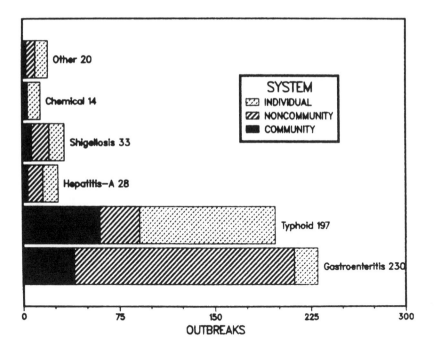

FIGURE 59. Etiology of waterborne outbreaks occurring in water systems using untreated ground water (1920—1980).

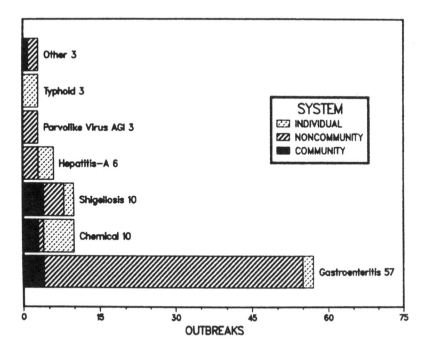

FIGURE 60. Etiology of waterborne outbreaks occurring in water systems using untreated ground water (1971—1980).

ice, buckets, hoses, tank truck, and drinking fountain, and suspected deliberate contamination with chemicals (Table 26).

Some 178 waterborne outbreaks occurred in water systems serving parks, campgrounds,

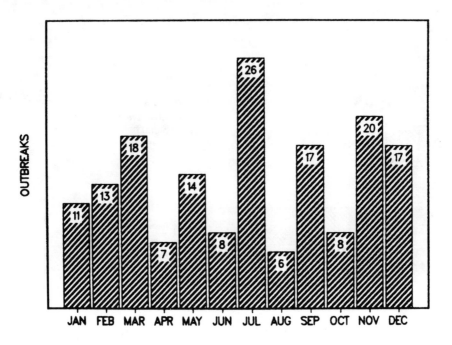

FIGURE 61. Seasonal distribution of waterborne outbreaks caused by a treatment deficiency in community water systems (1920—1980).

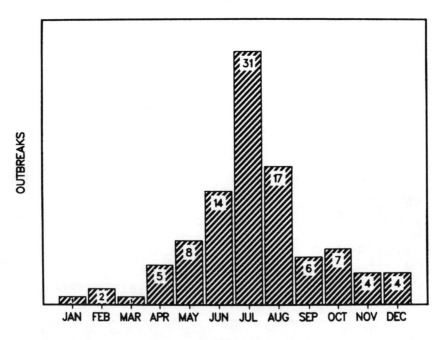

FIGURE 62. Seasonal distribution of waterborne outbreaks caused by a treatment deficiency in noncommunity water systems (1920—1980).

and recreational areas (Table 27), 149 outbreaks in water systems serving hotels, motels, and rooming houses (Table 28), 85 outbreaks in water systems serving industrial facilities (Table 29), and 77 outbreaks in water systems serving schools (Table 30). In water systems serving visitors and travelers to parks, campgrounds, recreational areas, hotels, motels, and rooming houses, outbreaks were caused primarily by contamination of untreated well water

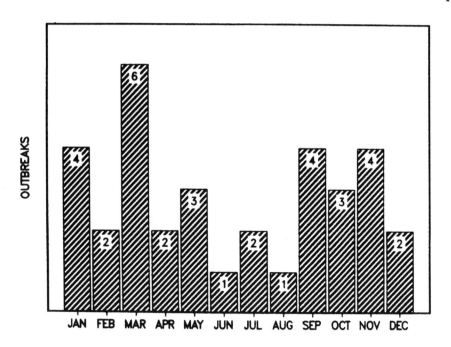

FIGURE 63. Seasonal distribution of waterborne outbreaks caused by a treatment deficiency in filtered water systems (1920—1980).

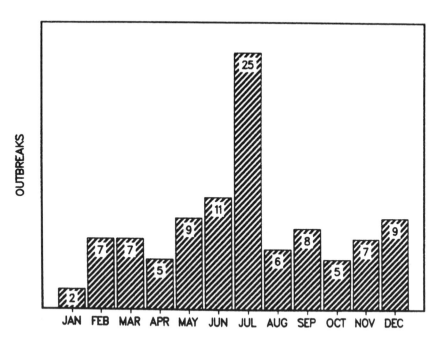

FIGURE 64. Seasonal distribution of waterborne outbreaks caused by inadequate or interrupted disinfection in surface water systems treated only by disinfection (1920—1980).

(43%) and inadequate or interrupted disinfection (21%). Outbreaks in these water systems showed a distinct seasonal variation with most outbreaks occurring in July (Figures 68 and 69). Cross-connections and backsiphonage (45%) and contamination of untreated well water (24%) caused most outbreaks at industrial facilities. Outbreaks at schools occurred primarily

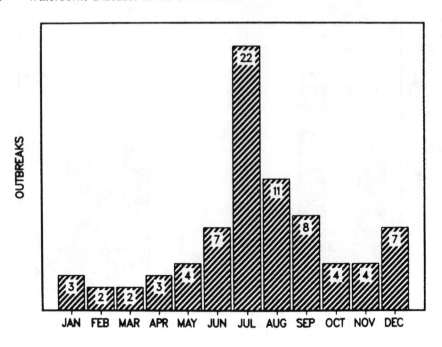

FIGURE 65. Seasonal distribution of waterborne outbreaks caused by inadequate or interrupted disinfection in ground water systems (wells) treated only by disinfection (1920—1980).

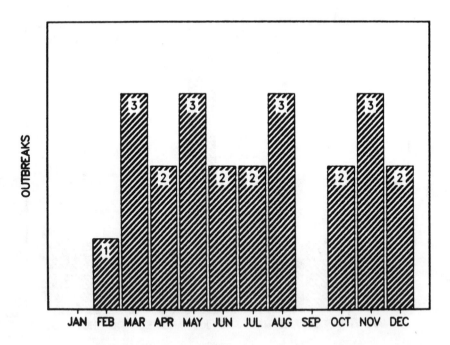

FIGURE 66. Seasonal distribution of waterborne outbreaks caused by inadequate or interrupted disinfection in ground water systems (springs) treated only by disinfection (1920—1980).

because of contaminated untreated well water (40%) and cross-connections and backsiphonage (16%). No clear seasonal distribution was observed for waterborne outbreaks in industrial facilities (Figure 70). Most waterborne outbreaks in schools occurred at the beginning of the school season in September and October (Figure 71). An increased number of outbreaks also occurred toward the end of the school season in April and May.

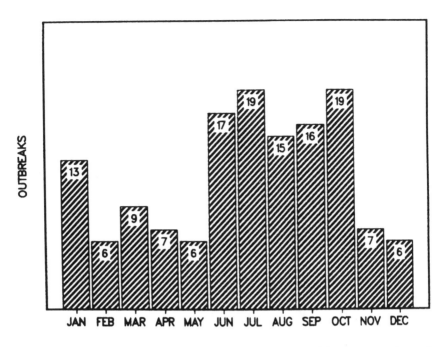

FIGURE 67. Seasonal distribution of waterborne outbreaks caused by cross-connections and backsiphonage (1920—1980).

Table 26
WATERBORNE OUTBREAKS CAUSED BY MISCELLANEOUS REASONS
(1920—1980)

	Type of water system							
	Community		Non-community		Individual		All	
Deficiency	OB[a]	Cases	OB[a]	Cases	OB[a]	Cases	OB[a]	Cases
Use of water not intended for drinking	2	12	14	1023	6	263	22	1,298
Contaminated ice	2	18	2	30	3	1,060	7	1,108
Contaminated buckets, hoses, tank truck	1	9	1	11	4	89	6	109
Contaminated drinking fountain	2	72	2	25	—	—	4	97
Ingestion of water while swimming	—	—	—	—	4	70	4	70
Suspected deliberate contamination with chemicals	1	2,000	—	—	2	18	3	2,018
Contaminated bottled water	—	—	2	32	1	3	3	35
Faulty distilling devices	—	—	1	14	—	—	1	14
Insufficient data	37	4,665	21	2,331	24	536	82	7,532
Total	45	6,776	43	3,466	44	2,039	132	12,281

[a] Number of outbreaks.

Table 27
CAUSE OF OUTBREAKS IN WATER SYSTEMS FOR PARKS, CAMPGROUNDS, RECREATION AREAS (1920—1980)

Cause	Outbreaks
Contaminated, untreated well water	61
Inadequate, interrupted disinfection	46
Contaminated, untreated surface water	24
Contaminated, untreated spring water	17
Cross-connections, backsiphonage	11
Contaminated storage facilities	3
Miscellaneous, insufficient data	16
Total	178

Table 28
CAUSE OF OUTBREAKS IN WATER SYSTEMS SERVING HOTELS, MOTELS, ROOMING HOUSES (1920—1980)

Cause	Outbreaks
Contaminated, untreated well water	80
Inadequate, interrupted disinfection	24
Contaminated, untreated surface water	15
Cross-connections, backsiphonage	10
Contaminated, untreated spring water	8
Miscellaneous, insufficient data	12
Total	149

IV. WATERBORNE OUTBREAKS (1971—1980)

A. Occurrence and Distribution of Outbreaks

During 1971 through 1980, 320 waterborne outbreaks and 77,989 cases of illness were reported by 43 states and Puerto Rico. Three deaths were associated with these outbreaks. No outbreaks were reported by Delaware, Kansas, and Nevada during this period. Community water systems accounted for 123 waterborne outbreaks, 160 outbreaks occurred in noncommunity water systems, and 37 outbreaks occurred in individual water systems during 1971 through 1980 (Figure 72). Although most outbreaks occurred in noncommunity systems (50%), most illness (76%) resulted from outbreaks in community systems; 23.5% of the cases of illness resulted from outbreaks in noncommunity systems and 0.5% of the cases of illness resulted from outbreaks in individual systems. From 1971 through 1980, outbreaks in community systems affected an average of 482 persons compared with 115 persons per outbreak in noncommunity systems and 10 persons per outbreak in individual systems.

A seasonal distribution of outbreaks was apparent in noncommunity water systems, with the largest number of outbreaks occurring during the summer months; these outbreaks affected primarily travelers, campers, restaurant patrons, and visitors to recreational areas (Figure 73). Outbreaks occurring among residents using individual water systems, which are primarily ground water systems, also showed a slightly increased occurrence during the summer months (Figure 74). Little seasonal variation was observed for outbreaks in community water systems; however, more outbreaks occurred in the month of July than any other month (Figure 75).

B. Causes of Outbreaks

The major causes of waterborne outbreaks and waterborne disease during this 10-year period were inadequate or interrupted treatment of water sources, use of contaminated, untreated ground water, and contamination of the distribution system (Figure 76). Specific sources of contamination were examined within each of these water supply deficiencies (Tables 31 to 34). Well water sources used without treatment accounted for 92 outbreaks (Table 31). Overflow or seepage of sewage, primarily from septic tanks or cesspools, was responsible for 38% of the outbreaks and 58% of the illness caused by use of contaminated, untreated well water. This included outbreaks where contaminants traveled through limestone or fissured rock. Chemical contamination accounted for 11% of the outbreaks and 2% of the cases of illness in untreated well water systems. Data were insufficient to establish a source of contamination for 35% of these outbreaks, emphasizing the need for better en-

Table 29
CAUSE OF OUTBREAKS IN WATER SYSTEMS FOR INDUSTRIAL FACILITIES (1920—1980)

Cause	Outbreaks
Cross-connections, backsiphonage	38
Contaminated, untreated ground water	20
Contaminated, untreated surface water	9
Contaminated mains, service lines, plumbing	5
Use of water not intended for drinking	4
Treatment deficiencies	3
Contaminated storage	2
Other	4
Total	85

Table 30
CAUSE OF OUTBREAKS IN WATER SYSTEMS SERVING SCHOOLS (1920—1980)

Cause	Outbreaks
Contaminated, untreated well water	31
Cross-connections, backsiphonage	12
Inadequate control of chemical feed or filtration	7
Inadequate, interrupted disinfection	6
Contaminated, untreated spring water	4
Contaminated mains, service lines, plumbing	4
Contaminated storage facilities	3
Miscellaneous, insufficient data	10
Total	77

gineering investigations if contamination problems for untreated ground water sources are to be understood. A few communities, individuals, and establishments still use surface water and spring water without treatment, and most of the outbreaks caused by use of contaminated, untreated surface water and spring water also had insufficient data to establish the source of contamination.

Most of the outbreaks (87%) and most of the cases of illness (85%) caused by treatment deficiencies occurred as the result of inadequate or interrupted disinfection (Table 32). In some areas of the country, surface water sources are treated only by disinfection. There were 39 outbreaks and 18,966 cases of illness caused by inadequate or interrupted chlorination of surface water; 1 outbreak of 900 cases of illness resulted from interruption of ultraviolet (UV) light disinfection.

Well water systems usually depend on having water of relatively good quality, and disinfection is often-times provided primarily to protect against possible contamination of the distribution system. In this situation, unexpected contamination of the source could completely overwhelm the disinfection provided. For ground water systems using a source known to be frequently or intermittently contaminated with bacteria, continuous disinfection is necessary to ensure potability until the sources of contamination are located and removed, improvements made in source protection, or alternative water sources developed. Outbreaks are caused by inadequate disinfection of the contaminated source water or interrupted disinfection due to equipment malfunction or an insufficient supply of the disinfectant. Inad-

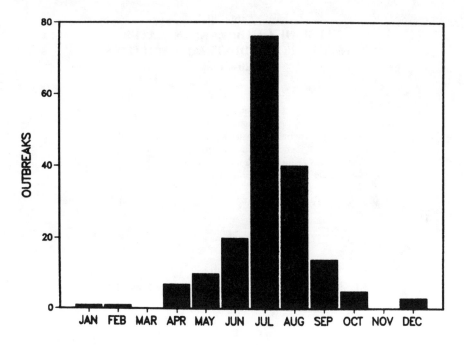

FIGURE 68. Seasonal distribution of waterborne outbreaks in campgrounds, parks, and recreational areas (1920—1980).

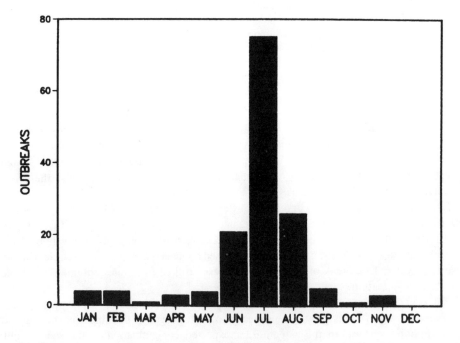

FIGURE 69. Seasonal distribution of waterborne outbreaks in hotels, motels, and rooming houses (1920—1980).

equate or interrupted disinfection of well water was responsible for 54% of the outbreaks and 48% of the cases of illnesses caused by treatment deficiencies in water systems where disinfection was the only treatment. Most of these outbreaks in well water systems were associated with the use of chlorine for disinfection, but four outbreaks were caused by

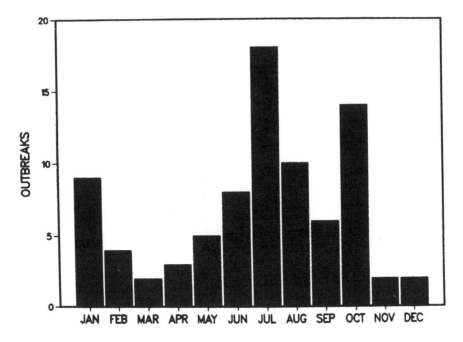

FIGURE 70. Seasonal distribution of waterborne outbreaks in industrial facilities (1920—1980).

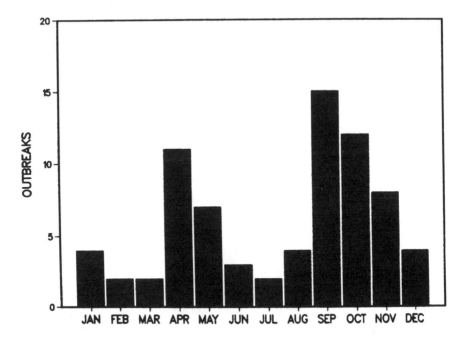

FIGURE 71. Seasonal distribution of waterborne outbreaks in schools (1920—1980).

inadequate or interrupted disinfection by iodine. These occurred in small noncommunity systems serving primarily a transient population.

The major causes of outbreaks in community systems (Figure 77) were treatment deficiencies (49%) and contamination of the distribution system (32%). Treatment deficiencies, such as inadequate filtration and interruption of disinfection, were responsible for most of the cases of illness related to community systems. Ineffective pretreatment and/or filtration

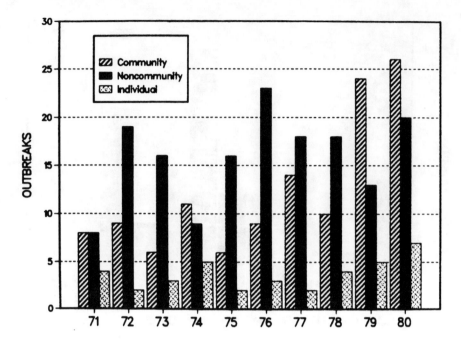

FIGURE 72. Waterborne outbreaks by type of system (1971—1980). (From Craun, G. F. et al., *Drinking Water and Human Health*, Bell, J. A. and Doege, T. C., Eds., American Medical Association, Chicago, 1984. With permission.)

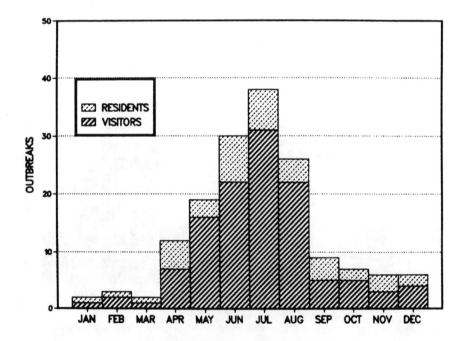

FIGURE 73. Seasonal distribution of waterborne outbreaks in noncommunity water systems (1971—1980). (From Craun, G. F. et al., *Drinking Water and Human Health*, Bell, J. A. and Doege, T. C., Eds., American Medical Association, Chicago, 1984. With permission.)

was documented in 6 outbreaks causing 7055 cases of waterborne illness. Almost all of the waterborne illness caused by distribution system deficiencies was due to cross-connections,

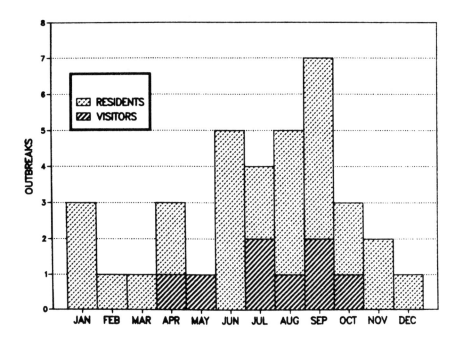

FIGURE 74. Seasonal distribution of waterborne outbreaks in individual water systems (1971—1980). (From Craun, G. F. et al., *Drinking Water and Human Health*, Bell, J. A. and Doege, T. C., Eds., American Medical Association, Chicago, 1984. With permission.)

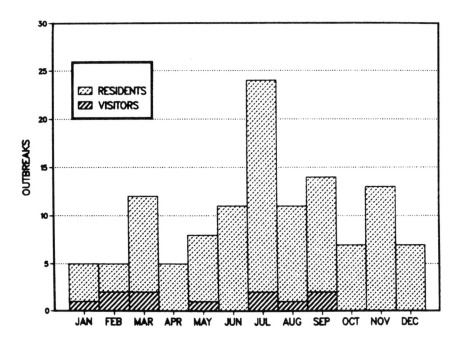

FIGURE 75. Seasonal distribution of waterborne outbreaks in community water systems (1971—1980). (From Craun, G. F. et al., *Drinking Water and Human Health*, Bell, J. A. and Doege, T. C., Eds., American Medical Association, Chicago, 1984. With permission.)

backsiphonage, contamination of water mains, and a contamination of distribution storage (Table 33), and most of these outbreaks occurred in community systems.

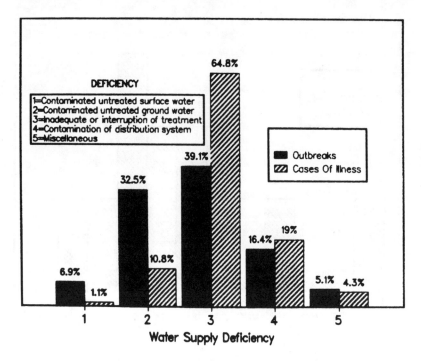

FIGURE 76. Causes of waterborne outbreaks in the U.S. (1971—1980). (From Craun, G. F. et al., *Drinking Water and Human Health*, Bell, J. A. and Doege, T. C., Eds., American Medical Association, Chicago, 1984. With permission.)

Contamination of ground water used without treatment and treatment deficiencies, primarily interrupted and inadequate disinfection, were responsible for almost all outbreaks (83%) and illness (80%) in noncommunity systems (Figure 78). The major cause of outbreaks (84%) and illness (85%) in individual water systems was use of contaminated, untreated surface water and ground water (Figure 79).

C. Etiologic Agents

An etiologic agent was determined in 45% of the outbreaks; the most commonly identified pathogen was *Giardia lamblia* (Tables 35 to 37). Chemical poisonings involving arsenic, chlordane, chromate, copper, cutting oil, developer fluid (hydroquinone, paramethylamino phenol), ethyl acrylate, fluoride, fuel oil, furadan, lead, leaded gasoline, a mixture of lubricating oil and kerosene, nitrate, phenol, polychlorinated biphenyls, selenium, and an unidentified herbicide accounted for 12% of the outbreaks. A large outbreak of 2000 illnesses occurred due to chlordane contamination of the Pittsburgh, Pa. water supply in 1980.

Acute copper poisoning (6 outbreaks) with copper levels of 4, 12.5, 38.5, and 80 mg/ℓ illustrate problems associated with the leaching of metals from water distribution systems: (1) naturally aggressive water with a low pH in contact with copper plumbing caused high levels of copper to be dissolved into the drinking water; (2) pH adjustment of naturally aggressive water was interrupted, allowing copper to be leached from copper plumbing; (3) a defective check valve allowed carbon dioxide from a drink dispensing machine to flow into the drinking water system, causing the water to become aggressive to copper plumbing; and (4) high levels of copper from a drinking fountain were felt to be caused by low water utilization and no flushing of the feed line or electrolysis because of improper electrical grounds.

Seven outbreaks of acute fluoride poisoning were caused by improperly designed or malfunctioning fluoridation equipment and operational problems in the addition of fluoride.

Table 31
WATERBORNE OUTBREAKS IN UNTREATED WATER SYSTEMS (1971—1980)

Cause of outbreak	Outbreaks	Illnesses
Contaminated surface water		
Used for supplemental water source	3	539
Contamination on watershed	2	41
Overflow of sewage near water intake	1	25
Heavy rains	1	79
Insufficient data	18	249
Total	25	933
Contaminated ground water (springs)		
Surface runoff, heavy rains	3	126
Insufficient data	7	914
Total	10	1040
Contaminated ground water (wells)		
Overflow or seepage of sewage	30	2953
Chemical contamination	10	140
Surface runoff, heavy rains	6	339
Creviced limestone, fissured rock	5	1221
Contaminated by stream or river	4	74
Improper construction, faulty well casing	3	101
Flooding	2	540
Insufficient data	32	1847
Total	92	7216

From Craun, G. F. et al., *Drinking Water and Human Health*, Bell, J. A. and Doege, T. C., Eds., American Medical Association, Chicago, 1984. With permission.

Four outbreaks occurred at elementary schools having their own water system. In the largest outbreak, 201 students and 12 adults became ill minutes after consuming orange juice made from the water supply at the school, which had a fluoride concentration of 270 mg/ℓ. The fluoride feeder pump at the well site had malfunctioned, causing fluoride to be fed continuously even while the water pump was not operating.

Waterborne diseases resulting from outbreaks caused by the use of contaminated, untreated water, inadequate or interrupted disinfection, and cross-connections were identified to determine which diseases might be associated with these deficiencies (Tables 38 and 39). Giardiasis and gastroenteritis of undetermined etiology were the two diseases most often transmitted by untreated and disinfected surface water sources. The disease most often transmitted via well water systems was gastroenteritis of undetermined etiology. Acute chemical poisonings were identified in 11% of the outbreaks that occurred in untreated well water systems. *Shigella* was the most commonly identified pathogen transmitted via well water systems and was identified in 11% of the outbreaks in untreated well water systems and 11% of the outbreaks in well water systems with inadequate or interrupted disinfection. Outbreaks caused by cross-connections and backsiphonage were of various etiologies (Table 39). Gastroenteritis of undetermined etiology and acute chemical poisonings were reported most often, but giardiasis, viral gastroenteritis, hepatitis A, shigellosis, and salmonellosis were also transmitted in this manner.

Table 32
WATERBORNE OUTBREAKS CAUSED BY TREATMENT DEFICIENCIES (1971—1980)

Cause of outbreak	Outbreaks	Illnesses
Disinfection only treatment		
Overflow or seepage of sewage sufficient to overwhelm normal disinfection		
Surface water systems	2	2,740
Well water systems	3	137
Spring water systems	3	1,249
Total	8	4,126
Inadequate disinfection		
Surface water systems	23	15,070
Well water systems	16	12,319
Total	39	27,389
Interruption of disinfection		
Surface water systems	15	2,056
Well water systems	38	8,127
Spring water systems	5	1,123
Total	58	11,306
Disinfection with other treatment		
Inadequate disinfection	1	41
Interruption of disinfection	1	26
Inadequate control of filtration/pretreatment	6	7,055
Inadequate control of chemical feed	9	700
Total	17	7,822

From Craun, G. F. et al., *Drinking Water and Human Health*, Bell, J. A. and Doege, T. C., Eds., American Medical Association, Chicago, 1984. With permission.

Table 33
WATERBORNE OUTBREAKS CAUSED BY DISTRIBUTION DEFICIENCIES (1971—1980)

Cause of outbreak	Outbreaks	Illnesses
Distribution system		
Cross-connection, unspecified	19	5656
Backsiphonage	13	1428
Contamination or corrosion of household plumbing or service lines	6	56
Contamination of broken, leaking mains	2	65
Contamination of water mains during construction or repair	2	1423
Corrosion due to CO_2 from soft drink machine	1	74
Total	43	8702
Storage facilities		
Contamination of distribution storage	6	5634
Contamination of private storage tank	3	331
Contamination of cistern	1	5
Total	10	5970

From Craun, G. F. et al., *Drinking Water and Human Health*, Bell, J. A. and Doege, T. C., Eds., American Medical Association, Chicago, 1984. With permission.

Table 34
WATERBORNE OUTBREAKS CAUSED BY MISCELLANEOUS DEFICIENCIES (1971—1980)

Cause of outbreak	Outbreaks	Illnesses
Use of water not intended for drinking	4	46
Contaminated ice	3	41
Contaminated bottled water	2	5
Suspected deliberate contamination	2	2008
Contaminated drinking fountain	1	6
Insufficient data	6	1379
Total	18	3485

From Craun, G. F. et al., *Drinking Water and Human Health*, Bell, J. A. and Doege, T. C., Eds., American Medical Association, Chicago, 1984. With permission.

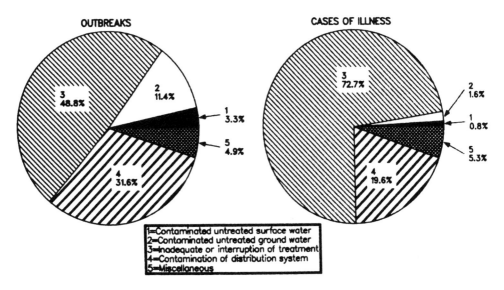

FIGURE 77. Causes of waterborne outbreaks and illness in community water systems (1971—1980). (From Craun, G. F. et al., *Drinking Water and Human Health*, Bell, J. A. and Doege, T. C., Eds., American Medical Association, Chicago, 1984. With permission.)

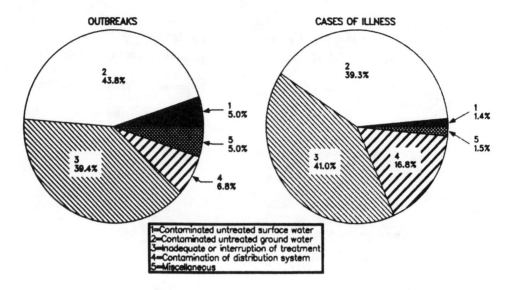

FIGURE 78. Causes of waterborne outbreaks and illness in noncommunity water systems (1971—1980). (From Craun, G. F. et al., *Drinking Water and Human Health,* Bell, J. A. and Doege, T. C., Eds., American Medical Association, Chicago, 1984. With permission.)

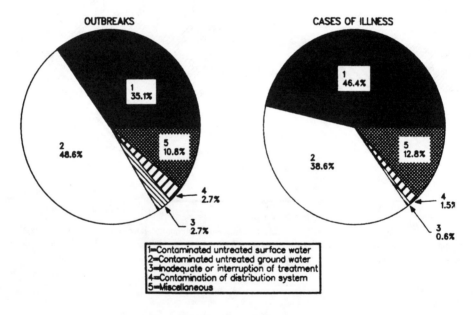

FIGURE 79. Causes of waterborne outbreaks and illness in individual water systems (1971—1980). (From Craun, G. F. et al., *Drinking Water and Human Health,* Bell, J. A. and Doege, T. C., Eds., American Medical Association, Chicago, 1984. With permission.)

Table 35
ETIOLOGY OF WATERBORNE OUTBREAKS IN COMMUNITY WATER SYSTEMS (1971—1980)

Disease	Outbreaks	Illnesses
Gastroenteritis, unidentified etiology	55	28,928
Giardiasis	22	17,090
Chemical poisoning	22	2,886
Shigellosis	10	3,788
Salmonellosis	5	1,075
Hepatitis A	5	130
Campylobacter diarrhea	2	3,800
Viral gastroenteritis	2	1,690
Total	123	59,387

From Craun, G. F. et al., *Drinking Water and Human Health*, Bell, J. A. and Doege, T. C., Eds., American Medical Association, Chicago, 1984. With permission.

Table 36
ETIOLOGY OF WATERBORNE OUTBREAKS IN NONCOMMUNITY WATER SYSTEMS (1971—1980)

Disease	Outbreaks	Illnesses
Gastroenteritis, unidentified etiology	110	10,783
Giardiasis	12	2,390
Shigellosis	12	1,401
Viral gastroenteritis	8	1,457
Hepatitis A	7	305
Chemical poisoning	7	645
Salmonellosis	2	72
Typhoid	1	210
Toxigenic *E. coli* diarrhea	1	1,000
Total	160	18,263

From Craun, G. F. et al., *Drinking Water and Human Health*, Bell, J. A. and Doege, T. C., Eds., American Medical Association, Chicago, 1984. With permission.

Table 37
ETIOLOGY OF WATERBORNE OUTBREAKS IN INDIVIDUAL WATER SYSTEMS (1971—1980)

Disease	Outbreaks	Illnesses
Gastroenteritis, unidentified etiology	12	134
Chemical poisoning	9	63
Giardiasis	5	72
Hepatitis A	4	28
Typhoid	3	12
Shigellosis	2	6
Salmonellosis	1	3
Campylobacter diarrhea	1	21
Total	37	339

From Craun, G. F. et al., *Drinking Water and Human Health*, Bell, J. A. and Doege, T. C., Eds., American Medical Association, Chicago, 1984. With permission.

Table 38
ETIOLOGY OF WATERBORNE OUTBREAKS OCCURRING IN UNTREATED OR DISINFECTED ONLY WATER SYSTEMS (1971—1980)

Disease	Outbreaks	Cases
Surface water sources		
Untreated		
Gastroenteritis	11	555
Giardiasis	10	312
Other	4	66
Inadequate disinfection		
Giardiasis	15	9,959
Gastroenteritis	8	4,661
Other	2	3,190
Interruption of disinfection		
Gastroenteritis	11	1,920
Giardiasis	2	52
Other	2	84
Well water sources		
Untreated		
Gastroenteritis	57	5,515
Chemical poisoning	10	140
Shigellosis	10	995
Hepatitis A	6	136
Viral gastroenteritis	3	338
Typhoid	3	12
Other	3	80
Inadequate disinfection		
Gastroenteritis	16	10,136
Viral gastroenteritis	1	120
Hepatitis A	1	50
Shigellosis	1	2,150
Interruption of disinfection		
Gastroenteritis	25	5,559
Shigellosis	5	1,563
Salmonellosis	3	291
Viral gastroenteritis	2	405
Hepatitis A	2	98
Typhoid	1	210

From Craun, G. F., et. al., *Drinking Water and Human Health,* Bell, J. A. and Doege, T. C., Eds., American Medical Association, Chicago, 1984. With permission.

Table 39
ETIOLOGY OF WATERBORNE OUTBREAKS CAUSED BY CROSS-CONNECTIONS AND BACKSIPHONAGE (1971—1980)

Disease	Outbreaks	Cases
Gastroenteritis	13	1208
Chemical poisoning	9	608
Giardiasis	3	2209
Viral gastroenteritis	2	1967
Hepatitis A	2	56
Shigellosis	2	286
Salmonellosis	1	750
Total	32	7084

From Craun, G. F., et al., *Drinking Water and Human Health*, Bell, J. A. and Doege, T. C., Eds., American Medical Association, Chicago, 1984. With permission.

ACKNOWLEDGMENTS

The assistance of Lon Winchester and Melda Hirth in typing this manuscript is gratefully acknowledged, as is the assistance of Art Hammonds in compiling summary data and Steve Waltrip in preparing figures and graphs.

The encouragement and assistance of my colleagues at the EPA and the CDC over the past 14 years in support of this effort are also acknowledged. Without the dedicated work of those public health officials in the field who recognize and investigate outbreaks, this report would not have been possible. Special gratitude is expressed to Abel Wolman for his review of this manuscript and suggestions. Dr. Wolman has provided the inspiration for many of us who have been involved in waterborne disease surveillance and water supply improvement.

REFERENCES

1. **Gorman, A. E. and Wolman, A.,** Water-borne outbreaks in the United States and Canada and their significance, *J. Am. Water Works Assoc.*, 31, 255, 1939.
2. **Gorman, A. E. and Wolman, A.,** Water-borne outbreaks in the United States and their significance, American Water Works Association, New York, 1948.
3. Committee on Public Works, Stream pollution control hearings before a subcommittee of the committee on public works, 80th Congress, 1st Session on S. 418, Washington, D.C., 1947.
4. **Eliassen, R. and Cummings, R. H.,** Analysis of waterborne outbreaks, 1938—45, *J. Am. Water Works Assoc.*, 40, 509, 1948.
5. **Weibel, S. R., Dixon, F. R., Weidner, R. B., and McCabe, L. J.,** Waterborne disease outbreaks, 1946—60, *J. Am. Water Works Assoc.*, 56, 947, 1964.
6. Division of Water Supply and Pollution Control, 228 waterborne disease outbreaks known to have occurred in the 15 year period 1946—60, Public Health Service, Cincinnati, Ohio, 1964.
7. Water Supply Program Division, Tabulation of 128 waterborne disease outbreaks known to have occurred in the United States, 1961—70, U.S. Environmental Protection Agency, Cincinnati, Ohio, 1971.
8. **Craun, G. F. and McCabe, L. J.,** Review of the causes of waterborne-disease outbreaks, *J. Am. Water Works Assoc.*, 65, 74, 1973.

9. **Craun, G. F.,** Waterborne disease outbreaks in the United States, *J. Environ. Health,* 41, 259, 1979.
10. **Craun, G. F.,** Outbreaks of waterborne disease in the United States, *J. Am. Water Works Assoc.,* 73, 360, 1981.
11. **Craun, G. F., Hammonds, A. F., and Waltrip, S. C.,** 1405 waterborne outbreaks reported in the United States 1920—80, U.S. Environmental Protection Agency, Cincinnati, Ohio, 1983.
12. **Craun, G. F., Waltrip, S. C., and Hammonds, A. F.,** Waterborne outbreaks in the United States 1971—81, in *Drinking Water and Human Health,* Bell, J. A. and Doege, T. C., Eds., American Medical Association, Chicago, Ill., 1984.
13. Foodborne Outbreaks Annual Summary 1972, Center for Disease Control, Department of Health, Education and Welfare, Publication No. (CDC)74-8185, 1973.
14. Foodborne and Waterborne Disease Outbreaks Annual Summary 1973, Center for Disease Control, Department of Health, Education and Welfare, Publication No. (CDC)75-8185, 1974.
15. Foodborne and Waterborne Disease Outbreaks Annual Summary 1974, Center for Disease Control, Department of Health, Education and Welfare, Publication No. (CDC)76-8185, 1976.
16. Foodborne and Waterborne Disease Outbreaks Annual Summary 1975, Center for Disease Control, Department of Health, Education and Welfare, Publication No. (CDC)76-8185, 1976.
17. Foodborne and Waterborne Disease Outbreaks Annual Summary 1976, Center for Disease Control, Department of Health, Education and Welfare, Publication No. (CDC)78-8185, 1977.
18. Foodborne and Waterborne Disease Surveillance Annual Summary 1977, Center for Disease Control, Department of Health, Education and Welfare, U.S. Government Printing Office 640-010-3610, 1979.
19. Water-Related Disease Outbreaks Surveillance Annual Summary 1978, Center for Disease Control, Health and Human Services Publication No. (CDC)80-8385, 1980.
20. Water-Related Disease Outbreaks Surveillance Annual Summary 1979, Center for Disease Control, Health and Human Services Publication No. (CDC)81-8385, 1981.
21. Water-Related Disease Outbreaks Surveillance Annual Summary 1980, Centers for Disease Control, Health and Human Services Publication No. (CDC)82-8385, 1982.
22. **Whipple, G. C.,** *Typhoid Fever. Its Causation, Transmission and Prevention,* John Wiley & Sons, New York, 1908.
23. Inventory of water supply facilities, *Eng. News-Rec.,* 123, 414, 1939.
24. **Weibel, S. R.,** A summary of census data on water treatment plants in the United States, *Public Health Rep.,* 57, 1679, 1942.
25. **Thoman, J. R.,** Statistical Summary of Water Supply and Treatment Practices in the United States, U.S. Public Health Service, Washington, D.C., 1953.
26. **Glass, A. C. and Jenkins, K. H.,** Statistical Summary of Municipal Water Facilities in the United States, U.S. Public Health Service, Publ. No. 1039, Washington, D.C. 1963.
27. **Lippy, E. C.,** Personal communication, 1983.
28. **Barker, W. H., Jr.,** Foodborne disease surveillance — Washington State, *Am. J. Public Health,* 64, 26, 1974.
29. **Weisman, J. B., Craun, G. F., Lawrence, D. N., Pollard, R. A., Saslaw, M. S., and Gangarosa, E. J.,** An epidemic of gastroenteritis traced to a contaminated public water supply, *Am. J. Epidemiol.,* 103, 391, 1976.
30. Annual Summary Reported Morbidity and Mortality in the United States, Health and Human Services Publ. No. (CDC)81-8241, 1981.
31. Water-Related Disease Outbreaks Annual Summary 1981, Centers for Disease Control, Health and Human Services Publication No. (CDC)82-8385, 1982.
32. **Saslow, M. S., Nitzkin, J. L., Feldman, R., Baine, W., Pfeiffer, K., and Pearson, M.,** Typhoid fever public health aspects, *Am. J. Public Health,* 65, 1184, 1975.
33. A waterborne epidemic of salmonellosis in Riverside, California, 1965: epidemiologic aspects, *Am. J. Epidemiol.,* 93, 33, 1971.
34. **Batik, O., Craun, G. F., Tuthill, R. W., and Kraemer, D. F.,** An epidemiologic study of the relationship between hepatitis A and water supply characteristics and treatment, *Am. J. Public Health,* 70, 167, 1980.
35. **Putnam, J. J. and Taylor, E. W.,** Is acute poliomyelitis unusually prevalent this season?, *Boston Med. Surg. J.,* 129, 509, 1893.
36. **Caverly, C. S.,** Notes of an epidemic of acute anterior poliomyelitis, *JAMA,* 26, 1, 1896.
37. **Mosley, J. W.,** Transmission of viral diseases by drinking water, in *Transmission of Viruses by the Water Route,* Berg, G., Ed., Interscience, New York, 1967, 5.
38. **Bancroft, P. M., Englehard, W. E., and Evans, C. A.,** Poliomyelitis in Huskerville (Lincoln) Nebraska, *JAMA,* 164, 836, 1957.
39. **Englehard, W. E. and Bancroft, P. M.,** Serological studies of the poliomyelitis epidemic in Huskerville, Neb., *J. Dis. Child.,* 97, 409, 1959.
40. **Mack, W. M., Lu, Y.-S., and Coohon, D. B.,** Isolation of poliomyelitis virus from a contaminated well, *HMSA Health Services Rep.,* 87, 271, 1972.

41. **Blacklow, N. R. and Cukor, G.,** Norwalk virus: a major cause of epidemic gastroenteritis, *Am. J. Public Health,* 72, 1321, 1982.
42. **Blacklow, N. R. and Cukor, G.,** Viral gastroenteritis, *N. Engl. J. Med.,* 304, 397, 1981.
43. **Greenburg, H. B., Valdesusa, J., Yolken, R. H., Gangarosa, E., Gary, W., Wyatt, R. G., Konno, T., Suzuki, H., Channock, R. M., and Kapikian, A. Z.,** Role of Norwalk virus in outbreaks of nonbacterial gastroenteritis, *J. Infect. Dis.,* 139, 564, 1979.
44. **Steinhoff, M. C.,** Viruses and diarrhea — a review, *Am. J. Dis. Child.,* 132, 302, 1978.
45. **Kapikian, A. Z.,** Viral diarrhea etiology and control, *Am. J. Clin. Nutr.,* 31, 2219, 1978.
46. **Blacklow, N. R., Dolin, R., Fedson, D. S., Dupont, H., Northrup, R. S., Hornick, R. B., and Chanock, R. M.,** Acute infectious nonbacterial gastroenteritis: etiology and pathogenesis, *Ann. Intern. Med.,* 76, 993, 1972.
47. **Adler, J. R. and Sickl, R.,** Winter vomiting disease, *J. Infect. Dis.,* 119, 688, 1966.
48. **Wenman, W. M., Hinde, D., Feltham, S., and Gurwith, M.,** Rotavirus infection in adults, *N. Engl. J. Med.,* 301, 303, 1979.
49. **Flewett, T. H. and Woode, G. N.,** The rotaviruses, brief review, *Arch. Virol.,* 57, 1, 1978.
50. **Kaplan, J. E., Gary, G. W., Baron, R. C., Singh, N., Schonberger, L. B., Feldman, R., and Greenberg, H. B.,** Epidemiology of Norwalk gastroenteritis and the role of Norwalk virus in outbreaks of acute nonbacterial gastroenteritis, *Ann. Intern. Med.,* 96, 756, 1982.
51. **Baron, R. C., Murphy, F. D., Greenberg, H. B., Davis, C. E., Bergman, D. J., Gary, W. G., Hughes, J. M., and Schonberger, L. B.,** Norwalk gastrointestinal illness — an outbreak associated with swimming in a recreational lake and secondary person-to-person transmission, *Am. J. Epidemiol.,* 115, 163, 1982.
52. **Koopman, J. S., Eckert, E. A., Greenburg, H. B., Strohm, B. C., Isaacson, R. E., and Monto, A. S.,** Norwalk virus enteric illness acquired by swimming exposure, *Am. J. Epidemiol.,* 115, 173, 1982.
53. **Gunn, R. A., Janowski, H. T., Lieb, S., Prather, E. C., and Greenberg, H. B.,** Norwalk virus gastroenteritis following raw oyster consumption, *Am. J. Epidemiol.,* 115, 348, 1982.
54. **Kaplan, J. E., Feldman, R., Campbell, D. S., Lookabaugh, C., and Gary, W.,** The frequency of a Norwalk-like pattern of illness in outbreaks of acute gastroenteritis, *Am. J. Public Health,* 72, 1329, 1982.
55. **Morens, D. M., Zweighaft, R. M., Vernon, T. M., Gary, G. W., Eslien, J. J., Wood, B. T., Holman, R. C., and Dolin, R.,** A waterborne outbreak of gastroenteritis with secondary person-to-person spread, association with a viral agent, *Lancet,* 1, 964, 1979.
56. **Craun, G. F.,** Disease outbreaks caused by drinking water, *J. Water Pollut. Control Fed.,* 51, 1751, 1979.
57. **Wilson, R., Anderson, L. J., Holman, R. C., Gary, G. W., and Greenburg, H. B.,** Waterborne gastroenteritis due to the Norwalk agent: clinical and epidemiologic investigation, *Am. J. Public Health,* 72, 72, 1982.
58. **Taylor, J. W., Gary, G. W., and Greenburg, H. B.,** Norwalk-related viral gastroenteritis due to contaminated drinking water, *Am. J. Epidemiol.,* 114, 584, 1981.
59. **Kaplan, J. E., Goodman, R. A., Schonberger, L. B., Lippy, E. C., and Gary, G. W.,** Gastroenteritis due to Norwalk virus: an outbreak associated with a municipal water system, *J. Infect. Dis.,* 146, 190, 1982.
60. **Lycke, E., Blomberg, J., Berg, G., Ericksson, A., and Madsen, L.,** Epidemic acute diarrhoea in adults associated with infantile gastroenteritis, *Lancet,* 2, 1056, 1978.
61. **Zamotin, B. A., Libiyainen, L. T., Bortnik, F. L., Chernitskaya, E. P., Enina, Z. I., Rossikhin, N. F., Veselov, V. I., Kharyutkin, S. A., Egerev, V. A., Nikulin, V. A., Dorozhkina, Z. D., Kostyukova, K. P., Dorozhkin, G. V., Shironina, V. G., Cherynl, V. S., Agapov, E. I., and Vlasov, V. I.,** Waterborne group infection of rotavirus etiology, *Zh. Microbiol. Epidemiol. Immunol.,* 11, 100, 1981.
62. **Sutmoller, F., Azeredo, R. S., Lacerda, M. D., Barth, O. M., Bereira, H. G., Hoffer, E., and Schatzmayr, H. G.,** An outbreak of gastroenteritis caused by both rotavirus and *Shigella sonnei* in a private school in Rio de Janeiro, *J. Hyg. (Camb.),* 88, 285, 1982.
63. **Merson, M. H., Morris, G. K., Sack, D. A., Wells, J. G., Feeley, J. C., Sack, R. B., Creech, W. B., Kapikian, A. Z., and Gangarosa, E. J.,** Travelers' diarrhea in Mexico. A prospective study of physicians and family members attending a congress, *N. Engl. J. Med.,* 294, 1299, 1976.
64. **Rosenburg, M. L., Koplan, J. P., Wachsmuth, I. K., Wells, J. G., Gangarosa, E. J., Guerrant, R. L., and Sack, D. A.,** Epidemic diarrhea at Crater Lake from enterotoxigenic *Escherichia coli*, *Ann. Intern. Med.,* 86, 714, 1977.
65. **Merson, M. H., Hughes, J. M., Wood, B. T., Yashuk, J. C., and Wells, J. G.,** Gastrointestinal illness on passenger cruise ships, *JAMA,* 231, 723, 1975.
66. **Lumish, R. M., Ryder, R. W., Anderson, D. C., Wells, J. G., and Puhr, N. D.,** Heat-labile enterotoxigenic *Escherichia coli* induced diarrhea aboard a Miami-based cruise ship, *Am. J. Epidemiol.,* 111, 432, 1980.
67. **Inoue, H.,** Outbreak of acute gastroenteritis due to waterborne infection of pathogenic *E. coli*, *J. Jpn. Assoc. Infect. Dis.,* 50, 336, 1976.

68. **Karoly, C.,** Fatal *Escherichia coli* 0124:K72:B17 infection caused by well water, *Orv. Hetil.*, 115, 1770, 1974.
69. **Craun, G. F.,** Health aspects of groundwater pollution, in *Groundwater Pollution Microbiology*, Bitton, G. and Gerba, C. P., Eds., John Wiley & Sons, New York, 1984, 135.
70. **Weber, G., Stanek, G., Massiczek, N., and Klenner, M. F.,** *Yersinia enterocolitica* in drinking water, *Z. Bakteriol. Mikrobiol. Hyg.*, 173, 209, 1981.
71. **Meadows, C. A. and Snudden, B. H.,** Prevalence of *Yersinia enterocolitica* in waters of the lower Chippewa river basin, Wisconsin, *Appl. Environ. Microbiol.*, 43, 953, 1982.
72. **Harvey, I., Greenwood, R., Pickett, M. J., and Mah, R. A.,** Recovery of *Yersinia enterocolitica* from streams and lakes of California, *Appl. Environ. Microbiol.*, 32, 352, 1976.
73. **Saari, T. N. and Jansen, G. P.,** Waterborne *Yersinia enterocolitica* in the midwest United States, *Contrib. Microbiol. Immunol.*, 5, 185, 1979.
74. **Highsmith, A. K., Feeley, J. C., Skaliy, P., Wells, J. G., and Woods, B. T.,** Isolation of *Yersinia enterocolitica* from well water and growth in distilled water, *Appl. Environ. Microbiol.*, 34, 745, 1977.
75. **Kapperud, G.,** *Yersinia enterocolitica* and *Yersinia*-like microbes isolated from mammals and water in Norway and Denmark, *Acta. Pathol. Microbiol.*, 85, 129, 1977.
76. **Schiemann, D. A.,** Isolation of *Yersinia enterocolitica* from surface and well waters in Ontario, *Can. J. Microbiol.*, 24, 1048, 1978.
77. **Shayegani, M., DeForge, I., McGlynn, D. M., and Rott, T.,** Characteristics of *Yersinia enterocolitica* and related species isolated from human, animal, and environmental sources, *J. Clin. Microbiol.*, 14, 304, 1981.
78. **Bartley, T. D., Quan, T. J., Collins, M. T., and Morrison, S. M.,** Membrane filter technique for the isolation of *Yersinia enterocolitica*, *Appl. Environ. Microbiol.*, 43, 829, 1982.
79. **Langeland, G.,** *Yersinia enterocolitica* and *Yersinia enterocolitica*-like bacteria in drinking water and sewage sludge, *Acta Pathol. Microbiol. Immunol. Scand.*, 91, 179, 1983.
80. **Lassen, J.,** *Yersinia enterocolitica* in drinking water, *Scand. J. Infect. Dis.*, 4, 125, 1972.
81. **Keet, E. E.,** *Yersinia enterocolitica* septicemia, *N.Y. State J. Med.*, 74, 2226, 1974.
82. **Craun, G. F.,** Microbiology — waterborne outbreaks, *J. Water Pollut. Control Fed.*, 48, 1378, 1976.
83. **Eden, K. V., Rosenburg, M. L., Stoopler, M., Wood, B. T., Highsmith, A. K., Skaliy, P., Wells, J. G., and Feeley, J. C.,** Waterborne gastroenteritis at a ski resort — isolation of *Yersinia enterocolitica* from drinking water, *Public Health Rep.*, 92, 245, 1977.
84. **Christiensen, S. G.,** Isolation of *Yersinia enterocolitica* 0:3 from a well suspected as the source of yersinosis in a baby, *Acta Vet. Scand.*, 20, 154, 1979.
85. **Blaser, M. J., Taylor, D. N., and Feldman, R. A.,** Epidemiology of *Campylobacter jejuni* infections, *Epidemiol. Rev.*, 5, 157, 1983.
86. **Blaser, M. J.,** *Campylobacter fetus* subspecies *jejuni:* the need for surveillance, *J. Infect. Dis.*, 141, 670, 1980.
87. **Blaser, M. J., LaForce, F. M., Wilson, N. A., and Wang, W. L. L.,** Reservoirs for human campylobacteriosis, *J. Infect. Dis.*, 141, 665, 1980.
88. **Vogt, R. L., Sours, H. E., Barrett, T., Feldman, R. A., Dickinson, R. J., and Witherell, L.,** Campylobacter enteritis associated with contaminated water, *Ann. Intern. Med.*, 96, 292, 1982.
89. **Mentzing, L. O.,** Waterborne outbreaks of campylobacter enteritis in central Sweden, *Lancet*, 2, 352, 1981.
90. **Palmer, S. R., Gully, P. R., White, J. M., Pearson, A. D., Suckling, W. G., Jones, D. M., Rawes, J. C. L., and Penner, J. L.,** Waterborne outbreak of campylobacter gastroenteritis, *Lancet*, 1, 287, 1983.
91. **Taylor, D. N., McDermott, K. T., Little, J. R., Wells, J. G., and Blaser, M. J.,** Campylobacter enteritis from untreated water in the Rocky Mountains, *Ann. Intern. Med.*, 99, 38, 1983.
92. **Hopkins, R. S., Olmsted, R., and Istre, G. R.,** Endemic *Campylobacter jejuni* infection in Colorado: identified risk factors, *Am. J. Public Health*, 74, 249, 1984.
93. **Deinhardt, F. and Gust, I. D.,** Viral hepatitis, *Bull. WHO*, 60, 661, 1982.
94. **Hejkal, T. W., Keswick, B., LaBelle, R. L., Gerba, C. P., Sanchez, Y., Dressman, G., Hafkin, B., and Melnick, J. L.,** Viruses in a community water supply associated with an outbreak of gastroenteritis and infectious hepatitis, *J. Am. Water Works Assoc.*, 74, 318, 1982.
95. **Mahoney, L. E., Friedman, G. T. H., Murray, R. A., Schulenburg, E. L., and Heidbreder, G. A.,** A waterborne gastroenteritis epidemic in Pico Riveria, California, *Am. J. Public Health*, 64, 963, 1974.
96. **Renteln, H. A. and Hinman, A. R.,** A waterborne epidemic of gastroenteritis in Madera, California, *Am. J. Epidemiol.*, 86, 1, 1967.
97. **Lobel, H. O., Bisno, A. L., Goldfield, M., and Prier, J. E.,** A waterborne outbreak of gastroenteritis with secondary person-to-person spread, *Am. J. Epidemiol.*, 89, 384, 1969.
98. **Mosely, J. W.,** Waterborne infectious hepatitis, *N. Engl. J. Med.*, 261, 703, 1959.
99. **Garibaldi, R. A., Murphy, G. D., and Wood, B. T.,** Infectious hepatitis outbreak associated with cafe water, *HSMHA Health Rep.*, 87, 164, 1972.

100. **Rosenberg, M. L., Koplan, J. P., and Pollard, R. A.**, The risk of acquiring hepatitis from sewage-contaminated water, *Am. J. Epidemiol.*, 112, 17, 1980.
101. **Prince, A. M.**, Non-A, non-B hepatitis virus, *Ann. Rev. Microbiol.*, 37, 217, 1983.
102. **Khuroo, M. S.**, Study of an epidemic of non-A, non-B hepatitis, *Am. J. Med.*, 68, 818, 1980.
103. **Black, R. E., Craun, G. F., and Blake, P. A.**, Epidemiology of common-source outbreaks of shigellosis in the United States, 1961—1975, *Am. J. Epidemiol.*, 108, 47, 1978.
104. **Craun, G. F. and Gunn, R. A.**, Outbreaks of waterborne disease in the United States: 1975—76, *J. Am. Water Works Assoc.*, 71, 422, 1979.
105. **Davis, C. and Ritchie, L. S.**, Clinical manifestations and treatment of epidemic amebiasis occurring in occupants of the Mantestsa apartment building, Tokyo, Japan, *Am. J. Trop. Med.*, 28, 817, 1948.
106. Centers for Disease Control, Giardiasis in travelers, *Morbid. Mortal. Weekly Rep.*, 19, 455, 1970.
107. **Walzer, P. D., Wolfe, M. S., and Schultz, M. G.**, Giardiasis in travelers, *J. Infect. Dis.*, 124, 235, 1971.
108. **Brodsky, R. E., Spencer, H. C., Jr., and Schultz, M. G.**, Giardiasis in travelers to the Soviet Union, *J. Infect. Dis.*, 130, 319, 1974.
109. **Moore, G. T., Cross, W. M., McGuire, D., Mollohan, C. S., Gleason, N. N., Healy, G. R., and Newton, L. H.**, Epidemic giardiasis at a ski resort, *N. Engl. J. Med.*, 281, 402, 1969.
110. **Shaw, P. K., Brodsky, R. E., Lyman, D. O., Wood, B. T., Hibler, C. P., Healy, G. R., MacLeod, K. I., Stahl, W., and Schultz, M G.**, A community wide outbreak of giardiasis with evidence of transmission by municipal water supply, *Ann. Intern. Med.*, 87, 426, 1977.
111. **Syrontynski, S. and Reamson, T. A.**, Unpublished work, 1977.
112. **Lippy, E. C.**, Waterborne disease: occurrence is on the upswing, *J. Am. Water Works Assoc.*, 72, 57, 1981.
113. **Kirner, J. C., Littler, J. D., and Angelo, L. A.**, A waterborne outbreak of giardiasis in Camas, Washington, *J. Am. Water Works Assoc.*, 70, 35, 1978.
114. **Dykes, A. C., Juranek, D. D., Lorenz, R. A., Sinclair, S., Jakubowski, W., and Davies, R.**, Municipal waterborne giardiasis: an epidemiologic investigation. Beavers implicated as possible reservoir, *Ann. Intern. Med.*, 92, 165, 1980.
115. **Lippy, E. C.**, Tracing a giardiasis outbreak at Berlin, New Hampshire, *J. Am. Water Works Assoc.*, 70, 512, 1978.
116. **Lopez, C. E., Dykes, A. C., Juranek, D. D., Sinclair, S. P., Conn, J. M., Christie, R. W., Lippy, E. C., Schultz, M. G., and Mires, M. H.**, Waterborne giardiasis: a community wide outbreak of disease and a high rate of asymptomatic infection, *Am. J. Epidemiol.*, 112, 495, 1980.
117. **Wright, R. A., Spencer, H. C., Brodsky, R. E., and Vernon, T. M.**, Giardiasis in Colorado: an epidemiologic study, *Am. J. Epidemiol.*, 105, 330, 1977.
118. **Weiss, H. B., Winegar, D. A., Levy, B. S., and Washburn, J. W.**, Giardiasis in Minnesota, 1971—75, *Minn. Med.*, 60, 815, 1977.
119. **Frost, F. and Harter, L.**, Giardiasis in Washington State, Office of Environmental Health Programs, Washington State Department of Social and Health Services, 1980.

Chapter 6

RECENT STATISTICS OF WATERBORNE DISEASE OUTBREAKS (1981—1983)

Gunther F. Craun*

TABLE OF CONTENTS

I. Occurrence and Distribution of Outbreaks 162

II. Etiologic Agents 162

III. Causes of Outbreaks 167

References... 168

---

* This chapter was written by Gunther F. Craun in his private capacity. No official support or endorsement by the Environmental Protection Agency or any other agency of the Federal Government is intended or should be inferred.

I. OCCURRENCE AND DISTRIBUTION OF OUTBREAKS

In 1981, 1982, and 1983, 112 outbreaks and 28,791 cases of illness occurring in potable water systems were reported in the U.S; 14 outbreaks and 394 cases of illness were caused by consumption of contaminated water from nonpotable sources[1-3] (Table 1). Results of microbiological analyses of water samples were available for 71 water systems experiencing an outbreak. Either coliform organisms or pathogens were found in the water system demonstrating water contamination in addition to epidemiologic evidence for the waterborne transmission of disease in 59 outbreaks.

In 1983, 47 waterborne outbreaks and 21,124 cases of illness were reported (Table 2). This is the largest number of waterborne outbreaks reported in any year since 1942, and the most illness resulting from waterborne outbreaks since 1940. The largest outbreak occurred when an estimated 11,400 persons became ill after consuming contaminated well water during a religious festival in 1983; some 20,000 persons attended the festival at a Pennsylvania campground which normally served only 168 persons.

As in past years, the occurrence of waterborne outbreaks during 1981 to 1983 was greater during the summer months (Table 3).

Community water systems accounted for 52% of the outbreaks and 48% of the cases of illness (Table 4). Noncommunity water systems accounted for 29% of the outbreaks and 50% of the cases of illness; 19 % of the outbreaks and 2% of the cases of illness occurred in individual water systems. Although most outbreaks occurred in community water systems, these data suggest that noncommunity water systems have a greater potential than other water systems for waterborne disease. Based on estimates that community water systems serve 180 million people, noncommunity water systems serve 20 million people, and individual water systems serve 30 million people, the incidence of waterborne disease during this period for noncommunity water systems was 243 cases of illness per 1 million persons served per year compared with 26 cases per 1 million persons per year for community systems and 7 cases per 1 million persons per year for individual systems.

Outbreaks caused by consumption of nonpotable water included an outbreak of giardiasis in a group of hikers in a backcountry area and an outbreak of gastroenteritis in a road work crew, both caused by drinking untreated water from streams. There were also two outbreaks of "sewage poisoning", which occurred after factor workers drank from an unmarked spigot used for sampling partially treated sewage effluent and after power-plant workers drank from an unlabeled tap delivering untreated water from the Potomac River. A giardiasis outbreak occurred in police and fire department divers working in the Hudson River near New York City. Shigellosis outbreaks associated with swimming in lakes in Oklahoma and Illinois and following the use of water slides in California also occurred. A giardiasis outbreak in Washington and a gastroenteritis outbreak in Iowa were associated with swimming in a community pool. A gastroenteritis outbreak was associated with swimming in an Iowa lake; two swimming-associated outbreaks in Minnesota parks were caused by Norwalk agent. A gastroenteritis outbreak was also associated with the use of river water in a gunite spray operation.

II. ETIOLOGIC AGENTS

An etiologic agent was determined in 60% of the outbreaks (Table 5). The most frequently identified pathogen in these outbreaks was *Giardia lamblia*. Water samples for *Giardia* cysts were collected during 21 of the giardiasis outbreaks, and cysts were recovered from water samples in 18 outbreaks. Most of the waterborne outbreaks of giardiasis occurred in water systems where surface water was disinfected only. Three giardiasis outbreaks occurred when filtration facilities were by-passed or malfunctioned, and five outbreaks occurred in filtered

Table 1
WATERBORNE OUTBREAKS REPORTED IN THE U.S.[1-3]

State	Month	Cases	Etiology	Type[a]	Remarks
Alabama	5	55	Norwalk agent	C	Community
	9	16	Benzene	I	
Alaska	1	2	Copper	C	Restaurant
	11	35	Gastroenteritis	C	Community
Arizona	5	24	Gastroenteritis	NC	Truck stop
	6	326	Gastroenteritis	C	Subdivision
California	4	61	Gastroenteritis	NC	Restaurant
	4	25	Norwalk agent	C	
	4	170	Norwalk agent	C	Marine Corps depot
	8	40	Shigella	I	Water slide
	11	6	Hepatitis A	C	Indian reservation
Colorado	1	10	Giardia	C	Ski resort
	1	4	Giardia	C	
	1	11	Giardia	C	
	1	17	Giardia	C	
	2	17	Giardia	C	Ski resort
	3	1761	Rotavirus	C	
	3	4	Giardia	C	Ski resort
	4	8	Giardia	C	Ski resort
	4	16	Gastroenteritis	C	
	5	10	Giardia	C	
	6	8	Gastroenteritis	C	
	6	11	Giardia	C	
	6	8	Giardia	C	
	7	72	Giardia	C	Family reunion
	7	27	Gastroenteritis	C	
	7	30	Giardia	C	
	8	28	Giardia	NC	Ranch resort
	8	36	Gastroenteritis	C	Trailer park
	8	110	Giardia	C	
	8	578	Gastroenteritis	C	
	9	32	Giardia	NC	Ski resort
	9	32	Giardia	NC	Camp
	10	11	Giardia	C	
	10	7	Giardia	I	Hikers
	11	38	Giardia	C	Ski resort
	12	14	Giardia	C	
	12	18	Giardia	C	
Connecticut	8	80	Gastroenteritis	NC	Park
Florida	3	3	Giardia	C	Trailer park
	5	871	Campylobacter	C	
	6	52	Gastroenteritis	C	Apartment
	10	7	Gastroenteritis	NC	Factory
Georgia	1	500	Norwalk agent	C	
	6	10	Hepatitis A	I	Day-care center
	7	35	Hepatitis A	C	Trailer park
Illinois	6	6	Shigella	I	Swimming
	6	32	Gastroenteritis	I	Swimming
	9	81	Campylobacter	C	Subdivision
Indiana	5	400	Gastroenteritis	C	
Iowa	6	14	Gastroenteritis	NC	Camp
	8	92	Gastroenteritis	NC	Camp
	9	60	Gastroenteritis	I	Swimming pool
Idaho	11	44	Giardia	C	
	11	71	Giardia	C	

Table 1 (continued)
WATERBORNE OUTBREAKS REPORTED IN THE U.S.[1–3]

State	Month	Cases	Etiology	Type[a]	Remarks
Kansas	8	100	Gastroenteritis	NC	Restaurant
Kentucky	9	150	Hepatitis A	I	Untreated wells
	11	58	Hepatitis A	C	
Maine	8	25	Gastroenteritis	C	Trailer park
	10	31	Fluoride	NC	School
	10	50	Gastroenteritis	I	
Maryland	8	72	Gastroenteritis	NC	Condominium
Minnesota	7	38	Norwalk agent	I	Swimming
	7	38	Norwalk agent	I	Swimming
Montana	7	100	Giardia	C	
New York	7	12	Giardia	I	Divers
	8	400	Gastroenteritis	NC	Camp
Nevada	8	342	Giardia	C	
New Hampshire	5	7	Giardia	C	
	10	13	Giardia	C	
New Mexico	8	100	Giardia	C	
North Carolina	9	153	Shigella	NC	School
Oklahoma	1	400	Salmonella	C	
	6	49	Shigella	I	Swimming
Oregon	2	40	Gastroenteritis	NC	Lodge
	9	9	Giardia	C	
Pennsylvania	1	3	Lead	I	
	2	16	Yersinia	I	
	4	9	Gastroenteritis	NC	Motel
	4	35	Gastroenteritis	NC	Motel
	4	16	Gastroenteritis	NC	
	5	51	Gastroenteritis	NC	Restaurant
	6	48	Gastroenteritis	NC	Country club
	6	11,400	Gastroenteritis	NC	Religious festival
	6	5	Gastroenteritis	I	Gunite spray
	7	71	Gastroenteritis	NC	Camp
	7	97	Gastroenteritis	NC	Camp
	7	30	Gastroenteritis	NC	Restaurant
	7	6	Gastroenteritis	I	Work crew
	7	31	Gastroenteritis	NC	Camp
	8	25	Gastroenteritis	NC	Recreational area
	8	200	Gastroenteritis	NC	Resort
	9	11	Gastroenteritis	I	Camp
	10	366	Giardia	C	
	10	146	Gastroenteritis	NC	Recreational area
	10	135	Giardia	C	
	11	84	Lead	C	
	11	9	Copper	C	School
Puerto Rico	6	500	Gastroenteritis	NC	
South Dakota	3	1	Nitrate	I	Farm
Tennessee	6	8	Hepatitis A	I	Church
Texas	8	17	V. cholera 01	NC	Oil rig
	8	3400	Gastroenteritis	C	
	9	65	Gastroenteritis	NC	Camp
Utah	1	41	Giardia	C	
	8	1272	Giardia	C	
	8	12	Gastroenteritis	NC	Camp

Table 1 (continued)
WATERBORNE OUTBREAKS REPORTED IN THE U.S.[1-3]

State	Month	Cases	Etiology	Type[a]	Remarks
Vermont	1	3	Copper	C	
	2	22	Gastroenteritis	NC	Subdivision
	7	105	Gastroenteritis	C	
	7	340	Gastroenteritis	NC	Camp
	10	22	Giardia	C	
	10	22	Giardia	C	
Virginia	4	4	Giardia	I	
	6	4	Giardia	I	
	6	16	Gastroenteritis	NC	Power plant
Washington	3	68	Gastroenteritis	C	
	5	78	Giardia	I	Swimming pool
	7	750	Salmonella	C	Restaurant
	7	253	Shigella	NC	Camp
	9	79	Gastroenteritis	C	Trailer park
West Virginia	6	30	Gastroenteritis	C	
	7	19	Shigella	C	Trailer park
	8	1000	Gastroenteritis	C	
Wisconsin	7	12	Shigella	I	Family reunion
	9	25	Giardia	C	

[a] Type of water system: C = community, NC = noncommunity, and I = individual.

Table 2
WATERBORNE OUTBREAKS BY YEAR (1981—1983)

Year	Outbreaks	Cases of illness
1981	35	4,450
1982	44	3,611
1983	47	21,124

Table 3
SEASONAL DISTRIBUTION OF WATERBORNE OUTBREAKS (1981—1983)

Month	Outbreaks Number	%
January	10	7.9
February	4	3.2
March	5	4.0
April	9	7.1
May	8	6.3
June	18	14.3
July	19	15.1
August	20	15.9
September	12	9.5
October	11	8.7
November	8	6.4
December	2	1.6
Total	126	100

Table 4
**WATERBORNE OUTBREAKS BY
TYPE OF SYSTEM (1981—1983)**

Water system	Outbreaks	Cases of illness
Community	66	13,979
Noncommunity	36	14,550
Individual	24	656
Total	126	29,185

Table 5
**ETIOLOGY OF WATERBORNE OUTBREAKS
(1981—1983)**

Disease	Outbreaks	Cases of Illness
Gastroenteritis, undetermined etiology	50	20,346
Giardiasis	42	3,169
Chemical poisoning	8	149
Shigellosis	7	532
Hepatitis A	6	267
Viral gastroenteritis, Norwalk agent	6	826
Salmonellosis	2	1150
Campylobacteriosis	2	952
Viral gastroenteritis, rotavirus	1	1761
Cholera	1	17
Yersiniosis	1	16
Total	126	29,185

water supplies because of inadequate treatment. Six giardiasis outbreaks occurred in water systems using untreated well or spring water, and one occurred because of inadequate chlorination of spring water. Two giardiasis outbreaks occurred when water mains became contaminated during breaks or repair.

Eight outbreaks were caused by chemicals (lead, copper, fluoride, nitrate, and benzene) with five outbreaks caused by lead or copper being leached from plumbing materials by corrosive water. One of the outbreaks of copper poisoning occurred when the backsiphonage of corrosive water containing carbon dixoide from a soda-mixing dispenser caused copper to be leached from water piping in a building. The benzene outbreak was caused after private, individual wells were contaminated by the underground seepage of chemicals from spillage of a railway car which exploded in 1965.

Five outbreaks of hepatitis A occurred in water systems using untreated water from wells or springs; the remaining outbreak occurred when contaminated, untreated surface water was distributed after a chlorinator became inoperable. Four gastroenteritis outbreaks in potable water systems were found to be caused by Norwalk agent; two resulted from contamination of the distribution system, and of the remaining two, one resulted from inadequate chlorination and the other from interrupted chlorination. The largest waterborne cholera outbreak in the U.S. during this century occurred in Texas in 1981; *V. cholera* 01 was found to be responsible for 17 cases of severe diarrhea after a potable water system on an oil rig was contaminated with sewage via a cross-connection. The waterborne transmission of rotavirus was demonstrated for the first time in the U.S. when 1761 cases of viral gastroenteritis occurred in a Colorado resort community in March 1981 after the malfunction of chlorination facilities.

Table 6
CAUSES OF WATERBORNE OUTBREAKS
(1981—1983)

Cause	Outbreaks	Cases of illness
Untreated water		
Contaminated surface water	6	743
Contaminated springs	9	264
Contaminated wells	28	2119
Treatment deficiencies		
Surface water sources	31	4,004
Spring water sources	5	659
Well water sources	17	18,004
Distribution deficiencies		
Corrosion of plumbing	5	101
Contamination during water main breaks	5	1600
Cross-connection	3	102
Storage	2	411
New plumbing	1	750
Miscellaneous		
Ingestion during swimming	8	313
Water not intended for drinking	4	68
Insufficient data	2	47
Total	126	29,185

Campylobacter jejuni, first identified as the cause of a waterborne outbreak in the U.S. in 1978, was responsible for an outbreak of 81 cases of illness in an Illinois community in 1981 when water became contaminated after a main break. *C. jejuni* was isolated from stool specimens of several ill persons and from a water sample collected 6 weeks after the outbreak from a house which had remained unoccupied. An outbreak of 871 cases of campylobacter gastroenteritis occurred in Florida in 1983 as a result of water system contamination by wild birds. Investigation revealed that local birds were carriers of *Campylobacter* and likely introduced the organisms into an open settling tank during water treatment. Analysis indicated no contamination of the well water source, but coliforms were found in the distribution system. Prechlorination facilities for the settling tank failed 2 weeks prior to the outbreak, and postchlorination facilities failed 1 to 2 days prior to the outbreak. Two outbreaks of campylobacteriosis which occurred in Washington during 1982 were suspected to have been waterborne, but insufficient data prevented their inclusion in this report.

A waterborne outbreak of 15 cases of yersiniosis was documented in Pennsylvania in 1982 when tap water from a private, untreated well was found to contain *Y. enterocolitica;* ill patients had eaten bean sprouts which had been soaked and rinsed with this water.

III. CAUSES OF OUTBREAKS

The use of untreated, contaminated ground water and deficiencies in water treatment, primarily inadequate or interrupted disinfection, were responsible for the most (71%) outbreaks and the most (86%) illness (Table 6). One outbreak in a ground water system resulted from inadequate or interrupted iodination; the remaining outbreaks caused by inadequate and interrupted disinfection occurred in water systems where chlorine was the disinfectant. Untreated, contaminated well water caused 28 (22%) outbreaks and 2119 (7%) illnesses; 17

(13%) outbreaks and 18,004 (62%) cases of illness were caused by inadequate or interrupted treatment of contaminated well water. Deficiencies in the treatment of surface water were responsible for 31 (25%) outbreaks and 4004 (14%) illnesses, and distribution deficiencies were responsible for 16 (13%) outbreaks and 3424 (12%) illnesses.

Although waterborne outbreaks are not a major cause of infectious disease in the U.S. a significant number of cases of waterborne disease continue to occur. This residual number of outbreaks can be prevented and waterborne disease can be reduced in the U.S. by increased surveillance of smaller community water systems and noncommunity water systems. Water system surveillance should emphasize engineering activities which (1) locate and remove sources of contamination for ground water systems, (2) ensure that continuous, effective disinfection is provided for ground waters which are subject to intermittent contamination or where contamination cannot be prevented, and (3) apply the proper treatment of surface waters to prevent the waterborne transmission of giardiasis.

REFERENCES

1. Water-Related Disease Outbreaks Annual Summary 1981, Centers for Disease Control, Health and Human Services Publication No. (CDC) 82-8385, 1982.
2. Water-Related Disease Outbreaks Annual Summary 1982, Centers for Disease Control, Health and Human Services Publication No. (CDC) 83-8385, 1982.
3. Water-Related Disease Outbreaks Annual Summary 1983, Centers for Disease Control, Health and Human Services Publications No. (CDC) 84-8385, 1984.

Section III: Investigation of Waterborne Outbreaks

Chapter 7

EPIDEMIOLOGIC PROCEDURES FOR INVESTIGATION OF WATERBORNE DISEASE OUTBREAKS

Frank L. Bryan

TABLE OF CONTENTS

I. Introduction 172

II. Getting Reports of Illness or Detecting Indicators of Waterborne Disease 172

III. Deciding Whether to Make Investigations 173

IV. Taking Action Before an Investigation 173

V. Investigating Outbreaks 174
    A. Verifying Diagnoses 174
        1. Getting Case Histories 174
        2. Collecting Specimens and Water Samples 175
        3. Defining Cases 175
    B. Making Epidemiologic Associations 175
    C. Determining Whether an Outbreak is Occurring 178
    D. Formulating or Modifying Hypotheses 178
    E. Searching for Additional Cases and Others at Risk and Getting Information From Them 179
    F. Determining Mode of Contamination and Other Factors that Contributed to the Outbreak 179

VI. Analyzing Data and Testing Hypotheses 180
    A. Identifying the Etiologic Agent 180
        1. Defining the Syndrome 180
        2. Calculating Incubation Period and Median 182
        3. Detecting the Responsible Pathogen, Toxin, or Poison 183
    B. Making Associations among the Afflicted Group 183
        1. Making Time Associations 183
        2. Making Place Associations 183
        3. Making Person Associations 185
        4. Making Associations with a Water Supply 186
        5. Evaluating Statistical Significance of Associations 187
        6. Summarizing the Investigation 191

VII. Taking Action After the Investigation 191
    A. Submitting Report 191
    B. Controlling the Outbreak 192
    C. Preventing Outbreaks 192
    D. Initiating or Maintaining Programs to Ensure Safe Water 192

References 192

I. INTRODUCTION

Surveillance of waterborne disease consists of (1) obtaining reports of illnesses that might be waterborne or detecting other indicators of waterborne disease outbreaks, (2) formulating an hypothesis that an outbreak has occurred and investigating the situation, (3) collating and interpretating data collected during the investigation, and (4) communicating findings. Surveillance, therefore, becomes the primary basis for taking corrective action and making recommendations for prevention of waterborne diseases. Furthermore, as a result of epidemiological investigations and bacteriological and bacteriological or chemical studies of contaminated water supplies, patterns of a disease and modes of transmission can be clarified. The nature of an agent, its source, its reservoirs, and factors that led to its entry into a water supply or its survival of treatment processes can be either determined or confirmed so that decisions can be made to plan and initiate effective public health programs.

II. GETTING REPORTS OF ILLNESS OR DETECTING INDICATORS OF WATERBORNE DISEASE

A report of an illness or reports of illnesses of such a nature (usually enteric) that they might be waterborne must come to the attention of a person who is concerned with either surveillance of disease or water supplies before an investigation can begin. Such reports may come from a review of medical records by health officials or calls from physicians or medical care facilities; they also may come from the person who is ill or from a family member or friend. Prevalence reports of illnesses in a community, laboratory findings of isolations of waterborne pathogens from patients, or reports of bacteriological or chemical contamination of water may provide information that suggests that an outbreak of waterborne disease is occurring. Potential problems occur either at times of heavy rainfall and floods, when there is loss of pressure in water lines, or during other situations that threaten the safety of a water supply.

Routine examinations of water samples detect the number of certain microorganisms or of groups of microorganisms in a water supply and provide information about chemical or physical characteristics of water. Thus, they can give an indication of the hygienic quality of a water supply and provide a basis for determining the potential for a waterborne outbreak. The two most commonly used microbiological tests are the aerobic colony (standard plate) count (APC) and the coliform test. In general, counts of good-quality treated water at its source will have fewer than 200 APC per milliliter. After water from any source is examined routinely, marked changes in the number of aerobic colonies should be viewed with concern, and efforts should be made to discover the reason for the change.

The coliform test is used as an indicator of the degree of pollution. Some members of this group, such as *Escherichia coli,* are present in large numbers in the feces of man and other animals. Fecal or sewage pollution of water usually results in relatively high numbers of coliforms compared to specific pathogens in the water. Certain members of the coliform group, however, are found in soil, leaves, fruits, and grain, and, therefore, in runoff water, which may reach a drinking water supply from these sources. Some strains can multiply on decaying vegetation in water or on washers or packing in pumps. The fecal coliform test provides a better indication of fecal contamination than the coliform test, and the *E. coli* test is even a better test for this purpose. Whatever their source, the presence of coliforms in a treated water supply or higher than expected numbers of these bacteria from a protected water source serves as a warning that either the treatment was inadequate or that contamination occurred after treatment, and that pathogens may also be present. Presence of the coliform group, fecal coliform, *E. coli,* or even a high number of these organisms is not proof, however, that a treated water supply contains pathogens. On the other hand, pathogens could be present where there are few or no coliforms present.

Other microbial tests (such as fecal streptococci, sulfate-reducing bacteria, *Salmonella*, or *Pseudomonas aeruginosa* can also be used as indicators or as direct evidence of hazard. No matter which test is used, evaluation of the safety of water and its potential for being a vehicle for transmission of pathogens should be based upon both analysis of water samples and a sanitary survey of the watershed, source-protection, treatment, or distribution system — whichever is applicable to the situation. The bacteriological results should be compatible with observed sources of contamination and treatment failures found during the survey.

III. DECIDING WHETHER TO MAKE INVESTIGATIONS

Hypotheses can be formulated when recent reports of illnesses, which could possibly be waterborne, are reviewed. A few situations that will illustrate this point are (1) large numbers of people develop gastroenteritis throughout a community, and both sexes and all age groups have similar rates; (2) a few persons who live in a rural community develop either (a) gastroenteritis or (b) jaundice within a few days of each others' illness; (3) children develop chills, fever, muscular aches, and stiff necks 1 or 2 weeks after swimming in a farm pond; (4) people develop gastroenteritis while at or shortly after leaving a resort (such as a ski lodge); (5) backpackers develop diarrhea (pale, greasy, malodorous stools) 1 week to 1 month after hiking and after drinking water from springs or streams; (6) children develop a rash on body areas covered by bathing suits after playing on a water slide; (7) young adults develop a rash a few days after bathing in a hot tub; (8) several persons vomit a few minutes after drinking water from a fluoridated water supply; (9) an infant less than 4 months of age develops a bluish coloration of skin; (10) children develop conjunctivitis after swimming in a swimming pool at a motel. Many other situations could be given, but these will provide examples of the way hypotheses can be formed immediately upon hearing or reading reports or as soon as certain initial associations are made between cases and potential exposures.

Examples 1, 2, 4, 5, 8, and 9 suggest the possibility of the illness resulting from ingestion of a contaminated water supply. One of several bacterial, viral, or parasitic diseases could be responsible for the illnesses in examples 1, 2a, and 4. Hepatitis A is a possibility for example 2b. Giardiasis and the likelihood that a spring or stream was contaminated would be suspected in example 5. A malfunctioning chemical feeder would come under suspicion in example 8. Well water containing an excessively high level of nitrites would be suspected in example 9.

Examples 3, 6, 7, and 10 suggest that the illnesses resulted from contact with contaminated, improperly treated, or improperly cleaned tubs or other water reservoirs. Example 3 suggests leptospirosis resulting from contact with water that was polluted by animal urine. The rash, mentioned in examples 6 and 7, could be due to *Pseudomonas aeruginosa*. Conjunctivitis (example 10) could have resulted from either improperly adjusted pH of the water or the presence of *Chlamydia* spp. in the water.

These reports and epidemiologic associations, and the hypotheses that follow, should indicate that further investigations are needed to confirm or refute the initial hypotheses and to provide further direction for the investigation. In many reports of illnesses, however, either multiple hypotheses (e.g., waterborne, foodborne, person-to person contact) may be formed at this phase of disease surveillance, or additional cases must be identified before hypotheses can be formed. Therefore, some of the afore-mentioned associations or hypotheses or their subsequent modifications await further investigation.

IV. TAKING ACTION BEFORE AN INVESTIGATION

After an hypothesis is formed, preparation needs to be made for the pursuant investigation of either illness reports or water-associated problems. Responsibility needs to be assigned to a person who will either make the investigation or will head the team that does this. A team might include chemists, epidemiologists, engineers, geologists, medical technicians, microbiologists, nurses, physicians, sanitarians, and others. If a team is to be used, the members must be assembled or alerted and either given direction or provided with appropriate information to carry out their parts of the investigation.

Forms will be needed to interview patients and others at risk, to record information about specimens and samples that are collected, and to record observations made during field investigations. Such forms will have to be either gathered or developed. Example forms for these purposes are illustrated in a booklet prepared by a Committee of the International Association of Milk, Food and Environmental Sanitarians.[1]

If the evidence suggests that an outbreak is or has occurred, the director of the supporting laboratory should be contacted. Discussions should concern the kind and quantity of clinical specimens and water samples to be collected, proper methods for collecting, preserving, and shipping them, and their estimated arrival time. Specimen and sample containers should be obtained from the laboratory, if they are not already available in a kit of materials, to enhance investigations.

If a surveillance protocol has already been developed, many of the preparatory steps mentioned above would have already been done. In spite of this, some sort of action usually has to be taken before field investigations are begun.

V. INVESTIGATING OUTBREAKS

At least four principal steps, each having several parts, are essential for the successful investigation of waterborne disease outbreaks. These are (1) verification of diagnosis, (2) formation of time, place, and/or person associations between cases, leading to determination of occurrence of an outbreak and formulation of or modification of an hypothesis or hypotheses, (3) finding additional cases and other persons at risk, with subsequent interviews, and (4) determination of the mode of contamination of the water and other factors contributing to the outbreak.

A. Verifying Diagnoses

After a report of an alleged illness or illnesses has been received, the diagnosis must be verified by obtaining a thorough case history from a few of the persons who are ill and by examining appropriate specimens from them. This is true whether the cases have been reported by a physician, a hospital, or the patients themselves.

1. Getting Case Histories

A person's experience with an illness is ascertained by a personal interview, a telephone conversation, or (sometimes, in large investigations) by a questionnaire. A standard format should be used to guide investigators and to record information.[1] The case history should identify information about the patient, such as name, address, telephone number, age, sex, occupation, and ethnic group. It should describe signs, symptoms, time of onset, severity, duration, and treatment of the illness up to the present time. It should list supplies from which water was consumed and foods eaten (if initial hypotheses include both possible vehicles) during the past 3 days along with time and place of ingestion. It should list water sources or impounded water contacted during wading, swimming, bathing, playing, or for any other purpose during the past 6 weeks if diseases likely to be spread by contact with water are suspected. Other information, of course, may be sought, depending on the nature of the illness and the hypotheses formed at this time. Other persons who may have been at risk because of having associations with ill persons during these periods should also be identified and interviewed.

2. Collecting Specimens and Water Samples

Only a few waterborne diseases can be diagnosed by information on signs and symptoms alone (Table 1). Diagnosis should be confirmed by laboratory tests for the presence of infectious agents or toxic substances in appropriate specimens from ill persons, and by clinical examination or tests for signs of infection. Specimens should be obtained at the time of the case-history interview or as soon as possible thereafter, because signs of infection may be present for only a few days after onset and some pathogens remain in the intestinal tract for just a short period of time.

In general, choice of specimens should be determined by signs and symptoms. (Table 1 lists specimens that should be collected when certain diseases are under suspicion.) Vomitus is collected if the person is vomiting. Either stools or rectal swabs are collected if the person has, or recently has had, diarrhea and a bacterial or parasitic agent is suspected. Blood (10 cm^3) should be collected during the acute phase of the illness and 2 to 4 weeks thereafter if a patient has a generalized infection with fever. Specific procedures for collecting specimens are provided by Bryan et al.[1] and Isenberg et al.,[2] and procedures for testing those specimens are provided by Lennette et al.[3] Methods for analyzing water samples are contained in Chapter 10.

If a water supply from an individual household is already under suspicion as a possible vehicle, separate samples of water from a tap, from water kept in a bottle in a refrigerator, and from ice from refrigerator trays should be collected. (See Bryan et al.[1] for procedures.) If the suspect vehicle is a commercially bottled water, the original bottle or container should be obtained if at all possible. A code number and place of bottling can be useful in continuing the investigation or in making a recall, if necessary.

3. Defining Cases

Before new cases are considered to be involved in the outbreak, they should comply with a prescribed definition, which is based upon the syndrome and laboratory findings. Cases can be divided into three categories. A suspected case is usually defined as a person with an illness that is clinically compatible with the disease under consideration, but neither serological nor other laboratory confirmation of the diagnosis has been made. For example, in an outbreak of gastroenteritis, a case may be defined as a person who developed diarrhea within a specified period of time. Diarrhea, also, will have to be defined, perhaps as three or more loose, watery stools during a 24-hr period. Alternately, the definition may be such that two or more signs and symptoms must be present or either one or the other are present. A presumptive case is a person with an illness that is clinically compatible with the disease under consideration, and laboratory evidence (e.g., elevated hemagglutination inhibition antibody titer) of infection has been made, but the etiologic agent has not been isolated. A confirmed case is a person with an illness that is clinically compatible with the disease under consideration and, additionally, there is either serological evidence of a fourfold or greater rise or fall in antibody titers (e.g., hemagglutination inhibition or complement fixation) or the etiologic agent has been isolated from (or otherwise identified in) a specimen from the patient. (See Table 1 for guidelines for confirmed waterbone disease outbreaks.)

B. Making Epidemiologic Associations

From information collected from persons initially contacted during the investigation (e.g., the person making the report or complaint, the first few cases interviewed), a determination should be made as to whether the cases had any time, place, or person associations with each other. Time associations usually concern illnesses with onsets that occur within a few hours or days of each other. They may also concern time that the cases attended a social event or outing. Place associations usually denote that the cases had drunk or otherwise contacted water at the same place, lived at the same place, worked at the same place, went

Table 1
GUIDELINES FOR CONFIRMATION OF WATERBORNE DISEASE OUTBREAKS

Etiologic agent	Clinical syndrome	Epidemiologic criteria
Bacterial		
Campylobacter jejuni	Incubation period usually 2 to 5 days. GI syndrome: majority of cases with diarrhea	Isolation of *C. jejuni* from epidemiologically implicated water and/or isolation of *C. jejuni* from stools of ill persons
Escherichia coli	Incubation period 6 to 36 hr. GI syndrome: majority of cases with diarrhea	Demonstration of organisms of same serotype in epidemiologically incriminated water and stools of ill persons and not in stools or controls, or isolation of organisms of the same serotype which have been shown to be enterotoxigenic or invasive by special laboratory techniques from stools of most persons
Salmonella typhi	Incubation period 7 to 28 days. High fever, tender and distended abdomen, rose spots, malaise	Isolation of *S. typhi* from epidemiologically implicated water, and/or isolation from *S. typhi* from stools of ill person. Fourfold rise in antibody titer in convalescent serum. Isolation of some phase type in stools of patients and from epidemiologically implicated water
Salmonella (other than *S. typhi*)	Incubation period 6 to 48 hr. GI syndrome: majority of cases with diarrhea	Isolation of *Salmonella* from epidemiologically implicated water, and/or isolation of *Salmonella* from stools or tissues of ill persons
Shigella	Incubation period 12 to 47 hr. GI syndrome: majority of cases with diarrhea	Isolation of *Shigella* from epidemiologically implicated water, and/or isolation of *Shigella* from stools of ill persons
Yersinia enterocolitica and *Y. pseudotuberculosis*	Incubation period 3 to 7 days. GI syndrome: majority of cases with diarrhea	Isolation of *Yersinia* spp. from epidemiologically implicated water, and/or isolation of *Yersinia* spp. from stools of ill persons, or significant rise in bacterial agglutinating antibodies in acute and early convalescent sera
Others	Clinical and laboratory data appraised in individual circumstances	
Chemical		
Fluoride	Incubation period usually less than 1 hr. GI illness: usually nausea, vomiting, and abdominal pain	Demonstration of high concentration of fluoride ion in epidemiologically incriminated water

Heavy metals (e.g., cadmium, copper)	Incubation period 5 min to 8 hr (usually less than 1 hr). Clinical syndrome compatible with heavy metal poisoning — usually GI syndrome and often metallic taste	Demonstration of high concentration of metallic ion in epidemiologically incriminated water
Others	Clinical and laboratory data appraised in individual circumstances	
Parasitic		
Entamoeba histolytica	Incubation period usually 2 to 4 weeks. Variable: GI syndrome from acute fulminating dysentery with fever, chills, and bloody stools to mild abdominal discomfort with diarrhea	Demonstration of *E. histolytica* cysts in epidemiologically incriminated water, and/or demonstration of *Entamoeba histolytica* trophs or cysts in stools of affected persons
Giardia lamblia	Incubation period 1 to 4 weeks. GI syndrome: chronic diarrhea, cramps, fatigue, and weight loss	Demonstration of *Giardia* cysts in epidemiologically incriminated water, and/or demonstration of *Giardia* trophs or cysts in stools or duodenal aspirates of ill persons
Others	Clinical and laboratory data appraised in individual circumstances	
Viral		
Enterovirus	Incubation period is variable. Syndrome: variable; poliomyelitis, aseptic meningitis, herpangina, etc.	Isolation of virus from epidemiologically implicated water and/or isolation of virus from the stool of ill persons
Hepatitis A virus	Incubation period 14 to 28 days. Clinical syndrome compatible with hepatitis — usually including jaundice, GI symptoms, dark urine	Liver function tests compatible with hepatitis in affected persons who drank the epidemiologically incriminated food or water
Norwalk agent and Norwalk-like agents	Incubation period 16 to 72 hr. GI syndrome: vomiting, watery diarrhea, abdominal cramps; duration approximately 30 hr	Demonstration of virus particles in stool of ill persons by immune electron microscopy, or significant rise in antiviral antibody in paired sera
Rotavirus	Incubation period 24 to 72 hr. GI syndrome: vomiting, watery diarrhea, abdominal cramps	Demonstration of the virus in the stool of ill persons, or significant rise in antiviral antibody in paired sera blood
Others	Clinical and laboratory data appraised in individual circumstances	

Modified from Water-Related Disease Outbreaks Surveillance Annual Summary 1980, Centers for Disease Control, Atlanta, Ga., 1982.

to the same school, had visited the same place, used water from the same source, or used the same brand of bottled water. Person associations refer to common experiences among the ill persons (e.g., being of the same age, sex, ethnic group, occupation, or religion, users of bottled water, being frequent bathers in hot tubs or whirlpools, or being frequent swimmers).

Hypotheses based upon associations that were made during the initial report may have to be modified or even rejected after information obtained during interviews of cases is reviewed. Causal associations (and criteria to judge them) are often made because of one or more of the following:

1. The association is strong. Higher rates of disease are observed in groups who are exposed to a vehicle, factor, or environment that contains or is likely to contain the etiologic agent than are observed in groups without such an exposure. The clinical disease is more severe, the incubation period is shorter, and a higher incidence of disease is observed in persons who are exposed to higher doses of an infectious or toxigenic agent. Minimum effect, longer incubation periods, and a low incidence are observed with low doses of the agent.
2. There is a consistency of occurrence. For example, association is observed repeatedly with different study populations, with different methods of study, by different investigators, and the contributory factors have been seen during investigations of previous outbreaks.
3. There is specificity of association. For example, the same manifestations occur when the agent infects different persons.
4. There is a temporal association. Exposure to the etiologic agent (or vehicle containing the agent) precedes the onset of illness and accounts for an appropriate incubation period or latency period.
5. The association is biologically or chemically plausible. The association must make sense in relation to existing information on causes of disease.

An association can be causal, coincidental, or related only through a third, often apparently unrelated, variable. No matter how valid an association between cases and an attribute appear, it is not necessarily conclusive evidence of a causal relationship. The association must be tested by further laboratory and field investigations.

C. Determining Whether an Outbreak is Occurring

An outbreak is an incident in which two or more persons have the same disease, have similar clinical features, or have the same pathogen (thus meeting the case definition), and there is a time, place, or person association among these cases. A waterborne outbreak is traceable to either ingestion of or contact with contaminated water. A single case of a disease, however, may be considered as an incident of waterborne disease if there is evidence that the ill person either ingested or contacted contaminated water and there is no other plausible causal factors. An example is the single case of infantile methemoglobinemia which occurred because of the ingestion of water from a well that contained a high concentration of nitrite.

D. Formulating or Modifying Hypotheses

As a result of making causal (time, place, and person) associations and inferring that an outbreak has occurred from the accumulated data, an hypothesis to explain the most likely cause, vehicle, and distribution of cases, unknown at this stage of the investigation, is formulated. Almost any set of observations will be compatible with at least one hypothesis, and one or a few of the more feasible are chosen for testing. An hypothesis can be formulated even without belief, so that its conformace with facts can be tested. The hypothesis is the starting point of a detailed investigation, and it directs the course of an investigation to produce data with which to test the validity of the hypothesis itself.

E. Searching for Additional Cases and Others at Risk and Getting Information From Them

After an hypothesis has been formulated, materials for performing the investigation must be assembled and personnel who will assist in the investigation must be contacted, if these have not already been done. Persons already known to be ill or exposed to situations associated by the hypothesis (or a representative sample) should be interviewed as previously described.

A search should be made for additional cases and for other persons who were at risk because they shared something in common with those at risk but did not become ill (e.g., lived in the same household, drank from the same water supply, contacted the same impounded water, attended the same event, traveled to the same places). These persons can sometimes be identified by questioning ill persons and persons in charge of group activities. Ill persons can also be identified by contacting physicians and hospitals to determine whether they have treated persons who might have the disease in question. A review of diseases logs often kept by health departments might also uncover other cases. Surveys of households in the same neighborhood of the afflicted is another approach in finding additional cases.

F. Determining Mode of Contamination and Other Factors that Contributed to the Outbreak

Discovery of the means by which the water supply became contaminated and other factors that contributed to the water becoming the vehicle for infectious or toxigenic agents is the most important part of a waterborne outbreak investigation from the standpoint of prevention and control. It is the overriding purpose for making an investigation in the first place.

Determination of the mode of contamination of water supplies or water courses calls for at least four types of activities. These are (1) reviewing records, (2) interviewing persons responsible for quality of suspect water, (3) conducting a sanitary survey of the watershed, well, treatment facilities, water storage facilities, and/or plumbing systems, and (4) collecting and testing samples of water. Although more information on much of this is described in Chapters 10 and 11, a brief review of these activities is necessary to continue the epidemiologic and investigative thought processes.

If a surface water supply or watershed is under investigation, a survey should be conducted for potential sources of pollution such as out-fails of sewage, overflowing septic tanks, and drainage from feed lots or barnyards. Also, access of animals to and the presence of dead animals in the water course should be determined. Whether the water is used for drinking without treatment or used as a secondary supply should be ascertained. Furthermore, large samples of the water might be collected and tested for specific pathogens.

If a ground water is under investigation, well logs, as well as soil and geological information, should be reviewed. The slope of the land and characteristics of subsurface formations should be determined. Methods of excreta disposal in the area must also be investigated. Sewage, for example, can travel long distances through creviced limestone or fissured rock. The well or spring location should be evaluated for the likelihood of surface or underground pollution reaching the water. The depth and construction of the well, type of pump, and adequacy of its operation, and other information about the water system should be determined. The survey should include the possibility of overflow or seepage of sewage (especially from septic tanks), abandoned wells, or buried or surface wastes. Evidence or unusually heavy rains and flooding or drought conditions should also be gathered. Samples might be collected and tested for pathogens, indicator organisms, or poisonous substances.[4] Fluorescein dye or other tracers (in appropriate soils) might be used to determine the manner by which contamination traveled from privy, septic tank, absorption system, sewer, drain, industrial waste lines, or other sources to reach the underground water source.

If treated water is under investigation, appropriate records of water quality (e.g., coliforms and aerobic colony count, chlorine residual, turbidity, chlorine demand, jar tests, and chemical analysis) must be reviewed. This review should cover a period equal to the incubation period of the suspect diseases. Raw water samples and water at particular points of processing, as well as the treated water, should be collected and tested to assess the effectiveness of treatment. Adequacy of chlorine concentration and contact time and whether there was any interruption of chlorination should be evaluated. Adequacy of filtration and prefiltration treatment should also be reviewed. Certain equipment such as chlorinators and fluoridators should be inspected to determine whether they are functioning properly.

If water storage facilities might be involved, the likelihood of pollution from sewage, surface water, or animals should be evaluated. Free available residual chlorine levels should be tested. Samples should be collected and pathogens sought.

If distribution systems are to be investigated, records on recent repairs and locations of water and sewer lines should be reviewed. If there were recent repairs to lines, an inquiry as to whether the mains and plumbing systems were subsequently disinfected should be made. A survey is essential for cross-connections and situations such as negative pressure and submerged inlets that might have led to backsiphonage. This should include determining whether the water supply line to a building at which the cases resided was turned off, if there was a recent fire in the neighborhood, or if there was any other event that could explain negative line pressure. Recordings of pressure at upper stories of buildings might also be useful to evaluate the possibility of backsiphonage. Dye tests might be done to reproduce likely events that may have contributed to backsiphonage.

If an infection resulted from swimming in a pool, the free chlorine residual and pH should be tested. The pool wall construction and filtration units should also be inspected. If schistosomiasis is under investigation, a survey for snails and the likelihood of animal or human waste reaching the water course should be made.

As an investigation proceeds, more specific hypotheses will be formulated and tested with subsequently gathered information and analyses. Factors that commonly contribute to waterborne disease outbreaks are reviewed by Craun and McCabe[5] and are summarized in Table 2. These should be sought if appropriate for the situation under investigation. Specific techniques and forms to use when determining the source and mode of contamination of a water supply or water course are given by Bryan et al.[1]

VI. ANALYZING DATA AND TESTING HYPOTHESES

The data collected during interviews of patients, field investigations, and laboratory analyses must be tested to confirm or refute hypotheses. Certain calculations or graphs are made to aid in identifying the etiologic agent, in making associations among the afflicted persons, and in determining the source of contamination and circumstances that contributed to the outbreak.

A. Identifying the Etiologic Agent

Several tests are routinely made to identify etiologic agents. These include (1) determination of the frequency of signs and symptoms of the afflicted group, (2) determination of incubation periods of each case and the median incubation period of the group, and (3) isolation of a pathogen from several persons who are ill, and, whenever possible, isolation of the same pathogen from the epidemiologically implicated water.

1. Defining the Syndrome

Signs and symptoms that present as a result of an infection, intoxication, or poisoning will depend upon a number of variables that are associated with the host and the parasite. These include resistance of the host, virulence and invasiveness of the infecting strain of

Table 2
CONTRIBUTORY FACTORS FOR WATERBORNE DISEASES[5]

Surface water
  Use of untreated surface water
  Contamination of watershed by human or animal sources
  Use of contaminated surface water for supplementary source
  Overflow of sewage or outfall near water intake
  Flooding
  Dead animals in stream or reservoir
  Live animals in stream or reservoir
Ground water
  Overflow or seepage of sewage into well or spring
  Surface contamination near well or spring
  Contamination through creviced limestone or fissured rock
  Flooding
  Chemical or pesticidal contamination
  Seepage from abandoned well
  Contamination of raw-water transmission line or suction pipe
  Improper well construction
Inadequate treatment of water
  Inadequate concentration or contact time of chlorination or other disinfection process
  Interruption of chlorination
  Inadequate filtration
  Inadequate prefiltration treatment
Storage deficiency
  Unprotected storage tanks or reservoirs
  Contamination of cistern or individual storage facility with surface water or sewage
  Improper or no disinfection of new storage facility
Distribution deficiency
  Backsiphoning
  Cross-connections
  Contamination of mains during construction or repair
  Water main and sewer in same trench or inadequately separated
  Improper or no disinfection of mains of plumbing
Water contact problems
  Snails in water
  Puncture injuries or wounds
  Diving, underwater swimming, water skiing
  Inadequacy of water-holding facilities
  Swimming or wading in parasite-infested waters
  Improper pH adjustment
  Improper chlorination
  Improper filtration
  Rough pool wall construction
Other factors
  Use of water not intended for drinking
  Contaminated buckets and other containers
  Contaminated drinking fountains
  Deliberate contamination
  Contaminated ice

the pathogen, and the quantity of pathogens, toxins, or poisons ingested or contacted. Consequently, individual reactions to invading pathogens or poisons will not be identical.

A tabulation of the percentage of each symptom experienced by ill persons or each sign seen by a physician (as illustrated in Table 3) will help classify the outbreak into general categories. The majority of the cases summarized in Table 3 had diarrhea, but other GI signs and symptoms were common. The disease could be caused by one of several bacterial or viral agents (see Table 1). Only one case reported bloody diarrhea; therefore, agents

Table 3
PERCENTAGE OF CLINICAL MANIFESTATIONS OF AFFECTED PERSONS

Sign or symptom	Number ill (n = 288)	%
Diarrhea	212	93
Abdominal cramps	176	77
Nausea	164	72
Vomiting	155	68
Chills	109	48
Headache	107	47
Muscle ache	103	45
Fever	93	41
Blood by diarrhea	1	0.3

Table 4
FREQUENCY DISTRIBUTION OF INCUBATION PERIODS[a]

Incubation period range	Frequency	Cumulative frequency
0—4	0	0
5—9	0	0
10—14	0	0
15—19	3	3
20—24	7	10
25—31	11	21
30—34[b]	9	30[c]
35—39	8	—[d]
40—44	7	
45—49	1	
50+	0	
Total	46	

[a] Middle-most number $= \dfrac{n+1}{2}$

$$\dfrac{46+1}{2} = \dfrac{47}{2} = 23.5$$

[b] Median falls with this range.
[c] Middle-most number falls between 22 and 30.
[d] Middle-most number exceeded.

(some Shigellae and invasive *Escherichia coli*) that cause colon infections would be questioned. Nausea and vomiting were quite common which would bring enterotoxigenic strains under suspicion. Laboratory confirmation would be required; biologic tests would probably be necessary to determine whether enterotoxigenic strains were responsible.

2. Calculating Incubation Period and Median

The incubation (latent) period of a disease is the time between exposure of a susceptible person to an etiologic agent and the appearance of the first sign or symptom. This span of time depends upon each particular host-parasite relationship and the ease with which the infecting organism or the toxin gains access to the tissue in which its primary multiplication or toxic reaction takes place. The range and median of the incubation period, however, cannot be determined until the time of ingesting the implicated water is known, but gathering information about specific times of drinking water are often difficult to determine.

The median is the value that divides a set of values ordered from shortest to longest into two equal parts. For sets of a small number of values this can readily be determined. If the median is to be found for a large number of incubation periods, a frequency distribution table can be constructed. In such a table, either each incubation period is listed or several groups of them are listed in order of increasing duration sequentially, with the shortest first, and their frequencies are cumulated until the middle-most number is reached. The middle-most number is the number of incubation periods plus one divided by two. The middle-most number is located in the cumulative frequency column, and the median is the corresponding number in the column of sequential incubation periods. Table 4 shows a frequency distribution of incubation periods; the middle-most number is 22.5, and the median falls within the 30- to 34-hr range.

The duration that signs and symptoms last may also be useful in defining an illness. For example, the mean duration of gastroenteritis caused by Norwalk agent, along with signs and symptoms and incubation period and the inability to isolate bacterial pathogens from stool specimens of patients, is useful in diagnosis. This is particularly so because this agent cannot be grown in vitro and can only be laboratory confirmed in very specialized laboratories.

3. Detecting the Responsible Pathogen, Toxin, or Poison

Data on the frequency of signs and symptoms and the median incubation period suggest a few etiologic agents and sometimes a source, and thus aid in directing the investigation and dictating laboratory tests. A laboratory test is needed to either confirm a case or to provide information for presumptive diagnosis. There is strong evidence to support a hypothesis that an outbreak is waterborne if a specific pathogen (e.g., *Shigella*) is isolated from several persons and there are time, place, or person associations among these persons that link them to either drinking from a common water supply or contacting water. There is additional evidence if the pathogen was also isolated from the epidemiologically implicated water. The strength of association is increased even more if the same strain (e.g., serotype) of pathogen is isolated from ill persons and from the epidemiologically implicated water. (See Table 1 for criteria for confirmation of waterborne disease outbreaks.)

B. Making Associations among the Afflicted Group

1. Making Time Associations

Associations of the time that illness occurred are important patterns of a disease outbreak. Information about time of onset of illness should be available from any set of case histories that are obtained during an investigation of an outbreak, and thus an epidemic curve can be constructed as one of the bases for forming and testing hypotheses. An epidemic curve is a graphic representation of a number of cases according to distribution of times of onset of illness. The primary reason for constructing an epidemic curve is to see whether a cluster of cases has characteristics of either a common-source or a propagated outbreak. A common-source epidemic is spread by a vehicle (such as water) that was shared by the affected people. In a common-source outbreak, when many persons are exposed more or less simultaneously, the relative uniformity of the incubation period for a specific disease results in a single cluster of cases in time. As illustrated in Figure 1, cases occur rapidly after the first onset, reach a peak, and then decline. Furthermore, the duration of the outbreak does not exceed the maximum incubation period of the disease. A common-source, single-event outbreak is typified by onsets of cases that follow an event such as a particular meal at which a common vehicle (source), such as a particular food or water, is ingested by persons who subsequently become ill. The initial part of a common-source, multiple-event epidemic curve resembles the common-source, single-event curve, but cases continue to occur at a high rate for a period equal to several incubation periods of the disease. Among the possible causes of this continuation of new cases is a water supply that is continually contaminated.

Propagated outbreaks are caused by agents that are transmitted from a human or animal reservoir by direct or indirect contact with susceptible persons. Onsets are insidious. The rapidity with which propagated outbreaks reach a peak and their duration depend upon the infectivity of the agent, the lenght of the incubation period of the disease, the initial proportion of susceptible persons in a population, and the degree of crowding and intimacy of contact. Propagated outbreaks sometimes follow common-source outbreaks; this is more apt to occur when only small numbers of the etiologic agent are capable of causing illness.

By defining the pattern of time of onset of illness for those afflicted and tracking back the average incubation period of the disease from the peak of the curve, the expected time of exposure can be determined (see Figure 1). This can provide valuable information about a likely vehicle and about events that may have contributed to contamination of the vehicle. The type of curve presented also provides evidence to help prove or refute hypotheses.

2. Making Place Associations

Associations among ill persons with place of residence or attendance at a common event

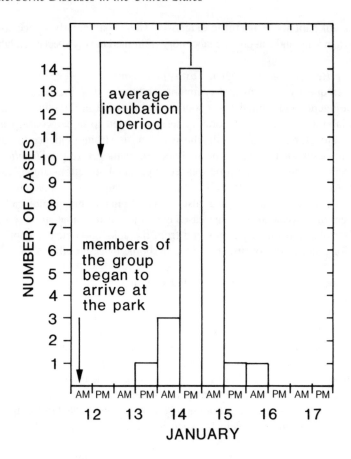

FIGURE 1. Number of cases of gastroenteritis by time of day and date of onset (epidemic curve).

can either lead to hypotheses or lend support to those already formed about a possible vehicle. These associations are usually expressed as incidence rates of the population at risk within some boundary or area of known census. An incidence rate is the number of new cases of a specified disease that are reported during a given time period in relation to the size of the population being studied, multiplied by a constant (frequently 100, 1000, 10,000, or 100,000).

Table 5 compares rates of gastroenteritis in three groups (residents of a community served by a water supply, students and teachers at a school served by the same water supply, and residents of a county surrounding but excluding the community and school populations). Higher incidence rates are evident for those persons using water supply B (community A and school C); lower rates are seen in persons whose households are not connected to water supply B (county D, excluding community A and school C). Thus, indirect associations between ill persons and a water supply can be made by the associations between ill persons and their place of residence.

Geographic representation can be made of place of residence or incidence (attack) rates of the afflicted as illustrated in Figure 2. Attack rates varied from 0 to 8.4% in different subdivisions. Rates were lower in sections on the west side of the community (A, B, D, G, and K) and higher in sections on the east side of the community (C, E, F, H, I, and L). Such illustrations should lead to investigation of the reason for the differences. When this was done, it was found that multiple water sources supplied the community. Subdivisions C, E, F, H, I, and L (where rates were higher) were supplied primarily by a spring (likely

Table 5
RATE OF ILLNESS BY MONTH AND PLACE OF RESIDENCE AND SCHOOL

	Number of cases			
Month	Community A on water supply B (pop. 3332)	School C in county on water supply B (pop. 4601)	County D excluding community A and school C; not on water supply B (pop. 78,024)	Total (pop. 85,957)
January	0	2	12	14
February	4	2	6	12
March	107	38	16	161
April	26	5	19	50
May	4	1	8	13
June	0	0	7	7
Total	141	48	68	257
Rate/1000	42.3	10.4	0.9	3.0

Modified from Mosley, J. W., Schrack, W. D., Denshaun, T. W., and Matter, L. D., *Am. J. Med.*, 26, 555, 1959.

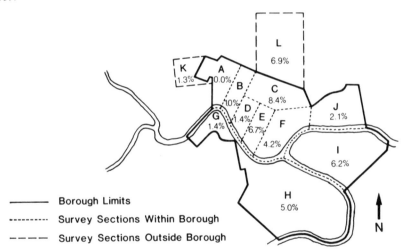

FIGURE 2. Attach rates by residence. (Modified from Mosley, J. W. et al., *Am. J. Med.*, 26, 555, 1959.)

to have been contaminated). Subdivisions A, B, D, G, and K derived their water from a reservoir. Households in subdivision J derived their water entirely from private supplies.

3. Making Person Associations

Variation of disease among persons is sometimes related to attributes acquired before birth (sex, race), at birth (age, birth order), or acquired later (religion, occupation, education, economic status). People can be categorized into appropriate groups based upon these attributed and if population figures are known for these groups, rates can be calculated. In community-wide outbreaks of waterborne diseases there is usually little difference in rates between the two sexes and between different age groups. An exception might be infants, who are not exposed to tap water, but who received sterilized milk and water. When differences are seen, a particular water supply, place visited, or events attended may be

Table 6
RATES OF PERSONS INFECTED WITH *GIARDIA LAMBLIA* BY AGE GROUP

Age group (years)	No. patients examined	No. infected[a]	Infection rate (%)
0—9	122	46	38
10—19	45	27	60
20—29	61	26	43
30—39	37	14	38
40—49	37	18	49
50—59	50	31	62
60+	77	34	44
All ages	429	196	46

[a] Identified in stool

Modified from Lopez, C. E. et al., *Am. J. Epidemiol.*, 112, 495, 1980.

Table 7
FOOD/BEVERAGE-SPECIFIC ATTACK RATES

Food/beverage	Ate				Did not eat				Difference in percentage
	Ill	Not ill	Total	Attack rate	Ill	Not ill	Total	Attack rate	
Ham sandwich	58	30	88	66	1	0	1	100	−34
Potato salad	49	27	76	64	10	3	13	77	−13
Beans	38	17	55	69	21	13	34	62	+7
Cake	48	28	76	63	11	2	13	85	+22
Water	59	14	73	81	0	16	16	0	+81

suggested, depending upon the characteristics of the afflicted group. Table 6 shows relative high infection rates among persons of different age groupings who were exposed to *Giardia lamblia* in a community water supply. There was no difference in infection rates by sex (data not presented).

4. Making Associations with a Water Supply

In addition to the indirect water-illness associations described under time, place, and person associations, direct associations can be made. Vehicle-specific attack rates should also be calculated whenever water or a beverage containing ice is either served at a common meal or ingested at a common place that is suspected of being the likely time and place of exposure. A tabulation of these rates (as shown in Table 7) provides an easy way to compare the percentages of ill persons who ingested each food and beverage with those who did not during the meal under investigation. These attack rates are calculated by dividing the number of ill persons who ingested a particular food or beverage by the total number (both ill and well) who ingested the same food or beverage, and multiplying the quotient by 100. The same calculation is performed for the total number who did not ingest the particular food or beverage, and the rates for the two groups are compared for differences.

An attack rate table is useful in identifying the food or beverage that is responsible for an outbreak. The food or beverage that has the highest attack rate (percentage ill) for persons who ingested it and the lowest attack rate for persons who did not ingest it (the greatest difference between the two rates) becomes suspect (as is water in Table 7). This suspicion

Table 8
NUMBER OF GLASSES OF WATER AND UNTREATED WATER-CONTAINING BEVERAGES INGESTED PER DAY

Number of glasses per day	Ill	Not ill	Total	Attack rate
0	4	61	65	6
1 or 2	23	134	157	15
3 or 4	51	131	182	28
5 +	65	133	198	33

is strenghtened if this item was also ingested by the vast majority of persons affected. Some persons who did not ingest the suspect vehicle may have nevertheless become ill. Possible explanations are that some persons may have become ill from other causes, some may have forgotten that they ingested water, or, on rare occasions, some may have exhibited psychological rather than real symptoms. Also, the tables often include some persons who ingested the implicated vehicle but did not become ill. Plausible explanations for this is that organisms or toxins are not always evenly distributed in the vehicle, some persons eat or drink more than others, some persons are more resistant to illness than others, and some persons will not admit that they were ill.

Attack rates for persons who ingested varying amounts of water, unheated water-containing beverages, ice, or ice-containing beverages will often provide additional evidence that water is or is not the responsible vehicle of diseases caused by microorganisms. In general, the larger the quantity of infectious agents ingested (thus, the larger the dose), the greater the chances are that illness will result. Table 8 shows a way to compare rates for groups of persons ingesting different quantities of water. This table shows that the attack rates increased as the consumption of water increased. If water was not involved, the rates would have been similar for persons who ingested large amounts of hot coffee, hot tea, beer, wine, or carbonated soft drinks which might have precluded them from drinking large amounts of water. If water were the vehicle, attack rates among persons ingesting these hot or low-pH beverages would be inversely proportionate (low) to those who ingested large amounts of unheated, contaminated water or contaminated ice. Such data, however, do not always show a positive association between illness and ingestion of large quantities of water for several reasons. Small quantities of some pathogens can cause illness, particularly when contaminated water is ingested during nonmeal times. Unequal distribution of contaminants in water, intermittent contamination, and varying susceptibility and resistance of individuals can sometimes account for the lack of associations.

5. Evaluating Statistical Significance of Associations

Two statistical methods which are often used to test for the association between illnesses and water exposure are Chi-square (χ^2) and Fisher's exact. Programs are available for computers to make such calculations, but a brief review of the means of making these calculations without computers is given to provide a basic understanding of these tests.

The χ^2 method compares an observed distribution with an expected distribution based on marginal proportions. The proportion of persons who have an attribute, such as having drunk water, would be expected to be the same in groups of both ill and well persons, if there is no association between water and illness (null hypothesis).

The level of statistical significance used to reject or fail to reject the null hypothesis must be chosen by the investigator. Frequently, a 5% significance level ($P < 0.05$) is used, or in other words, chance could account for the event 1 time in 20 occurrences or 5 times in 100 occurrences when the null hypothesis holds.

Table 9
GENERAL TABLE FOR DATA WHICH IS USED IN THE χ^2 AND FISHER'S EXACT FORMULAE

Attribute	Condition present	Condition absent	Total
Positive	a	b	a + b
Negative	c	d	c + d
Total	a + c	b + d	a + b + c + d = n

Table 10
RELATIONSHIP OF ILL PERSONS AND WATER INGESTION AT A COMMON PLACE (OBSERVED VALUES)

Attribute; ingestion history of suspected vehicle	Condition present (Ill)	Condition absent (Well)	Total
Drank water	76	29	105
Did not drink water	2	24	26
Total	78	53	131

Although χ^2 can be used to compare several proportions, the most common situation for outbreak data are two dichotomies — ill, not ill, and water exposure, no water exposure. A table with two columns (ill, not ill) and two rows (with attribute, without attribute) is designated as a 2 × 2 contingency table. A 2 × 2 contingency table with symbols used in the formula for the χ^2 calculation is shown in Table 9. The cells of the observed values are entered into the table and columns and rows are totaled. An example of a 2 × 2 table of observed data collected during an investigation is shown in Table 10. Because the χ^2 is an approximation of an exact method, it should not be used when the expected values in any cell are small (<5). Rather, the Fisher's exact test should be used.

In the χ^2 method, observed values are compared with expected values so an expected table should be developed. This is done by calculating the expected proportion of persons who would fall into each contingency (i.e., ill and drank water, ill and did not drink water, remained well and drank water, or remained well and did not drink water) if there were no differences between the groups. The calculation is done by first dividing the total number of ill persons by the number of persons in the universe (sample). This quotient is then miltiplied by the total number of persons with an attribute (e.g., drank water) to give a product of the expected rate for the persons who had the attribute and acquired the condition (drank water and ill). This value is entered into the appropriate cell of the 2 × 2 contingency expected table, and the remaining cells are filled by subtracting this value from the marginal totals. To complete Table 11, the proportion of persons who would be expected to be ill would be the result of 78 divided by 131. The expected rate for those drinking water who became ill would be this value (0.595) times 105, which equals 62.5. The remaining values for the other cells are calculated by subtracting 62.5 from a + b and from a + c and the result of either from the other appropriate marginal total (b + d; c + d) (see Table 11). Because only one calculated value is needed in order to obtain the other values in a 2 × 2 contingency table, there is one degree of freedom. In such cases χ^2 can be calculated from observed and expected values by using the following formula: $\chi^2 = \Sigma \, (|\, O - E\, | - \frac{1}{2})^2\, |\, E$. Because χ^2 is based on the normal approximation to the binomial, a continuity correction called Yates' correction is frequently used. The approximation continuity correction consists of reducing the absolute value of O − E by $\frac{1}{2}$.

Table 11
EXPECTED DISTRIBUTION IF THERE IS NO DIFFERENCE BETWEEN RATES FOR DATA IN TABLE 9

Attribute; ingestion history of suspected vehicle	Condition present (Ill)	Condition absent (Well)	Total
Drank water	62.5	42.5	105
Did not drink water	15.5	10.5	26
Total	78	53	131

For ease of calculation, the χ^2 formula can be shown algebraically, similar to the following formula:

$$\chi^2 = \frac{n\,[|ad - bc| - n/2]^2}{(a + b)(c + d)(a + c)(b + d)}$$

Data from an observed table is used to replace the symbols shown in Table 9 to make the χ^2 calculation. From data in Table 10, for example, the χ^2 value would be:

$$\chi^2 = \frac{131\,|76 \times 24 - 29 \times 2| - 131/2]^2}{(105)(26)(78)(53)} = 33.6$$

The probability (P) of an event or more extreme event occurring by chance alone is obtained from a χ^2 table for one degree of freedom as given in Table 12. In the example just given, the χ^2 value is 33.6, and the null hypothesis would be rejected. The probability ($P<0.000001$) of difference between an observed distribution and the expected distribution by this much or more happening by chance alone would be less than 1 in 1 million. Therefore, the difference in these rates is statistically significant, and the ingestion of contaminated water could have been responsible for the illness and the event that was associated with this difference.

Whenever any of the expected values is less than five, Fisher's exact test, instead of the χ^2 test, is used to determine statistical significance of the data. For making the Fisher's exact calculation, the same general table (Table 9) is used as is used for the χ^2 test. The formula for the Fisher's exact test is

$$P = \frac{(a + c)!\,(b + d)!\,(c + d)!\,(a + b)!}{n!\,a!\,b!\,c!\,d!}$$

Factorial (!), which is a value multiplied by all possible lower whole numbers (e.g., 5! = 5 × 4 × 3 × 2 × 1), is used in this formula. Zero factorial equals one.

When applying Fisher's exact test to data in a 2 × 2 contingency table, the value P must be calculated for the observed data and for data from all other 2 × 2 tables that can be developed from the same marginal totals, giving a greater difference in attack rates in the same direction. This is necessary because the true value of "P" would be the probability associated with getting a difference that was as great as or greater than the observed values by chance; (therefore, $P = P1 + P2 + ...Px$). The marginal totals (a + b, c + d, a + c, b + d, n) must remain fixed, but the internal numbers (a, b, c, d) can be changed to determine whether a greater difference occurs. If any of the internal cell numbers is zero or become zero as these numbers are changed, a more extreme 2 × 2 table is not possible.

Table 12
χ^2 TABLE FOR ONE DEGREE OF FREEDOM

χ^2	2.71	3.84[a]	6.64	7.88[b]	10.83	15.14	19.51	23.93
P	0.1	0.05	0.01	0.005	0.001	0.0001	0.00001	0.000001

[a] Any χ^2 value 3.84 or greater is considered statistically significant because chance occurs 5 or less times in 100.
[b] Any χ^2 value 7.88 or greater is considered highly statistically significant because chance occurs 5 or less times in 1000.

Table 13
2 × 2 CONTINGENCY TABLE: OBSERVED VALUES

Attribute; status of water ingestion	Condition present (Ill)	Condition absent (Well)	Total	Attack rate
Drank water	11	8	19	58
Did not drink water	1	7	8	13
Total	12	15	27	71

Table 14
EXPECTED VALUES IF THERE IS NO DIFFERENCE BETWEEN RATES FOR TABLE 13

Attribute; status of water ingestion	Condition present (Ill)	Condition absent (Well)	Total
Drank water	8.4	10.6	19
Did not drink water	3.6[a]	4.4[a]	8
Total	12	15	27

[a] Cells less than 5, χ^2 test cannot be used to test probability, but Fisher's exact test can.

Table 15
2 × 2 CONTINGENCY TABLE: MORE EXTREME TABLE OF TABLE 13

Attribute; status of water ingestion	Condition present (Ill)	Condition absent (Well)	Total	Attack rate
Drank water	12	7	19	63
Did not drink water	0	8	8	0
Total	12	15	27	63

A 2 × 2 contingency observed table in which values within some cells are small is shown in Table 13. When the expected distribution is calculated and filled into a 2 × 2 contingency table (Table 14), two cells (b and c) are below five. Therefore, the χ^2 test cannot be used and the Fisher's exact test is called for to calculate the probability of whether or not the proportions are different. From data in Tables 13 and 15, which have a greater difference in attack rate in the same direction, and have the marginal values fixed, the probability of the difference being by chance alone would be:

$$P_1 = \frac{12!15!8!19!}{27!11!1!7!8!} = 0.0348 \quad P_2 = \frac{12!15!8!19!}{27!12!0!8!7!} = 0.0029$$

$$P_1 + P_2 = P \therefore 0.0348 + 00.0029 = 0.0377$$

To make this calculation, first cancel factorials whenever possible (e.g., 19!/27! = 1/27 × 26 × 25 × 24 × 23 × 22 × 21 × 20). Then cancel as many of the remaining numbers as possible. The probability that the difference would occur by chance alone in this example would be approximately 4 times in 100. Because a 5% significance level was selected, there is a statistical difference between the proportions. Thus, the null hypotheses would be rejected and strength is given to the hypothesis that water is the vehicle.

6. Summarizing the Investigation

Data collected during an investigation must either confirm or at least support the hypothesis that the outbreak was waterborne and preclude the possibility of transmission from a source other than water, such as food or person-to-person spread. Finding the eitologic agent in the implicated water may be required to fully confirm the hypothesis, but this is often difficult to accomplish. The finding of indicator organisms or certain chemical substances in the water may demonstrate that contamination occurred, but it does not necessarily mean that the suspected etiologic agents was present. A reasonable explanation, with supporting evidence whenever possible, must also be made to account for the way in which the etiologic agent reached the water, the manner by which it survived any treatment provided, and whether there was treatment failure.

The source of the etiologic agent also should be traced. It may be human excrement, manure, toxic industrial wastes, or pesticides. Such substances can reach the water through cross-connections, backsiphonage, seepage, surface runoff, flooding, or by other means. Whenever possible, the source, mode of contamination, and means by which the contaminant survived any treatment should be demonstrated by appropriate field tests, or it should at least be supported by observations.

VII. TAKING ACTION AFTER THE INVESTIGATION

After investigations have been completed, action must be taken to accomplish the purposes of surveillance of waterborne diseases. These are to gather information on the ecology of etiologic agents and the epidemiology of the diseases they cause, to control the current outbreaks to prevent future outbreaks, and to provide guidance for improving public health programs, water supplies, and water sources.

A. Submitting Report

After data have been analyzed and interpreted, a report of the investigation should be sent to state or district health authorities. These and similar reports of other investigations should be collated by these agencies and transmitted to the national agency responsible for waterborne disease surveillance. Persons concerned with water supply and with public health should make every effort to ensure complete investigation and reporting of waterborne disease. Without reliable complete information, trends in waterborne disease incidence and causal factors of the disease are difficult to determine. Over a period of time, a data bank can be developed about the epidemiology of waterborne diseases and the ecology of agents that cause them.

B. Controlling the Outbreak

Control can be accomplished, based upon an hypothesis of which strong supporting evidence is found during the investigation, by warning the public of the situation and by recommending certain precautionary actions, such as to boil water. Action can also be taken to chlorinate a water supply or to increase the chlorine dosage if the supply is already chlorinated. If spring or surface water is contaminated, official agencies can post signs that state that the public is not to drink or swim, as appropriate. Whether or not any preliminary control actions have been taken, control measures should be taken immediately after proving that an outbreak is waterborne. Adequate treatment of the water should be applied, and needed repairs in the system made. After the appropriate changes have been made and the water has been tested and found safe, emergency measures, such as a "boil water order", can be suspended.

C. Preventing Outbreaks

Surveys should be conducted to determine whether the situations that contributed to the outbreak are occurring in other water supplies or impounded bodies of water. If so, several measures should be taken. Appropriate treatment should be put into effect and repairs to the systems made. Operators of the facilities should be informed about the problem and their skills improved by training. Periodic inspections and surveys should be made to determine whether the faulty conditions have been corrected or are allowed to be reintroduced. Legal action may even be necessary to ensure compliance with official standards and acceptable sanitary practices. The public should be informed of any potential or actual harm that may result from ingesting or contacting contaminated water. Critical control points of treatment processes should be monitored periodically to determine whether the water is within limits of standards set for microorganisms, inorganic or organic chemicals, turbidity, and radionuclides. Most waterborne diseases are preventable, but prevention requires constant vigilance by personnel in water treatment facilities and officials of health and regulatory agencies to ensure that the hazards are understood and that questionable construction or practices are avoided.

D. Initiating or Maintaining Programs to Ensure Safe Water

Information gathered during investigation of outbreaks or subsequent surveys should be used as a basis for setting public health program priorities, training staff and persons who are responsible for water treatment and quality control, developing and passing laws and regulations, and educating the public. Dynamic water safety programs must be based on contemporary problems and their solutions.

REFERENCES

1. **Bryan, F. L., Anderson, H. W., Baker, K. J., Craun, G. F., Duel, W., Lewis, K. H., McKinley, T. W., Robinson, R. A., Swanson, R. C., and Todd, S. C. D.,** *Procedures to Investigate Waterborne Illness*, International Association of Milk, Food and Environmental Sanitarians, Ames, Iowa, 1979.
2. **Isenberg, H. D., Washington, J. A., II, Balows, A., and Sonnenwirth, A. C.,** Collection, handling, and processing of specimens, in *Manual of Clinical Microbiology*, 3rd ed., Lennette, E. H., Balows, A., Hausler, W. J., Jr., and Truant, J. P., Eds., American Society for Microbiology, Washington, D.C., 1980.
3. **Lennette, E. H., Balows, A., Hausler, W. J., Jr., and Truant, J. P., Eds.,** *Manual of Clinical Microbiology*, 3rd ed., American Society for Microbiology, Washington, D.C., 1980.
4. **Rand, M. C., Greenbury, A. E., and Taras, M. J., Eds.,** *Standard Methods for the Examination of Water and Wastewater*, 14th ed., American Public Health Association, Washington, D.C., 1976.

5. **Craun, G. F. and McCabe, L. J.,** Review of the causes of waterborne disease outbreaks, *J. Am. Water Works Assoc.,* 68, 420, 1973.
6. **Mosley, J. W., Schrack, W. D., Denshaun, T. W., and Matter, L. D.,** Infectious hepatitis in Clearfield County, Pennsylvania. I. A probable waterborne epidemic, *Am. J. Med.,* 26, 555, 1959.
7. **Lopez, C. E., Dykes, A. C., Juranek, D. D., Sinclair, S. P., Conn, J. M., Christie, R. W., Lippy, E. C., Schultz, M. G., and Mires, M. H.,** Waterborne giardiasis: a community-wide outbreak of disease and a high rate of asymptomatic infection, *Am. J. Epidemiol.,* 112, 495, 1980.
8. Water-Related Disease Outbreaks Surveillance Annual Summary 1980, Centers for Disease Control, Atlanta, Ga., 1982.

Chapter 8

METHODS TO IDENTIFY WATERBORNE PATHOGENS AND INDICATOR ORGANISMS*

T. H. Ericksen and A. P. Dufour

TABLE OF CONTENTS

I.	Introduction	196
II.	Isolation and Enumeration Methods for Indicator Bacteria	196
	A. Total Coliforms	196
	B. Fecal Coliforms	197
	C. *Escherichia coli*	198
	D. Fecal Streptococci	198
	E. Enterococci	199
III.	Isolation and Enumeration Methods for Pathogens and Opportunistic Pathogens	200
	A. *Salmonella*	200
	B. *Shigella*	201
	C. *Yersinia enterocolitica*	201
	D. *Campylobacter*	202
	E. *Pseudomonas aeruginosa*	203
	F. *Aeromonas*	204
	G. *Mycobacterium*	204
	H. *Vibrio*	205
	I. *Legionella* Species	206
	J. *Chromobacterium*	206
	K. *Leptospira*	207
	L. Enterotoxigenic *Escherichia coli*	208
	M. *Giardia, Entamoeba histolytica,* and *Balantidium coli*	208
IV.	Summary	209
References		210

---

* The research described in this paper has been reviewed by the U.S. Environmental Protection Agency and approved for publication. Mention of trade names or commercial products does not constitute endorsement or recommendation for use.

I. INTRODUCTION

The microbiological methods used to enumerate and isolate bacteria from water samples obtained from aquatic environments and potable water fall into two categories. One group of methods measures indicator bacteria to signal the potential presence of pathogens or undesirable microorganisms in water. The classic bacterial indicators are those that indicate the presence of fecal contamination. Early microbiologists were aware of the difficulties associated with measuring or detecting enteric pathogens in water. They reasoned that by isolating from water an innocuous bacterium, consistently found only in feces, they could assume the presence of fecal material and hence, the possible presence of enteric pathogens. Indicator bacteria are usually harmless, occur in high densities in their natural environments, and are easily cultured on relatively simple bacteriological media. Most of the enteric pathogens that were impossible or difficult to isolate by the early microbiologists can now be cultured, but the philosophy of using indicators has not changed. The reason for this is that pathogens usually occur sporadically in the water environment and the isolation of one pathogen will not necessarily predict the presence of another pathogen. Furthermore, many enteric viral pathogens, such as rotavirus and Norwalk agent, are still difficult to culture or are unculturable. Thus, the concept of using indicator bacteria is as useful today as it was near the turn of the century when Escherich[1] first proposed their use.

The second category involves methods which isolate pathogens or opportunistic pathogens from aquatic environments and potable water systems. The methods in this category usually are modifications of methods developed in clinical laboratories for isolating pathogens from patient specimens. The isolation of pathogens from water is frequently difficult and cumbersome because of interferences by natural competing flora and the fastidious nutrient requirements of the pathogens. Enrichment procedures must be used in some cases before the organisms are transferred to harsh selective media. The need to enrich obviously limits the ability of some methods to quantify the pathogens in the sample.

Methodology for viral isolation will not be addressed in this chapter since many of the important viral pathogens are difficult or impossible to culture from water samples. Furthermore, the sporadic nature of the occurrence of viral pathogens and the logistical complexity of routine monitoring virtually precludes measuring the culturable viruses on a systematic basis.

II. ISOLATION AND ENUMERATION METHODS FOR INDICATOR BACTERIA

A. Total Coliforms

The coliforms[2] are a group of bacteria defined as Gram-negative, aerobic or facultatively anaerobic, nonspore-forming rods that ferment lactose in 48 hr at 35°C with gas production by the most probable number (MPN) methodology. Coliforms were originally believed to indicate the presence of fecal pollution, and they are still used for that purpose. Coliform testing also is prescribed for determining the sanitary integrity of potable water in distribution systems, even though some coliforms have been found to be widely distributed in nature and are not associated with the intestinal tract of warm-blooded animals.[3] This finding has limited the use of coliforms as indicators of fecal pollution; in surface waters, however, they are still widely used as indicators of potable water quality in the U.S.

The methods[2] approved by the U.S. Environmental Protection Agency (EPA) for regulatory purposes and most commonly used to enumerate coliforms are the lactose broth or lauryl sulfate tryptose (LTB) broth MPN method and the membrane filter technique employing m-Endo broth or agar.

Another membrane filter method, the mC method,[4] has been reported for enumerating coliforms and their component genera from seawater. Confirmation of typical coliform colonies on this media from marine water samples was 95%. A comparison of marine water samples analyzed by this method and the coliform MPN and m-Endo methods showed that in 14 of 20 samples, recovery of coliforms by the mC method was greater than with the m-Endo method. The mC and MPN method were comparable in their recovery of coliforms. A new method, using m-T7 medium,[5] has been reported to recover higher numbers of injured coliforms than currently recommended methods. The new method recovered an average of 43% more verified coliforms from drinking water samples than the m-Endo technique.

LeChevallier et al.[6] have proposed new verification tests for coliforms from membrane filter methods. These tests would substitute for the subculturing of colonies to lactose broth and subsequent transfer of gas-positive culture tubes to brilliant green bile broth. These investigators indicated that the new verification procedure increases the confirmation rate by 87% over the currently accepted methods for membrane filter coliform colonies obtained from drinking water.

Total coliforms are used to measure the quality of potable water, shellfish harvesting waters, and, in at least 23% of the states, to determine the quality of recreational waters. The finding of coliforms in these different types of waters is interpreted as a degradation of the quality of the water, and therefore, a potential health hazard to users or to consumers of shellfish harvested from the water. This interpretation has severe limitations, however, because one member of the coliform group, *Klebsiella*, has been shown to grow in industrial and agricultural wastes which are unrelated to fecal contamination.

B. Fecal Coliforms

Fecal coliforms[2] are a subgroup of the total coliform group which can be differentiated by their ability to grow at an elevated temperature (44.5 ± 0.2°C). This group of organisms is measured with the EC broth MPN procedure or the mFC membrane filter method. The fecal coliforms can comprise four genera of bacteria — *Escherichia, Klebsiella, Citrobacter,* and *Enterobacter* — some of which are capable of thermotolerant (44.5°C) growth. The thermotolerant characteristic separates the fecal coliforms, which may be found in the GI tract of warm-blooded animals, from the other coliforms.

Rose et al.[7] proposed a two-layer method for enumerating fecal coliforms which yielded higher recovery than existing fecal coliform membrane filter procedures. This procedure was evaluated using raw waste water, chlorinated waste water, fresh surface water, and marine waters. Fecal coliform colony verification was 92%, and comparison of the two-layer method with existing methods showed that the new method was more efficient for recovery of fecal coliforms in these waters. Presswood and Strong[8] reported a possible toxic effect of rosolic acid used in mFC medium. They observed that the use of mFC medium without rosolic acid resulted in higher counts 71% of the time compared to mFC containing rosolic acid, in water samples from unchlorinated and chlorinated sewage, and from fresh surface waters. Verification of typical blue fecal coliform colonies from the modified medium had a fecal coliform verification rate of about 94%.

Lin[9] has described a two-step membrane filter method utilizing a phenol red lactose broth pre-enrichment technique followed by transfer of the filter to mFC agar incubated at 44.5°C. This method was very effective for detecting fecal coliforms in chlorinated sewage effluents. He showed that this method was comparable to the EC MPN method when both were used to assay 126 chlorinated water samples. Green et al.[10] reported that recovery of fecal coliforms from chlorinated sewage effluents by the mFC technique was improved by first incubating the medium at 35°C for 5 hr followed by incubation for 18 hr at 44.5°C.

Stuart et al.[11] compared a two-layer membrane filter medium (IM-MF) for injured fecal coliforms with the EC MPN procedure and the standard mFC method using various types of water samples. The IM-MF method recovery of fecal coliforms was equal to or greater than the MPN technique recoveries in 50% of the samples tested and was always greater than the mFC method. The IM-MF method exceeded the standard method in 30 of 33 trials using stream water samples.

LeChevallier et al.[12] evaluated m-T7 agar as a fecal coliform medium using sewage effluent samples. The m-T7 medium was compared to the standard mFC method and the two-layer enrichment procedure of Rose et al.[7] Higher fecal coliform counts were obtained with the m-T7 medium than with the other methods tested. Verification of isolated fecal coliforms averaged 89% for M-T7 and 83% for mFC. The highest fecal coliform isolation rates were observed with sewage effluents when m-T7 was preincubated at 37°C for 8 hr followed by transfer to 44.5°C for 12 hr.

The presence of fecal coliforms in water is indicative of fecal contamination and the possible presence of enteric pathogens. Thus, finding fecal coliforms in samples of drinking water, recreational water, or shellfish harvesting water indicates that water or shellfish are contaminated by fecal pollution or that consuming the water, swimming in the water, or consuming shellfish harvested from the water may result in acute illness.

C. *Escherichia coli*

E. coli is an unquestionable inhabitant of the GI tract of man and other warm-blooded animals. It is unable to show sustained growth in water or other extraintestinal environments. This organism, which is a member of the well-known fecal coliform group, is easily distinguished from other genera of the fecal coliform group by the urease test. *E. coli* does not possess this enzyme, whereas the enzyme is possessed by other coliforms, such as *Klebsiella, Enterobacter,* and *Citrobacter.* This characteristic was used by Dufour et al.[13] to develop a membrane filter method that was 90% specific for *E. coli* from marine, estuarine, and fresh water samples. The method (mTEC) used a lactose base medium which is incubated at 44.5°C to select for thermotolerant coliforms, and those coliforms able to grow at this temperature were differentiated using an *in situ* urease test.

The finding of *E. coli* in water samples assayed with the mTEC procedure is clear evidence of fecal contamination of the water from which the sample was taken. Furthermore, at fresh water beaches, the density of *E. coli* in the water has been shown to be directly related to the incidence of swimming-associated illness.[14] This relationship is limited to fresh water, since Cabelli[15] has reported that this indicator organism does not relate very well to swimming-associated illness in marine waters.

D. Fecal Streptococci

The fecal streptococci[2] are comprised of *Streptococcus faecalis, S. faecalis* subsp. *liquefaciens, S. faecalis* subsp. *zymogenes, S. faecium, S. bovis, S. equinus,* and *S. avium.* All of these streptococci are members of Lancefield's Group D Streptococci. The fecal streptococci have long been proposed as indicators of fecal contamination, since most of them are found in the GI tracts of man, domestic and feral animals, and birds. However, there are reports that some of the fecal streptococci may be found on vegetation and insects.[2]

The fecal streptococci do not reproduce in the environment, but may survive for prolonged periods. They are less dominant in the feces of many animals, birds, etc. than other GI bacterial flora, such as members of the coliform group.

Three methods used to enumerate the fecal streptococci group are the azide dextrose (AD) MPN multitube method, the KF membrane filter technique, and the Pfizer Selective Enterococcus (PSE) agar pour plate procedure.

The AD multitube MPN method involves incubation of AD broth, to which measured quantities of the water sample have been added, at 35°C for 24 to 48 hr. Inoculum from tubes showing turbidity after 24 and/or 48 hr is streaked onto PSE agar plates. The plates exhibiting typical colonies with esculin hydrolysis are positive for fecal streptococci. MPN tables are used to estimate the number of streptococci in the water sample. The direct plate count methods may be used for samples with high turbidity. KF agar, PSE agar, or esculin-azide agar can be used as pour plate media for turbid water samples. The membrane filter method involves filtration of the water sample followed by incubation of KF streptococcus agar medium at 35°C for 48 hr.

Typical streptococci isolated with these methods may be confirmed by the following tests: presence of catalase, growth in BHI broth incubated at 45°C for 48 hr, and growth in 40% bile broth medium incubated at 35°C for 72 hr. The fecal streptococci may be differentiated further by the methods described by Facklam,[16] Gross et al.,[17] or Waitkins et al.[18]

Pavlova et al.[19] compared five media, (M-enterococcus agar, KF agar, azide-sorbitol agar, thallous-acetate agar, and PSE medium) for the isolation and enumeration of fecal streptococci from animal and bird feces, raw sewage, and foods. These investigators determined that KF and PSE agar were essentially equal in their ability to enumerate fecal streptococci, but PSE required only 24 hr of incubation. Brodsky and Schiemann[20] compared PSE and KF media membrane filter procedures for the recovery of fecal streptococci from sewage effluent. They reported that PSE medium was highly specific for enterococci (86%) and fecal streptococci (90%) and that KF medium was less selective for fecal streptococci (83%). KF broth, PSE agar, mSD agar, Slanetz-Bartley broth (35°C), and Slanetz-Bartley (SB) broth (4 hr at 30°C and then 44 hr at 44°C) were compared for the enumeration of fecal streptococci from sewage by Stanfield et al.[21] They reported that KF was the least selective method tested. It is suggested that PSE or Slanetz-Bartley medium at 35°C is best for the enumeration of Group D streptococci from sewage.

Dutka and Kwan[22] compared eight media for the recovery of fecal streptococci from unchlorinated sewage effluent, sewage lagoon effluent, and fresh surface waters. The media compared were m-Enterococcus (mEnt) agar, KF agar, PSE agar, mE plus EIA agar, and AD-EVA broth. Several of these methods involved using different incubation temperatures. These investigators reported that an incubation temperature of 44.5°C did not enhance the selectivity of any of the media. The most efficient membrane filter methods tested were PSE agar, KF agar, and mE agar (41°C for 48 hr). The AD-EVA MPN procedure at 35°C was as efficient as the most efficient membrane filter method. Pagel and Hardy[23] compared PSE agar, KF agar, mEnt agar, mE-EIA, and SB media for the recovery of fecal streptococci from natural sources. They concluded that PSE agar was the least inhibitory, KF agar had the lowest specificity, and the best overall medium was mEnt agar.

The MPN and pour plate methods are limited to the use of relatively small volumes of water samples. The membrane filter methods, on the other hand, allow large volumes of water samples to be tested, and they are more precise. The literature indicates that without confirmation, all of the media are limited in their ability to accurately enumerate fecal streptococci.

The presence of fecal streptococci in water is an indication of fecal pollution and the potential presence of GI pathogens, but this interpretation is limited by similar considerations noted for finding coliforms in water.

E. Enterococci

The enterococci are a subgroup of the Group D fecal streptococci and include the following genera: *Streptococcus faecalis, S. faecium,* and *S. avium*.[24,25] The members of this group commonly inhabit the intestinal tract of humans and warm-blooded animals, and therefore, they have long been used to detect fecal contamination of water. Several media have been developed to isolate and enumerate enterococci using membrane filter methods. Slanetz and Bartley[26] developed mEnt agar for detecting enterococci in water, sewage, and feces. Daoust and Litsky[27] modified the PSE agar developed by Isenberg et al.[28] for use as a membrane filter method, and Levin et al.[29] developed mE agar for isolating enterococci from marine waters. Stanfield et al.[21] compared the growth of various species of streptococci on these three media and found none of them to be specific for enterococci. The mEnt agar supported the growth of *Streptococcus bovis, S. equinus, S. mitis,* and *S. salivarius,* as well as the enterococci strains. The *S. bovis* and *S. equinus* also were able to grow on PSE agar in addition to the enterococci. The only nonenterococcus species able to grow on mE medium was *S. equinus.* Although none of the media are specific for enterococci, in practice very few fecal streptococci that are not enterococci are enumerated.

Pagel and Hardy[23] compared the specificity of these three media using raw sewage, primary sewage effluents, and storm sewer effluents. *S. bovis* was identified twice on PSE agar and once on mEnt agar and not at all on mE agar. *S. salivarius* was identified once on mEnt agar. Stanfield et al.[21] found the mE method to be the most specific of the three methods, followed by mEnt agar and then PSE agar, whereas Pagel and Hardy[23] found the PSE method to be most specific, followed by the mE and mEnt methods. Dutka and Kwan[22] examined various types of water samples using many methods, including the three methods described above. They ranked the methods according to their ability to isolate enterococci and found that it followed the order: mE agar, PSE agar, and mEnt agar. The results of these three studies[21-23] indicate that, overall, mE medium probably has the highest specificity for enterococci. The ability of this medium to selectively isolate enterococci from water samples and waste water effluents was a major reason for its use during the health effects recreational water quality studies conducted by Cabelli et al.[30,31] The organisms enumerated by this method showed the best relationship to swimming-associated gastroenteritis in marine and fresh water swimmers. The finding of these enterococci in fresh or marine surface waters, as measured by the mE method, can be directly translated to a rate of swimming-associated gastroenteritis using the relationship reported by Cabelli[15] and Dufour.[14]

III. ISOLATION AND ENUMERATION METHODS FOR PATHOGENS AND OPPORTUNISTIC PATHOGENS

A. *Salmonella*

A great deal of emphasis has been placed on the presence of *Salmonella* species in water environments because of their historic relationship to waterborne and water contact disease. Although the incidence of illness caused by *Salmonella* has dramatically decreased in the U.S. since the mid-20th century because of good hygiene and sewage treatment practices (as well as the availability of antibiotics to treat patients with the disease), its presence in water must still be considered a threat to public health. Methods for the detection and isolation of this genus are important for identifying possible sources of the organism in waterborne disease outbreaks.

Most methods for isolation of this organism from water samples employ concentration techniques because of the low densities at which *Salmonella* occur in water. Filtration through membranes or diatomaceous earth, followed by incubation of the filter material or membranes in an enrichment medium, i.e., Dulcitol Slenite or Tetrathionate broth, and the subsequent streaking of inocula from the enrichment broth to several selective plating media, such as Brilliant Green Agar, Bismuth Sulfite Agar, Xylose Lysine Desoxycholate agar is the most commonly used method for isolation of *Salmonella.* Isolated typical *Salmonella* colonies are then identified by biochemical and serological techniques.

The methods described by Grunnet,[32] Cheng et al.,[33] and Levin et al.[34] have been developed to concentrate and enumerate *Salmonella* in water samples. Methods which use adsorbent pads suspended in water for long periods of time have also been used to concentrate *Salmonella* from water.[35]

The procedure described by Cheng et al.[33] is stated to give results in 48 hr and is based on the use of specific antisera in a mannitol medium for identification of *Salmonella.*

However, problems associated with serological flocculation have been noted with this technique.

The method described by Levin et al.[34] requires the use of fiberglass epoxy depth filters, followed by enrichment and selective plating. The use of these filters and their accompanying apparatus may be cumbersome and expensive; however, this method has been successful for enumerating low levels of *Salmonella* in marine waters.

A facile, definitive method for the isolation and enumeration of *Salmonella* in water samples is not available, but the methods described above have been quite adequate, especially for detecting *Salmonella* associated with waterborne outbreak episodes.

B. *Shigella*

The *Shigella* genus consists of four species:[36] *S. dysenteriae, S. flexneri, S. boydii,* and *S. sonnei.* This pathogen can be present in the intestinal tract of infected humans, and the transmission of shigellosis usually is fecal-oral in nature. This organisms is one of the leading causes of diarrhea in humans.[36] The presence of *Shigella* in any type of water represents an immediate human health risk to individuals using the water. This pathogen has been implicated in waterborne outbreaks[36] and in an outbreak of shigellosis among bathers swimming in contaminated river water.[37]

The methodology for the isolation of *Shigella* from water has been adopted from techniques primarily used for clinical samples, and the success of using these techniques in isolating this organism from natural waters has been varied. The organisms can be concentrated from water by a variety of techniques.[2] After concentration of the water sample, enrichment techniques are employed, followed by plating on selective media and biochemical or serological identification.

Several reports[38-40] have indicated that Gram-negative (GN) broth is the enrichment medium of choice for the isolation of *Shigella* from feces. These same investigators, including Morris et al.,[41] report that the best selective plating medium for isolating *Shigella* from enrichment broths is XLD agar.

Park and colleagues[42,43] introduced a novel technique for selecting *Shigella* from water samples. They incorporated 4-chloro-2-cyclopentylphenyl-β-D-galactopyranoside, an analogue of lactose, into lactose broth. Bacteria that utilize lactose, such as coliforms, hydrolyze the compound to galactose and a toxic moiety which subsequently kills them. *Shigella*, which do not utilize lactose, are not affected. These investigators reported that 42 of 48 *Shigella* strains were selectively enriched by this procedure. This selective enrichment procedure, followed by plating on a selective medium, could be employed as a modified MPN method for the isolation and enumeration of *Shigella* from water. This technique has not gained wide acceptance, probably because of the high cost of the lactose analogue.

Because of the sporadic nature of disease caused by *Shigella* species, it is not reasonable to monitor waters for this pathogen on a routine basis. However, in situations where an outbreak of shigellosis has occurred, it is mandatory that the suspected waters be examined to determine the source of the pathogen if at all possible. The detection of *Shigella* species in surface or potable waters, to which ill individuals have been exposed, should be interpreted as *prima facie* evidence of the source of the etiologic agent.

C. *Yersinia enterocolitica*

Y. enterocolitica has been implicated in many waterborne disease outbreaks.[44] *Yersinia* associated with disease and the environment recently have been divided into four species: *Y. enterocolitica, Y. kristensenii, Y. frederiksenii,* and *Y. intermedia,* based on DNA-relatedness and biochemical reactions.[44]

Methodology[46] for the isolation of *Y. enterocolitica* from water has utilized media developed for other purposes. Traditionally, isolation of *Yersinia* has been accomplished by incubation of the water sample itself or the membrane filter through which a water sample has been passed in mild buffer or Cooked Meat Media (CMM) at low temperature (4°C) for up to 21 days. The water sample or medium containing the membrane filter is sampled at 7, 14, and 21 days by plating portions onto an agar medium.

Several methods have recently been developed which are more rapid and facile. The membrane filter method of Bartley et al.[47] is based on the ability of *Y. enterocolitica* to ferment sorbitol, the absence of the enzymes, lysine decarboxylase and arginine decarboxylase-dihydrolase, and the presence of urease. Presumptive identification of *Y. enterocolitica* is made in approximately 48 hr. Confirmation[48] of the isolates is accomplished by biotyping and serotyping.

An enrichment broth for the isolation of *Y. enterocolitica* has been reported by Weagant and Kaysner.[49] This medium uses sorbitol as the primary differential substrate. This broth medium may be inoculated directly or water samples may be filtered and the filters placed into the medium. The inoculated medium is incubated at 22°C for 48 hr. Portions from tubes of enrichment broth showing growth are inoculated onto an agar medium to obtain colony isolates. Isolates are identified further by biochemical and serological test procedures.

A selective differential plating media (CIN) based on mannitol fermentation and the use of antibiotics and bile salts for the suppression of competing bacterial flora has been described by Schiemann.[50] Although developed for the isolation of *Y. enterocolitica* from feces, this medium may be useful for isolating this organism from water samples.

Head et al.[51] compared six *Y. enterocolitica* isolation media using 35 laboratory bacterial strains seeded into fecal specimens. They determined that CIN medium was the most efficient for inhibiting competing flora and supporting the growth of the target organism after 40 hr of incubation. The membrane filter method of Bartley et al.[47] isolated 33 strains of *Y. enterocolitica* from 15 of 27 river water samples. The frequency of isolation by the enrichment technique of Weagant and Kaysner[49] using natural samples of fresh and marine water recovered *Yersinia* in 22 of 24 (92%) and 95 of 127 (75%) samples, respectively. These investigators also reported the isolation of indigenous *Y. enterocolitica* from public drinking water systems by using epoxy-glass fiber filters followed by placing the inoculated filter into Peptone-Sorbitol-Bile Salts (PSB) enrichment media.

A thorough review of the literature on *Y. enterocolitica* has been conducted by Swaminathan et al.[52] Water has been implicated as the common source in several outbreaks[44] involving *Y. enterocolitica*. *Yersinia* has been isolated from numerous waters, but the strains found in water usually differ from those found in human disease. Therefore, the significance of finding *Y. enterocolitica* and *Y. enterocolitica*-like organisms in water is not known at this time. Unless organisms are isolated during a waterborne outbreak or the isolates are members of those serotypes known to cause disease in humans, it is unlikely that organisms isolated from aquatic environments pose a significant health risk for humans.

D. *Campylobacter*

C. jejuni recently has become recognized as one of the most common causes of bacterial gastroenteritis.[53] This organism is commonly found in the intestinal tracts of domestic and feral animals and in birds. These animals apparently act as the reservoir of this organism, which is transmitted to humans through direct contact or indirectly through contaminated water, milk, or food.[54]

C. jejuni produces gastroenteritis in healthy and impaired persons at relatively low levels[55] and has been implicated as the etiologic agent of gastroenteritis outbreaks associated with municipal water supplies in Sweden and the U.S. It is generally believed that this organism normally is found in waters which contain fecal coliforms, which suggests that fecally contaminated water is a risk factor of *Campylobacter* gastroenteritis.

The methods currently used for the isolation of this pathogen were initially developed for isolating *Campylobacter* from clinical fecal specimens. Using these methods, *Campylobacter* has been isolated from marine and fresh waters.[56-58] The methods involve filtration of the water sample, followed by enrichment in broth medium or placing the filter on a selective media, with subsequent incubation under microaerophilic conditions at 42°C. The organisms usually require further purification for identification purposes. One of the basic limitations of these methods is their inability to prevent the spreading and coalescing of colonies.

A selective medium for the isolation of *Campylobacter* was reported by Bolton and Robertson,[59] which they called Preston medium. Included in the report was an enrichment medium containing the same ingredients as Preston agar but without the agar. Preston medium was compared with Skirrow's medium for the isolation of *Campylobacter* from the feces of humans, cattle, pigs, sheep, chickens, and seagulls. The Preston medium isolated more *Campylobacter* than the Skirrow medium and if the Preston enrichment medium was used, considerably more *Campylobacter* isolates were found. These investigators also reported that the enrichment broth and Preston agar medium have been used successfully for $2\frac{1}{2}$ years for the isolation of *Campylobacter* from coastal seawater, drain swab effluents, and abattoir environmental specimens.

Ribeiro et al.[60] have reported the successful use of the Preston medium enrichment method for the isolation of *Campylobacter* from human feces and domestic pets. They also reported that membrane filtration of water sample followed by placing the membrane in Preston enrichment medium and subsequent subculture to a selective medium resulted in 12 positive out of 64 water samples examined. All of these water samples were negative using direct plating techniques.

Bolton et al.[61] have reported the use of Preston enrichment broth followed by subculture to Preston agar medium and reported that this method was capable of estimating as few as 10 *Campylobacter* per 100 mℓ of water. Using their method, 11 or 49 marine water samples were *Campylobacter* positive and counts ranged from 10 to 230 per 100 mℓ. Positive river water samples contained 10 to 36 *Campylobacter* colonies each in 7 of 44 samples. They also reported that by using a larger volume of water sample and membrane filters that an additional 24 of the "negative" samples were found to contain *Campylobacter*.

The Preston enrichment technique was tested further by Ribeiro and Price[62] and they reported that of 128 water samples tested, 34 were positive using this technique. Bolton et al.[63] compared Skirrow's, Butyle's, Blaser's Campy-BAP, and Preston media for *Campylobacter* isolation from human, animal, and environmental specimens. Preston enrichment broth followed by subculture to Preston agar gave the highest isolation rates.

Campylobacter isolates obtained by the Preston enrichment technique can be further identified by the procedures of Skirrow and Benjamin[64] and Kaplan.[65]

Mathewson et al.[66] evaluated three microporous filters and three pore sizes of a depth filter. They determined that a positive-charged depth filter gave the highest percent recovery using seeded sterile tap and surface waters. This type of methodology could possibly be used to enhance *Campylobacter* recoveries when using the filter enrichment MPN technique.

Another enrichment technique has been reported by Oosterom et al.[67] for detecting *Campylobacter* in sewage and river water samples. However, very little supporting information concerning this methodology is available.

The presence of *C. jejuni* in water indicates recent fecal contamination and risk of infection for individuals using the water.

E. *Pseudomonas aeruginosa*

P. aeruginosa has frequently been implicated as the causative agent for infections and illnesses associated with water contact. Two methods have been proposed for isolating this organism from water environments. Drake[68] proposed using a liquid medium in conjunction with the most probable number (MPN) technique. The medium has been used to quantify

P. aeruginosa in feces[69] and in river and sewage samples.[70] A modification of Drakes's method also has been recommended in the 15th edition of *Standard Methods for the Examination of Waters and Wastewaters*.[2] This method, however, has all of the shortcomings commonly identified with MPN procedures. Confirmation of questionable positive tubes is usually required, other microorganisms may interfere with the growth of *P. aeruginosa*, the volume of sample that can be assayed is limited, and the technique itself is very imprecise. Many of these shortcomings were eliminated by the membrane filter method developed by Levin and Cabelli[71] for quantifying *P. aeruginosa*. The medium used for this method is highly selective, and it incorporates a differential dye system that is specific for *P. aeruginosa*.

The significance of finding *P. aeruginosa* in surface waters is not clear. Foster et al.[72] attempted to relate the density of this organism to the incidence of ear infections among bathers. They were unable to demonstrate a relationship between *P. aeruginosa* densities and the rate of ear infections. Calderon and Mood[73] observed similar results in swimming pools. They found that the water quality of swimming pools as measured by the presence of *P. aeruginosa* was not associated with cases of otitis externa. Seyfried and Frazer,[74] on the other hand, did find a relationship between the presence of *P. aeruginosa* in swimming water and colonization of the outer ear and subsequent development of ear infections in swimmers.

It is unlikely, however, that *Pseudomonas* can be used to predict the incidence of otitis externa until a direct relationship between these two variables has been clearly demonstrated. *P. aeruginosa* is the causative agent in many opportunistic infections linked to water contact, and it is in those cases that the methods described above will be useful for identifying the source of the pathogen.

F. *Aeromonas*

A. hydrophila shares many characteristics with the genera that comprise the coliform group. The fact that many members of this genus ferment lactose with gas production has drawn much attention to this ubiquitous organism because it may inflate coliform densities and give a false indication of poor-quality water. It can easily be distinguished from coliforms because it possesses a cytochrome oxidase that can be detected with a simple test.

Shubert[75] developed a medium for *A. hydrophila* (DSF medium) in 1967. DSF medium, which contains sodium sulfite and dextrin, was used in conjunction with a membrane filter procedure. In 1973, Shotts and Rimler[76] developed a selective-differential medium (RS) to isolate *A. hydrophila*. Sodium desoxycholate and novobiocin are the selective agents and aeromonads are differentiated from other organisms that grow on this medium with an indicator system based on H_2S production, maltose fermentation, and ornithine decarboxylation. Fliermans,[77] in 1977, used the RS medium with membrane filters and found this technique to be effective. In 1979, Rippey and Cabelli[78] described a membrane filter enumeration method for *A. hydrophila* (mA). The medium employs trehalose as a fermentable carbohydrate, and ampicillin and ethanol as selective inhibitors. After 20 hr of incubation at 37°C, and *in situ* mannitol fermentation test followed by an *in situ* oxidase test is used to differentiate *A. hydrophila* from other competing organisms. This method was very successful in recovering *A. hydrophila* from fresh surface waters.

A. hydrophila is ubiquitous in fresh water environments, and therefore, detecting this organism in surface or potable water samples does not carry any special significance. The methods can be useful, however, for determining the source of the etiological agent of illness or infection.

G. *Mycobacterium*

The mycobacteria usually associated with water environments are *M. kansasii*, *M. xenopi*, *M. intracellulare*, and *M. marinum*. Isolating mycobacteria from water requires three steps.

The first step involves concentration of the sample, either by centrifugation or filtration. This step is usually followed by a decontamination process whereby sodium hydroxide,[79-81] or a mixture of sodium hydroxide and sodium hypochlorite[82] or cetylpyridinium chloride,[83] is added to the concentrate or directly to the water sample to eliminate unwanted competing bacteria. Bailey et al.[80] eliminated the decontamination step by applying their membrane filters directly onto plates containing selective medium.

The third step involves placing the concentrated sample or the membrane on a selective medium. Lowenstein-Jensen medium and Middlebrook 7H10 medium have both been used to isolate mycobacteria from water samples. These media have been modified by various investigatiors who added antibiotics to the base formulations. Petron and Vera[84] added cyclohexamide, lincomycin, and nalidixic acid to Lowenstein-Jensen or Middlebrook 7H11 media, and this modification was found to inhibit Gram-negative bacilli while not affecting the recovery of mycobacteria. Lorian and Maddock[85] incorporated colimycin, spiramycin, and amphotercin B into Middlebrook 7H10 agar and found this to be an effective combination of inhibitors for decreasing contamination of their plates. Mitchison et al.[86] developed a selective 7H11 agar which contained polymyxin B, carbenicillin, amphotercin B, and trimethoprim. McClatchy et al.[87] recommended that the concentration of carbenicillin in the selective 7H11 medium be reduced from 100 to 50 μg/mℓ for better growth of *M. intracellulare*, *M. kansasii*, and *M. scrofulaceum*.

The finding of mycobacteria in surface or potable waters is probably not cause for taking steps to eliminate these organisms in the absence of overt disease in the exposed population. In those situations where outbreaks or infections are observed, isolation of mycobacteria from the water may be indicative that water is the vehicle of transmission and that appropriate control measures be considered.

H. *Vibrio*

V. parahaemolyticus, *V. alginolyticus*, *V. anguillaruim*, and *V. cholera* are found in marine waters, where they frequently cause disease in individuals who come in contact with the water. Thiosulfate-Citrate-Bile-Salt-Sucrose (TCBS) agar[88] is the most common medium used to isolate *Vibrio* species. High pH (8.6), bile salts and citrate inhibit some of the background flora found in marine waters, but the medium is frequently overwhelmed by growth of organisms such as *Aeromonas* species. Thus, follow-up tests are required to confirm *Vibrio* species. Akiyama et al.[89] and Horie et al.[90] have developed methods for *V. cholera*, but these methods are not very efficient for measuring the low densities of *Vibrio* that are found in marine environments. Horie et al.[91] developed a membrane filter technique for enumerating *V. parahaemolyticus* in 1967. The medium employed arabinose, ammonium sulfate, sodium cholate, and a high pH (8.6). Baross and Liston,[92] Twedt and Novelli,[93] and Vanderzant and Nickelson[94] developed media for *V. parahaemolyticus*, but the utility of the media was limited because large volumes of water could not be tested. In 1976, Watkins et al.[95] developed a membrane filter enumeration technique for *V. parahaemolyticus*. Selection of *V. parahaemolyticus* was based on the ability of this organism to grow in 3% NaCl, at high pH (8.6), and at 40°C. *In situ* galactose and sucrose fermentation tests, along with the oxidase test, were used to differentiate other species and verify *V. parahaemolyticus* colonies.

Vibrio species are similar to other aquatic bacteria in that they are ubiquitous in marine environments, and therefore, no special significance can be placed on detecting them in marine waters. The advantage of being able to isolate *Vibrio* species from water samples is that the source of this organism can be identified when it is suspected of causing disease or infections.

I. Legionella Species

L. pneumophila is the most important of the *Legionella* species because of its frequent association with respiratory illness. Although many other species have been identified since 1976, when *L. pneumophila* was first discovered,[96] these other species have not gained the importance of *L. pneumophila*. Since *L. pneumophila* has this position of importance, most of the media developed to isolate *Legionella* species have used it as the target bacterium. Presently, three media are available for isolating *Legionella* species. All three use the charcoal, yeast extract, cysteine, ferric pyrophosphate agar base developed by Feeley et al.[97] Bopp et al.,[98] Wadowsky and Yee,[99] and Edelstein[100] modified the buffered basal medium by adding various antibiotic inhibitors. Bopp et al.[98] added cephalothin, colistin, vancomycin, and cyclohexamide to control or eliminate non-*Legionella* bacteria found in water samples. Wadowsky and Yee[99] used glycine, vancomycin, and polymyxin B to inhibit background flora. Edelstein[100] added anisomycin, bromthymol blue, and bromocresol purple to Wadowsky and Yee's modification of the basal medium. The efficiency of the media for eliminating background flora and recovering *Legionella* have been examined by Calderon and Dufour.[101] The modification developed by Edelstein was the most efficient of the three media for recovering *Legionella* species seeded into environmental water samples. Edelstein's modification was also the most efficient of the three media for isolating *Legionella* species from natural samples obtained from plumbing systems and from lakes and rivers. The background flora of water samples can be further reduced by acidifying the water sample to pH 2 for 10 min and then neutralizing the sample to pH 7.[97]

The three isolation media are usually used as plating media and when used in this manner, they have the disadvantage of not being able to assay large volumes of sample. The Centers for Disease Control[102] recommend a method which involves membrane filtration of water samples. This technique allows much larger volumes of sample to be examined. After a water sample has been passed through the filter, the membrane is placed grid-side down in a flask containing 10 mℓ of the sterile filtrate and the flask is then placed into a sonic bath to release the concentrated bacteria from the membrane. The bacterial suspension then can be inoculated onto any of the above media, or it can be acid treated and then inoculated onto the medium.

The significance of isolating *Legionella* species, especially *L. pneumophila*, from water samples is not clear. *Legionella* species appear to be ubiquitous in the aquatic environment, and therefore, their detection is of little consequence except when the water is to be used for cooling or other purposes, such as showering. If the water is to be used for either of these purposes, there should be some concern because this organism can be amplified in hot water tanks or heat exchange systems and then dispersed as an aerosol. Some investigators believe that detecting *Legionella* in water samples, in the absence of overt disease in the exposed population, is not justification for taking action to eliminate the organism. Others believe that immediate action should be taken to eliminate *Legionella* from the water delivery system. However, in situations where high-risk populations are exposed, such as intensive care wards in hospitals and hospital patients taking immunosuppressive drugs, there is no disagreement. There is general agreement that steps should be taken to eliminate *Legionella* if it is found in water to which these groups are exposed.

J. Chromobacterium

C. violaceum is found in soil and water, particularly in tropical and subtropical regions. This bacterium, which causes severe illness and even death, is the most important health-related species of the genus *Chromobacterium*.[103] The number of cases associated with this organism, fortunately, is very limited.[103]

Several media — Bennett's agar,[104] BAN,[105] Ryall and Moss selective media,[106] and *Aeromonas* membrane agar without indicator[107] — have been reported for the isolation of

Chromobacterium from water. Bennett's medium is nonselective and was originally developed for the production of aerial mycelia from streptomyces. BAN medium is Bennett's agar with selectivity provided by the addition of neomycin hydrochloride, cyclohexamide, and nystatin. The selective medium of Ryall and Moss contains one fourth strength nutrient agar and the selective agents colistin, sodium deoxycholate, and cyclohexamide. The membrane *Aeromonas* medium is that described by Rippey and Cabelli[78] without the pH indicator, but containing ampicillin and sodium desoxycholate as inhibitors. Identification of members of the genus *Chromobacterium* can be made using the cultural and biochemical tests of Leifson[108] and Sneath.[109,110]

Kobarger and May[107] compared Bennett's agar, Ryall and Moss medium, and membrane filter *Aeromonas* agar incubated at 25 and 35°C for the recovery of *Chromobacterium* species from foods, soil, and water. The *Aeromonas* medium incubated at 35°C gave the highest recovery of *C. violaceum*. They reported *C. violaceum* only from soil and water. The nonpathogenic *C. lividum* was recovered from foods, soil, and water.

The presence of *C. violaceum* in water appears to be limited to tropical and subtropical regions. This genus favors warm waters and its presence may be seasonal in nature. Based on the small number of reported cases of illness, it appears to be of limited health hazard potential.

K. *Leptospira*

The isolation of *Leptospira* from surface waters is extremely difficult because these organisms grow very slowly, and therefore, they can be overgrown by competing bacteria. This difficulty is overcome by prefiltering the sample. *Leptospira* are commonly found in sediments of ponds and streams, and therefore, the first step in isolating this organism is to prefilter the sample through Whatman® No. 1 filter paper to remove gross particulates. This step is followed by filtration through membranes of smaller porosity, the minimum pore size usually being 0.45 µm.

The filtrate may then be cultured using several methods. Braun et al.[111] proposed that 0.2-µℓ and 2.0-mℓ volumes be inoculated into tubes of Fletcher's semisolid medium containing 10% rabbit serum. The medium is incubated at 30°C for 6 weeks. Tubes are examined weekly using dark field illumination to determine if *Leptospira* are present. Medium in tubes that are positive for *Leptospira* is passed through a 0.22-µm membrane filter and the filtrate is inoculated once again into tubes of Fletcher's medium, which are incubated at 30°C. Cultures that are not contaminated are used to determine the serotype and pathogenicity of the isolated strain.

Baseman et al.[112] recommend placing a drop of the filtrate in the center of each plate of serum agar or SM agar, and these plates are incubated at 30°C. Growth appears in 7 to 9 days as thin veils moving to the periphery of the plate. *Leptospira* were transferred from the veils to other plates and streaked for isolation. Purified cultures may be serotyped and the pathogenicity of the strains determined.

Pathogenic and saprophytic isolates can be differentiated in a number of ways. Johnson and Rogers[113] showed that 8-azaguanine inhibits the growth of pathogenic *Leptospira* and allows the growth of saprophytic strains. Johnson and Harris[114] examined the growth characterisitics of pathogenic and saprophytic *Leptospira* at low incubation temperatures. They found that the minimal growth temperature for pathogenic *Leptospira* was between 13 and 15°C, whereas saprophytic *Leptospira* have minimal growth temperatures between 5 and 10°C. Thus, saprophytes can easily be differentiated from pathogens by incubating the cultures at 13°C in a medium containing 10% rabbit serum. The saprophytic strains grow at this temperature and the pathogenic strains cannot. Füsi and Csóka[115] found that the growth of pathogenic *Leptospira* is inhibited by the presence of 10 mg/ℓ of copper sulfate in the culture medium, whereas saprophytic strains grew well.

The finding of pathogenic *Leptospira* in water samples from "swimming holes" or other surface waters used for swimming is a clear indication that the water is not suitable for bathing. Furthermore, it also indicates that domestic or wild animals sharing the water may be infected with *Leptospira*.

L. Enterotoxigenic *Escherichia coli*

Enterotoxigenic *E. coli*[116,117] produce illness through the production of heat-labile (LT) and heat-stable (ST) enterotoxins, cellular invasiveness, and heavy colonization of the intestinal mucosa.

These *E. coli* are specifically related to fecal contamination of water by warm-blooded animals and when found in water, represent a health risk for individuals using the water.

Enterotoxigenic *E. coli* (ETEC) cannot be differentiated from other *E. coli* unless they are tested for their effect on Y-1 adrenal and Chinese hamster ovary tissues, or with the suckling mouse, rabbit ileal loop, or Sereny tests.[116] The development of specific genetic hybridization probes[118] may simplify the test for ETEC.

The methods available for the detection of ETEC include the MPN and membrane filter methods described in *Standard Methods for the Examination of Water and Wastewater*.[2] However, the isolates from these methods must first be identified as *E. coli* by biochemical testing and then tested by previously described methods to determine their enterotoxigenicity. These methods are time-consuming and expensive. Furthermore, since the enterotoxigenic mechanism may be plasmid controlled, these characteristics can be lost during the identification procedure.

Calderon and Levin[119] modified the *E. coli* membrane filtration method described by Dufour et al.[13] to isolate and enumerate ETEC from water. ETEC colonies on the modified medium were identified by transferring the membrane to Y-1 adrenal cell monolayers. *E. coli* producing LT toxin in approximately 2 days caused the Y-1 adrenal cells to round up. This method was successfully tested using fecal samples and polluted marine water samples. Although this method is the most facile one available, its usefulness is limited because it requires costly tissue culture techniques.

The detection of enteropathogenic *E. coli* in potable water indicates that there is a considerable risk for persons who drink the water and that some action to eliminate these organisms should be taken.

M. *Giardia*, *Entamoeba histolytica*, and *Balantidium coli*

Giardia are protozoan flagellates which cause diarrhea in humans and have been commonly identified in waterborne disease episodes.[120] The presence of this parasite in humans usually is detected by the examination of fecal specimens by the microscopic examination of flotation or sedimentation concentrates. Microscopy also can be employed to identify *Giardia* cysts in water samples after concentration from the water. This organism was recovered initially from potable water during an outbreak in Rome, N.Y.[121] Methodology for the successful concentration, isolation, and identification of *Giardia* from tap water using a depth filter are described by Jakubowski and Ericksen.[122] Holman et al.[123] compared 1- and 7-μm porosity yarn-wound depth filters for recovery of as few as 6000 *Giardia* cysts per test, and determined that the 1-μm porosity depth filter exhibited higher cyst recovery rates. These investigators also compared membrane filtration and centrifugation methods for cyst recovery from yarn-wound filter concentrates and reported that higher numbers of cysts were recovered by centrifugation.

A workshop on *Giardia* methodology was held at the U.S. Environmental Protection Agency Laboratory in Cincinnati, Ohio in September 1980. The recommended methodology for *Giardia* detection from potable water derived from this meeting has been described by Jakubowski.[124] The low efficiency of detection methods, relatively high cost of procedures, and lengthy time for examination limit the use of present *Giardia* methodology for routine monitoring. However, these methods can be employed in suspected waterborne disease outbreaks where *Giardia* is the suspected etiologic agent.

E. histolytica causes diarrhea. Chang and Kabler[125] suggested the use of membrane filter procedures for the concentration and identification of *E. histolytica* from nonturbid waters. This organism is identified by microscopic examination of the water concentrate. The identification of *E. histolytica* trophozoites and/or cysts requires expertise in parasitological identification.

Chang and Kabler[125] also recommended the cultivation of samples or sample concentrates into liver infusion medium. This methodology is presented in the 15th edition of *Standard Methods for the Examination of Water and Wastewater*.[2] Examination of water for *E. histolytica* probably should be limited to those waters involved in waterborne disease outbreaks where this protozoan is the suspected etiologic agent and should not be employed for routine monitoring.

B. coli is the only ciliate pathogenic for man and exhibits world-wide distribution.[126] A waterborne outbreak of 100 cases was reported in Micronesia in 1971.[127] This organism rarely is found in specimens from patients in the U.S. *B. coli* is isolated and identified from fecal specimens using clinical procedures. The protozoan can be cultivated in Barret and Yarbrough's medium.[126] Since this ciliate is the largest protozoan to infect humans, the use of membrane filtration procedures as outlined for *E. histolytica* of Chang and Kabler[125] might be appropriate for the concentration, isolation, and identification of this organism from nonturbid waters. Waters should be examined for this organism only if it is the etiologic agent in a waterborne outbreak and routine monitoring would be superfluous.

IV. SUMMARY

The literature describing the enumeration and isolation of bacteria is replete with methods used to measure water quality and detect pathogens, and new methods are continuously being developed. The limited number of methods described in this chapter were selected because they have been successfully used to monitor water quality or to detect pathogens associated with water contact diseases. Some are included because they have characteristics that fulfill a unique purpose, i.e., high-volume enumeration techniques for measuring bacteria that normally occur in water at very low densities. None of the methods are ideal in the sense that they are free of limitations, but any shortcomings they might have are not great enough so as to preclude their use as tools for measuring bacteria in water environments.

The bacterial indicator methods, if used properly, are an effective means of monitoring water quality and thereby protecting groups at risk against pathogens associated with fecal material from humans and other warm-blooded animals. The use of fecal indicator bacteria are most efficient when they are used to index the potential presence of pathogens, especially viral pathogens, that are unculturable or very difficult to culture. Norwalk virus, rotavirus, and hepatitis A virus fall into this category. However, one should be mindful that indicator bacteria are not effective in all situations. For instance, *Giardia* cysts, which contaminate water as a result of the deposition of human or animal fecal wastes, have been found in potable water in the absence of indicator bacteria that warn of the presence of fecal contamination.

The methods that isolate pathogens and opportunistic pathogens from water samples are most useful for identifying the source of these organisms during waterborne disease outbreaks. From time to time, proposals have been introduced to monitor environmental waters for the pathogens themselves. Two factors discourage the acceptance of proposals to routinely monitor for pathogens. First, pathogens occur sporadically in water and the presence of one pathogen is usually unrelated to the presence of other pathogens and, second, the isolation of bacterial pathogens is usually difficult because of their fastidious nutritional requirements. In the case of viruses, the logistically complex concentration and tissue culture methods discourage their use for routine monitoring. These factors preclude the use of pathogen isolation methods for any role in water microbiology other than tracing pathogens to their source, so that corrective measures may be taken to eliminate the health hazard.

REFERENCES

1. **Escherich, T.,** Die Darmbakterien des Naugenborenen und Sauglings, *Fortschr. Med.*, 3, 515, 1885.
2. *Standard Methods for the Examination of Water and Wastewater,* 15th ed., American Public Health Association, Wahington, D.C., 1980.
3. **Dutka, B. J.,** Coliforms are an inadequate index of water quality, *J. Environ. Health,* 36, 39, 1973.
4. **Dufour, A. P. and Cabelli, V. J.,** Membrane filter procedure for enumerating the component genera of the coliform group in seawater, *Appl. Microbiol.,* 29, 826, 1975.
5. **LeChevallier, M. W., Cameron, S. C., and McFeters, G. A.,** New medium for improved recovery of coliform bacteria from drinking water, *Appl. Environ. Microbiol.,* 45, 484, 1983.
6. **LeChevallier, M. W., Cameron, S. C., and McFeters, G. A.,** Comparison of verification procedures for the membrane filter total coliform technique, *Appl. Environ. Microbiol.,* 45, 1126, 1983.
7. **Rose, R. E., Geldreich, E. E., and Litsky, W.,** Improved membrane filter method for fecal coliform analysis, *Appl. Microbiol.,* 29, 532, 1975.
8. **Presswood, W. G. and Strong, D. K.,** Modification of M-FC medium by eliminating rosolic acid, *Appl. Environ. Microbiol.,* 36, 90, 1978.
9. **Lin, S. D.,** Membrane filter method for recovery of fecal coliforms in chlorinated sewage effluent, *Appl. Environ. Microbiol.,* 32, 547, 1976.
10. **Green, B. L., Clausen, E. M., and Litsky, W.,** Two-temperature membrane filter method for enumerating fecal coliform bacteria from chlorinated effluents, *Appl. Environ. Microbiol.,* 33, 1259, 1977.
11. **Stuart, D. G., McFeters, G. A., and Schillinger, J. E.,** Membrane filter technique for the quantification of stressed fecal coliforms in the aquatic environment, *Appl. Environ. Microbiol.,* 34, 42, 1977.
12. **LeChevallier, M. W., Jakanoski, P. E., Camper, A. K., and McFeters, G. A.,** Evaluation of m-T7 agar as a fecal coliform medium, *Appl. Environ. Microbiol.,* 48, 371, 1984.
13. **Dufour, A., Strickland, E., and Cabelli, V.,** Membrane filter method for enumerating *Escherichia coli, Appl. Environ. Microbiol.,* 41, 1152, 1981.
14. **Dufour, A.,** Health Effects Criteria for Fresh Recreational Waters, EPA-600/1-84-004, U.S. Environmental Protection Agency, Cincinnati, Ohio, 1984.
15. **Cabelli, V.,** Health Effects Criteria for Marine Recreational Waters, EPA-600/1-80-031, U.S. Environmental Protection Agency, Cincinnati, Ohio, 1983.
16. **Facklam, R. F.,** Recognition of group D streptococcal species of human origin by biochemical and physiological tests, *Appl. Microbiol.,* 23, 1131, 1972.
17. **Gross, K. C., Houghton, M. P., and Senterfit, L. B.,** Presumptive speciation of *Streptococcus bovis* and other group D streptococci from human sources by using arginine and pyruvate tests, *J. Clin. Microbiol.,* 1, 54, 1975.
18. **Waitkins, S. A., Ball, L. C., and Fraser, C. A. M.,** A shortened scheme for the identification of indifferent streptococci, *J. Clin. Pathol.,* 33, 47, 1980.
19. **Pavlova, M. T., Brezenski, F. T., and Litsky, W.,** Evaluation of various media for isolation, enumeration and identification of fecal streptococci from natural sources, *Health Lab. Sci.,* 9, 289, 1972.
20. **Brodsky, M. H. and Schiemann, D. A.,** Evaluation of Pfizer selective enterococcus and KF media for recovery of fecal streptococci from water by membrane filtration, *Appl. Environ. Microbiol.,* 31, 695, 1976.
21. **Stanfield, G., Irving, T. E., and Robinson, J. A.,** *Isolation of Faecal Streptococci from Sewage,* Technical Report TR83, Water Research Centre, Stevenage, England, 1978.
22. **Dutka, B. J. and Kwan, K. K.,** Comparison of eight media-procedures for recovering faecal streptococci from water under winter conditions, *J. Appl. Bacteriol.,* 45, 333, 1978.
23. **Pagel, J. E. and Hardy, G. M.,** Comparison of selective media for the enumeration and identification of fecal streptococci from natural waters, *Can. J. Microbiol.,* 26, 1320, 1980.
24. **Hartman, P. A., Reinhold, G. W., and Saraswat, D. S.,** Indicator organisms — a review. I. Taxonomy of the fecal streptococci, *Int. J. Syst. Bacteriol.,* 16, 197, 1966.

25. **Buchanan, R. E. and Gibbons, N. E.**, *Bergey's Manual of Determinative Bacteriology*, 8th ed., Williams & Wilkins, Baltimore, 1974, 490.
26. **Slanetz, L. W. and Bartley, C. H.**, Numbers of enterococci in water, sewage and feces determined by the membrane filter technique with an improved medium, *J. Bacteriol.*, 74, 591, 1957.
27. **Daoust, R. A. and Litsky, W.**, Pfizer selective enterococcus agar overlay method for the enumeration of fecal streptococci by membrane filtration, *Appl. Microbiol.*, 29, 584, 1975.
28. **Isenberg, H. G., Goldberg, D., and Sampson, J.**, Laboratory studies with a selective enterococcus medium, *Appl. Microbiol.*, 20, 433, 1970.
29. **Levin, M. A., Fischer, J. R., and Cabelli, V. J.**, Membrane filter technique for enumeration of enterococci in marine waters, *Appl. Microbiol.*, 30, 66, 1975.
30. **Cabelli, V. J., Dufour, A. P., McCabe, L. J., and Levin, M. A.**, A marine recreational water quality criterion consistent with indicator concepts and risk analysis, *J. Water Pollut. Control Fed.*, 55, 1306, 1983.
31. **Cabelli, V. J., Dufour, A. P., McCabe, L. J., and Levin, M. A.**, Swimming-associated gastroenteritis and water quality, *Am. J. Epidemiol.*, 115, 606, 1982.
32. **Grunnet, K.**, Quantitative *Salmonella* demonstration by pad technique, *Rev. Int. Oceanogr. Med.*, 34, 155, 1974.
33. **Cheng, C. M., Boyle, W. C., and Goepfert, J. M.**, Rapid quantitative method for *Salmonella* detection in polluted waters, *Appl. Microbiol.*, 21, 662, 1971.
34. **Levin, M. A., Fischer, J. R., and Cabelli, V. J.**, Quantitative large-volume sampling technique, *Appl. Microbiol.*, 28, 515, 1974.
35. **Spino, D. F.**, Elevated-temperature technique for the isolation of *Salmonella* from streams, *Appl. Microbiol.*, 14, 591, 1966.
36. **Hornick, R. B.**, Shigellosis, in *Infectious Diseases: A Modern Treatise of Infectious Process*, 2nd ed., Hoeprich, P. D., Ed., Harper & Row, Hagerstown, Md., 1977.
37. **Rosenberg, M. L., Hazlet, K. K., Schaefer, J., Wells, J. G., and Pruneda, R. C.**, Shigellosis from swimming, *JAMA*, 236, 1849, 1976.
38. **Dunn, C. and Martin, W. T.**, Comparison of media for isolation of Salmonellae and Shigellae from fecal specimens, *Appl. Microbiol.*, 22, 17, 1971.
39. **Taylor, W. I. and Schelhart, D.**, Isolation of Shigellae. VIII. Comparison of xylose lysine deoxycholate agar, hektoen enteric agar, *Salmonella-Shigella* agar and eosin methylene blue agar with stool specimens, *Appl. Microbiol.*, 21, 32, 1971.
40. **Taylor, W. I. and Schelhart, D.**, Effect of temperature on transport and plating media for enteric pathogens, *J. Clin. Microbiol.*, 2, 281, 1975.
41. **Morris, G. K., Koehler, J. A., Gangarosa, E. J., and Sharrar, R. G.**, Comparison of media for direct isolation and transport of Shigellae from fecal specimens, *Appl. Microbiol.*, 19, 434, 1970.
42. **Park, C. E., Rayman, M. K., Szabo, R., and Stankiewicz, Z. K.**, Selective enrichment of *Shigella* in the presence of *Escherichia coli* by use of 4-chloro-2-cyclopentylphenyl β-D-galactopyranoside, *Can. J. Microbiol.*, 22, 654, 1976.
43. **Park, C. E., Rayman, M. K., and Stankiewicz, Z. K.**, Improved procedure for selective enrichment of *Shigella* in the presence of *Escherichia coli* by use of 4-chloro-2-cyclopentylphenyl β-D-galactopyranoside, *Can. J. Microbiol.*, 23, 563, 1977.
44. **Schiemann, D. A.**, A synoptic review on *Yersinia enterocolitica*, *J. Environ. Health*, 44, 183, 1982.
45. **Kay, B. A., Wachsmuth, K., Gemski, P., Feeley, J. C., Quan, T. J., and Brenner, D. J.**, Virulence and phenotypic characterization of *Yersinia enterocolitica* isolated from humans in the United States, *J. Clin. Microbiol.*, 17, 128, 1983.
46. **Highsmith, A. K., Feeley, J. C., and Morris, G. K.**, *Yersinia enterocolitica*: a review of the bacterium and recommended laboratory methodology, *Health Lab. Sci.*, 14, 253, 1977.
47. **Bartley, T. D., Quan, T. J., Collins, M. T., and Morrison, S. M.**, Membrane filter technique for the isolation of *Yersinia enterocolitica*, *Appl. Environ. Microbiol.*, 43, 829, 1982.
48. **Sonnenwirth, A. C.**, *Yersinia*, in *Manual of Clinical Microbiology*, 2nd ed., Lennette, E. H., Spaulding, E. H., and Truant, J. P., Eds., American Society of Microbiology, Washington, D.C., 1974.
49. **Weagant, S. D. and Kaysner, C. A.**, Modified enrichment broth for isolation of *Yersinia enterocolitica* from nonfood sources, *Appl. Environ. Microbiol.*, 45, 468, 1983.
50. **Schiemann, D. A.**, Synthesis of a selective agar medium for *Yersinia enterocolitica*, *Can. J. Microbiol.*, 25, 1298, 1979.
51. **Head, C. B., Whitty, D. A., and Ratnam, S.**, Comparative study of selective media for recovery of *Yersinia enterocolitica*, *J. Clin. Microbiol.*, 16, 615, 1982.
52. **Swaminathan, B., Harmon, M. C., and Mehlman, I. J.**, A review: *Yersinia enterocolitica*, *J. Appl. Bacteriol.*, 52, 151, 1982.
53. **Skirrow, M. B.**, Campylobacter enteritis — the first five years, *J. Hyg. Camb.*, 89, 175, 1982.

54. **Norkrans, G. and Svedhem, A.,** Epidemiological aspects of *Campylobacter jejuni* enteritis, *J. Hyg. Camb.,* 89, 163, 1982.
55. **Robinson, D. A.,** Infective dose of *Campylobacter jejuni* in milk, *Br. Med. J.,* 282, 1584, 1981.
56. **Knill, M., Suckling, W. G., and Pearson, A. D.,** Environmental isolation of heat-tolerant *Campylobacter* in the Southampton area, *Lancet,* ii, 1002, 1978.
57. **Palmer, S. R., Gully, P. R., White, J. M., Pearson, A. D., Suckling, W. G., Jones, D. M., Rawes, J. C. L., and Penner, J. L.,** Water-borne outbreak of campylobacter gastroenteritis, *Lancet,* i, 287, 1983.
58. **Taylor, D. N., Brown, M., and McDermott, K. T.,** Waterborne transmission of campylobacter enteritis, *Microb. Ecol.,* 8, 347, 1982.
59. **Bolton, F. J. and Robertson, L.,** A selective medium for isolating *Campylobacter jejuni/coli, J. Clin. Pathol.,* 35, 462, 1982.
60. **Ribeiro, C. D., Gray, S. J., and Price, T. H.,** A new medium for isolating *Campylobacter jejuni/coli, J. Clin. Pathol.,* 35, 1036, 1982.
61. **Bolton, F. J., Hinchcliffe, P. M., Coates, D., and Robertson, L.,** A most probable number method for estimating small numbers of campylobacters in water, *J. Hyg. Camb.,* 89, 185, 1982.
62. **Ribeiro, C. D. and Price, T. H.** The use of Preston enrichment broth for the isolation of "thermophilic" campylobacters from water, *J. Hyg. Camb.,* 92, 45, 1984.
63. **Bolton, F. J., Coates, D. Hinchcliffe, P. M., and Robertson, L.,** Comparison of selective media for isolation of *Campylobacter jejuni/coli, J. Clin. Pathol.,* 36, 78, 1983.
64. **Skirrow, M. B. and Benjamin, J.,** Differentiation of enteropathogenic *Campylobacter, J. Clin. Pathol.,* 33, 1122, 1980.
65. **Kaplan, R. L.,** *Campylobacter,* in *Manual of Clinical Microbiology,* 3rd ed., Lenette, E., Ed., American Society of Microbiology, Washington, D.C., 1980.
66. **Mathewson, J. J., Keswick, B. H., and DuPont, H. L.,** Evaluation of filters for recovery of *Campylobacter jejuni* from water, *Appl. Environ. Microbiol.,* 46, 985, 1983.
67. **Oosterom, J., Vereijken, M. J. G. M., and Engels, G. B.,** *Campylobacter* isolation, *Vet. Q.,* 3, 104, 1981.
68. **Drake, C. H.,** Evaluation of culture media for the isolation and enumeration of *Pseudomonas aeruginosa, Health Lab. Sci.,* 3, 10, 1966.
69. **Sutter, U. L., Hurst, V., and Lane, C. W.,** Quantification of *Pseudomonas aeruginosa* in feces of healthy human adults, *Health Lab. Sci.,* 4, 245, 1967.
70. **Hoadley, A. W.,** On the significance of *Pseudomonas aeruginosa* in surface waters, *N. Engl. Water Works Assoc.,* 82, 99, 1968.
71. **Levin, M. A. and Cabelli, V. J.,** Membrane filter technique for enumeration of *Pseudomonas aeruginosa, Appl. Microbiol.,* 24, 864, 1972.
72. **Foster, D. H., Hanes, N. B., and Lord, S. M.,** A critical examination of bathing water quality standards, *J. Water Pollut. Control Fed.,* 43, 2229, 1971.
73. **Calderon, R. and Mood, E. W.,** An epidemiological assessment of water quality and "swimmers ear", *Arch. Environ. Health,* 37, 300, 1982.
74. **Seyfried, P. L. and Frazer, D. J.,** *Pseudomonas aeruginosa* in swimming pools related to the incidence of otitis externa infections, *Health Lab. Sci.,* 15, 50, 1978.
75. **Schubert, R. H. W.,** Das Vorkommen der Aeromonadan in Oberirdischen Gewässern, *Arch. Hyg.,* 150, 688, 1967.
76. **Shotts, E. G. and Rimler, R.,** Medium for the isolation of *Aeromonas hydrophila, Appl. Microbiol.,* 26, 550, 1973.
77. **Fliermans, C. B., Gorden, R. W., Hazen, T. C., and Esch, G. W.,** Aeromonas distribution and survival in a thermally altered lake, *Appl. Environ. Microbiol.,* 33, 114, 1977.
78. **Rippey, S. R. and Cabelli, V. J.,** Membrane filter procedure for enumeration of *aeromonas hydrophila* in fresh waters, *Appl. Environ. Microbiol.,* 38, 108, 1979.
79. **Beerwerth, W.,** Culture of mycobacteria from droppings of domestic animals and the role of these bacteria in epizootiology, *Prax. Pneumol.,* 25, 189, 1967.
80. **Bailey, R. K., Wyles, S., Dingley, M., Hesse, F., and Kent, G. W.,** The isolation of high catalase *Mycobacterium kansasii* from tap water, *Am. Rev. Respir. Dis.,* 101, 430, 1970.
81. **McSwiggan, D. A. and Collins, C. H.,** The isolation of *M. kansasii* and *M. xenopi* from water systems, *Tubercle,* 55, 291, 1974.
82. **Goslee, S. and Wolinsky, E.,** Water as a source of potentially pathogenic mycobacteria, *Am. Rev. Respir. Dis.,* 113, 287, 1976.
83. **DuMoulin, G. C. and Stottmeier, K. D.,** Use of cetylpridinium chloride in the decontamination of water for culture of mycobacteria, *Appl. Environ. Microbiol.,* 36, 771, 1978.
84. **Petran, E. I. and Vera, H. D.,** Media for selective isolation of mycobacteria, *Health Lab. Sci.,* 8, 225, 1971.

85. **Lorian, V. and Maddock, S.,** The effect of anticontamination agents in media for the isolation of mycobacteria, *Dis. Chest,* 50, 630, 1966.
86. **Mitchison, D. A., Allen, B. W., Carrol, L., Dickinson, J. M., and Aber, V. R.,** A selective oleic acid albumin agar medium for tubercle bacilli, *J. Med. Microbiol.,* 5, 165, 1972.
87. **McClatchy, J. K., Waggoner, R. F., Kane, W., Cernich, M. S., and Bolton, T. L.,** Isolation of mycobacteria from clinical specimens by use of selective 7H11 medium, *Am. J. Clin. Pathol.,* 65, 412, 1976.
88. **Kobayashi, T., Enomoto, S., Sakazaki, R., and Kuwahara, S.,** A new selective medium for pathogenic vibrios TCBS agar (Modified Nakanishi's Agar), *Jpn. J. Bacteriol.,* 18, 387, 1975.
89. **Akiyama, S., Takizawa, K., and O'Hara, Y.,** Application of teepol to isolation media for *Vibrio parahaemolyticus, Jpn. J. Microbiol.,* 18, 255, 1963.
90. **Horie, S., Saheki, K., and Okuzumi, M.,** Quantitative Enumeration of *Vibrio parahaemolyticus* in sea and estuarine waters, *Bull. Jpn. Soc. Sci. Fish.,* 33, 126, 1967.
91. **Horie, S., Saheki, K., Kozima, T., Nara, M., and Sekine, Y.,** Distribution of *Vibrio parahaemolyticus* in plankton and fish in the open sea, *Bull. Jpn. Soc. Sci. Fish.,* 30, 786, 1964.
92. **Baross, J. and Liston, J.,** Occurrence of *Vibrio parahaemolyticus* and related hemolytic vibrios in marine environments of Washington state, *Appl. Microbiol.,* 20, 179, 1970.
93. **Twedt, R. M. and Novelli, R. M. E.,** Modified selective and differential isolation medium for *Vibrio parahaemolyticus, Appl. Microbiol.,* 22, 593, 1971.
94. **Vanderzant, C. and Nickelson, R.,** Procedure for isolation and enumeration of *Vibrio parahaemolyticus, Appl. Microbiol.,* 23, 26, 1972.
95. **Watkins, W. D., Thomas, C. D., and Cabelli, V. J.,** Membrane filter procedure for enumeration of *Vibrio parahaemolyticus, Appl. Environ. Microbiol.,* 32, 679, 1976.
96. **McDade, J. E., Shepard, C. C., Frazer, D. W., Tsai, T. F., Redus, M. A., Dowdle, W. R., and the Laboratory Investigation Team,** Legionnaire's disease. Isolation of a bacterium and demonstration of its role in other respiratory diseases, *N. Engl. J. Med.,* 297, 1197, 1977.
97. **Feeley, J. C., Gibson, R. J., Gorman, G. W., Langford, N. C., Rasheed, J. K., Mackel, D. C., and Baine, W. B.,** Charcoal yeast extract agar: a primary isolation medium for the Legionnaire's disease bacterium, *J. Clin. Microbiol.,* 16, 437, 1979.
98. **Bopp, C. A., Sumner, J. W., Morris, G. K., and Wells, J. G.,** Isolation of *Legionella* spp. from environmental water samples by low-pH treatment and use of a selective medium, *J. Clin. Microbiol.,* 13, 714, 1981.
99. **Wadowsky, R. and Yee, R. B.,** Glycine-containing selective medium for isolation of Legionellaceae from environmental specimens, *Appl. Environ. Microbiol.,* 42, 768, 1981.
100. **Edelstein, P. H.,** Comparative study of selective media for isolation of *Legionella pneumophila* from potable water, *J. Clin. Microbiol.,* 16, 697, 1982.
101. **Calderon, R. L. and Dufour, A. P.,** Media for detection of *Legionella* spp. in environmental water samples, in *Legionella, Proceedings of the 2nd International Symposium,* Thornsberry, C., Balows, A., Feeley, J., and Jakubowski, W., Eds., American Society of Microbiology, Washington, D.C., 1984.
102. **Gorman, G. W. and Feeley, J. C.,** *Procedures for the Recovery of Legionella from Water, Developmental Manual,* Centers for Disease Control, Atlanta, 1982.
103. **Sinnott, J. T., Yangco, G., Feldman, D. H., and Gunn, R. A.,** Chromabaceriosis — Florida, *Morbid. Mortal. Weekly Rep.,* 29, 613, 1981.
104. **Jones, K. L.,** Fresh isolates of actinomycetes in which the presence of sporogenous aerial mycelia is a fluctuating characteristic, *J. Bacteriol.,* 57, 141, 1949.
105. **Keeble, J. R. and Cross, T.,** An improved medium for the enumeration of chromobacterium in soil and water, *J. Appl. Bacteriol.,* 43, 325, 1977.
106. **Ryall, C. and Moss, M. O.,** Selective media for the enumeration of *Chromobacterium* spp. in soil and water, *J. Appl. Bacteriol.,* 38, 53, 1975.
107. **Kobarger, J. A. and May, S. O.,** Isolation of *Chromobacterium* spp. from foods, soil, and water, *Appl. Environ. Microbiol.,* 44, 1463, 1982.
108. **Leifson, E.,** Morphological and physiological characteristics of the genus *Chromobacterium, J. Bacteriol.,* 71, 393, 1956.
109. **Sneath, P. H. A.,** Cultural and biochemical characteristics of the genus *Chromobacterium, J. Gen. Microbiol.,* 15, 70, 1956.
110. **Sneath, P. H. A.,** Identification methods applied to *Chromobacterium,* in *Identification Methods for Microbiologists. Part A,* Gibbs, B. M. and Skinner, F. A., Eds., Academic Press, London, 1966.
111. **Braun, J. L., Diesch, S. L., and McColloch, W. F.,** A method of isolating leptospires from natural surface waters, *Can. J. Microbiol.,* 14, 1011, 1968.
112. **Baseman, J. B., Hennebery, R. C., and Cox, C. D.,** Isolation and growth of leptospira on artificial media, *J. Bacteriol.,* 91, 1374, 1966.

113. **Johnson, R. C. and Rogers, P.,** Differentiation of pathogenic and saprophytic leptospires with 8-azaguanine, *J. Bacteriol.,* 88, 1618, 1964.
114. **Johnson, R. C. and Harris, V. G.,** Differentiation of pathogenic and saprophytic leptospires, *J. Bacteriol.,* 94, 27, 1967.
115. **Füsi, M. and Csóka, R.,** Differentiation of pathogenic and saprophytic leptospirae by means of a copper sulfate test, *Zentralbl. Bakteriol., Parasitenkd., Infektionskr. Hyg., Abt. 1, Orig.,* 179, 231, 1960.
116. **Sack, R. B.,** Enterotoxigenic *Escherichia coli:* identification and characterization, *J. Infect. Dis.,* 142, 279, 1980.
117. **Morris, G. K., Merson, M. H., Sack, D. A., Wells, J. G., Martin, W. T., Dewitt, W. E., Feeley, J. C., Sack, R. B., and Bessudo, D. M.,** Laboratory investigation of diarrhea in travelers to Mexico: evaluation of methods for detecting enterotoxigenic *Escherichia coli, J. Clin. Microbiol.,* 3, 486, 1976.
118. **Moseley, S. L., Hug, I., Alim, A. R. M. A., So, M., Samadpour-Motalebi, M., and Falkow, S.,** Detection of enterotoxigenic *Escherichia coli* by DNA colony hybridization, *J. Infect. Dis.,* 142, 892, 1980.
119. **Calderon, R. L. and Levin, M. A.,** Quantitative method for enumeration of enterotoxigenic *Escherichia coli, J. Clin. Microbiol.,* 13, 130, 1981.
120. **Craun, G. F.,** Waterborne disease — a status report emphasizing outbreaks in ground-water systems, *Ground Water,* 17, 183, 1979.
121. **Shaw, P. K., Brodsky, R. E., Lyman, D. O., Wood, B. T., Hibler, C. P., Healy, G. R., MacLeod, K. I. E., Stahl, W., and Schultz, M. G.,** A community wide outbreak of giardiasis with evidence of transmission by a municipal water supply, *Ann. Intern. Med.,* 87, 426, 1977.
122. **Jakubowski, W. and Ericksen, T. H.,** Methods for detection of *Giardia* cysts in water supplies, in *Waterborne Transmission of Giardiasis,* Jakubowski, W. and Hoff, J. C., Eds., U.S. Environmental Protection Agency, 600/9-79-001, Cincinnati, Ohio, 1979, 193.
123. **Holman, B., Frost, F., Plan, B., Fukutaki, K., and Jakubowski, W.,** Recovery of *Giardia* cysts from water: centrifugation vs. filtration, *Water Res.,* 17, 1705, 1983.
124. **Jakubowski, W.,** Detection of *Giardia* cysts in drinking water: state of the art, in *Giardia and Giardiasis: Biology, Pathogenesis, and Epidemiology,* Erlandsen, S. L. and Meyer, E. A., Eds., Plenum Press, New York, 1984.
125. **Chang, S. L. and Kabler, P. W.,** Detection of cysts of *Entamoeba histolytica* in tap water by use of membrane filter, *Am. J. Hyg.,* 64, 170, 1956.
126. **Mackie, T. T., Hunter, G. W., and Worth, C. B.,** *A Manual of Tropical Medicine,* W. B. Saunders, Philadelphia, 1954.
127. **McCabe, L. J. and Craun, G. F.,** Status of waterborne diseases in the U.S. and Canada, committee report, *J. Am. Water Works Assoc.,* 67, 95, 1975.

Chapter 9

ENGINEERING ASPECTS OF WATERBORNE OUTBREAK INVESTIGATION

Edwin C. Lippy

TABLE OF CONTENTS

I. Introduction 216

II. Evaluation of Disinfection (Priority No. 1) 218
    A. Concentration and Contact Time 218
    B. pH and Temperature 220
    C. Interferences 221

III. Evaluation of the Water System (Priority 2 if cases are not occurring) 223
    A. Source of Supply 224
    B. Treatment 225
    C. Distribution Network 225

IV. Sampling for the Agent (Priority 3 if cases are not occurring) 227

V. Communication 229

References 230

I. INTRODUCTION

In most cases, waterborne disease outbreak investigations are difficult to conduct. An investigation is made difficult because normally the outbreak has subsided by the time an investigation is started. If a water system is challenged by a pathogen and cannot respond to the challenge, an exposed susceptible person will generally become ill in 24 to 72 hr for most bacterial/viral agents, in 1 to 3 weeks for parasitic agents, or in 2 to 6 weeks for hepatitis A virus. The ill person may suffer with his illness several days before seeking medical treatment, at which time a doctor or nurse who has seen a number of patients with similar symptoms may recognize that an outbreak is in progress. They may report the unusual number of illnesses to government authorities who, before beginning an investigation, may require several days to determine whether the cases constitute an outbreak or are sporadic and whether the preliminary data indicate a possible common source. Thus, depending on the causative agent and respective incubation period, 1 or more weeks may elapse between the exposure and the beginning of an investigation. These delays contribute to failure in identifying the causative agent, especially from water samples, since the contaminant may be purged from the distribution system through normal usage of water, in reconstructing events that permitted contamination to occur, and in specifically identifying the event or defect that caused contamination.

Besides being hindered by delays in initiating an investigation, engineering studies are further hampered because clues and information developed by the epidemiologist through interviews with cases and controls do not become available in usable form for 3 to 5 days. Complaints of odorous or strange-tasting water, especially if time and place are identified, are clues which are helpful in the engineering investigation. Attack rates also can be very useful if they are stratified by a common zone of the distribution network. High attack rates in a specific zone may lead the engineer to suspect localized contamination introduced through a cross-connection, main break, open reservoir, or a well that serves as a source of supply for one area of the community.

Because of the delays inherent in beginning outbreak investigations and the time required to develop reliable epidemiological evidence to confirm that illness is transmitted via water, the engineer must, after a preliminary assessment of the situation, assume that drinking water is the cause and must act quickly. To err is human but failure to act in preventing further illness is rarely forgiven. Therefore, the engineering investigation should consider as the first order of priority an evaluation of the treatment currently provided for water to assure that it is capable of responding to challenge by the suspected agent. The most critical water treatment process in combating infectious agents is disinfection, and a thorough evaluation of disinfection practices should be undertaken immediately. In most situations where ground water is used as a source of supply, disinfection will be the only treatment. Where surface water is used, complementary processes, including mixing, coagulation-flocculation, sedimentation, and filtration, normally will be provided along with disinfection. The complementary processes are important for preparing water for disinfection, but disinfection is the primary treatment for inactivation of bacteria, viruses, and parasites.

After disinfection has been evaluated and adjusted to neutralize the suspected microbiological challenge, the engineer must choose his second priority. The choice is based upon the continued occurrence of illness. If new cases are seeking medical attention, and this reporting is apparent to physicians, health clinics, and hospitals, it may mean the agent is still present in the water system. The second priority should then be the collection of water samples to capture and identify the agent. If the same agent is identified from both water samples and clinical specimens from ill cases, the required water system improvements are implemented quickly and with few objections.

If there are no recent reports of illness, the engineer's second priority should be a thorough evaluation of the entire water system to determine weaknesses and strengths in facilities and operations (collection of water samples becomes the third priority). This information is not only useful to the engineer, but also is important to the epidemiologist in designing population surveys of households served by the water system. Knowledge of the treatment and distribution of water within a community and its environs is extremely important to the epidemiologist in selecting the survey sample. For example, one segment of the community may receive treated water while another may receive untreated water, and the survey should be designed accordingly. Also, it is to the engineer's advantage to receive as soon as possible survey data from the epidemiologist on the location of the cases. For example, if 85% of the illness occurred in the northeast quadrant of the distribution network, the engineering evaluation should focus on determining what is unusual about the treatment and delivery of water in that quadrant. The date cases occurred is important to establish the timing of contamination or an event that caused contamination. The epidemiologist, through a survey and/or questionnaire, determines the date of onset of illness, symptoms of illness, incubation period, diet histories, probable etiologic agent, and the time, person, and place associations to identify the vehicle of transmission. The collection and analysis of these data provides important clues for the engineer, particularly the graphing of cases of illness according to their date of onset (the epidemiological curve or epi-curve). From the epi-curve the engineer can derive an appropriate time of exposure by subtracting the incubation period for the illness from the median date of onset of all ill persons. This provides a good estimate of the critical time period for contamination of the system, and information should be obtained on how the operation, maintenance, and control of the water system relate to that time. It should be obvious that the engineer and epidemiologist are important to each other in sharing information obtained early in the investigation of the outbreak.

The fourth priority of the investigation is related to communication in that information collected and analyzed through the engineer's participation must be disseminated. He must communicate not only with the epidemiologist, but also with the water utility, local governing bodies, and the public both during the investigation and afterwards. During the investigation, communication to the utility and local government will primarily be verbal; afterwards it will be a detailed written report intended to present findings of the evaluation and recommendations for improvement of the water system. Both forms of communication are important and the principles of disseminating oral and written information should be followed. Communication with the public will be accomplished primarily through newspapers, television, and radio. In some instances it is more efficient to communicate findings to a central authority who has previously been designated as an official spokesperson.

The discussion which follows focuses on waterborne disease outbreaks caused by microbiological agents. Engineering aspects of outbreaks caused by chemical agents were not specifically addressed because (1) chemical poisonings constitute a small percentage of the total number of outbreaks, (2) normally, chemical agents are easily recognized because they impart color, taste, or odor which consumers detect and report, and (3) in most cases, these outbreaks are brought under control rapidly through identification of the contaminant and isolation of the source of contamination. The acute health effects of exposure to chemical agents also contribute to identification and early control of outbreaks. The incubation period for acute effects is brief for ingested chemicals. They normally affect the upper GI tract, causing vomiting and symptoms not associated with microbiological agents, such as tingling of the skin, irritation of eyes and other membranes, vertigo, etc. Current technology for analysis of water for organic and inorganic constituents through gas chromatography and atomic absorption spectrophotometry permits rapid identification of contaminants. Once the contaminant is identified, the source of contamination can generally be located quickly through association with use or manufacture of the chemical. Control of the source of contamination does not usually involve adjustments in water treatment but is directed towards isolation, containment, or cessation of discharge. If the contaminant cannot be controlled at the source, alternatives to adjustment in water treatment are normally chosen, and these alternatives include use of other sources of supply. Cross-connections, backsiphonage, corrosion of plumbing, and chemical-feed problems also have caused acute illness due to high concentrations of chemicals in the water system and should be considered during the investigation of an outbreak caused by chemical contaminants.

II. EVALUATION OF DISINFECTION (Priority No. 1)

The most important aspect of treatment is its capability to withstand challenge by a contaminant. Regardless of whether or not the epidemiologic study has confirmed that illness is related to drinking water, the engineer is responsible during the investigation to ensure that treatment is optimized, with the primary objective of preventing the transmission of disease to preserve and protect public health.

The unit process having the greatest impact on reducing microbial challenge and preventing waterborne disease transmission is disinfection. In the U.S. the most commonly used disinfectant is chlorine. The dosage of chlorine is normally easy to adjust and in some circumstances chlorine contact time can be increased by simply relocating the injection point. No other unit process can reduce microbial concentrations on the order of four or five orders of magnitude by simply adjusting the chemical dosage. Therefore, evaluating chlorination is the first order of business for the engineer.

The factors which influence the effectiveness of chlorination to inactivate bacteria, viruses, and parasites are (1) concentration, (2) contact time, (3) pH, (4) temperature, and (5) interfering substances. Each of these factors must be evaluated to determine disinfection capability under current operating conditions and the adjustments that are necessary to cope with the challenge of a pathogen.

A. Concentration and Contact Time

The product of chlorine concentration (C) in milligrams per liter and contact time (T) in minutes produces a CT relationship which is used in determining the requirements for inactivation of bacteria, virus, and protozoa. Since CT relationships are developed through laboratory research under carefully controlled conditions, applying the data to a real world situation requires engineering judgment. The CT values listed in Table 1 for virus were developed by White[1] and also can be applied for inactivation of bacterial agents. The viral CT values are based upon inactivation of coxsackie A2 virus, which are more resistant than polio virus or any pathogenic vegetative bacteria implicated as causative agents in waterborne disease outbreaks. Values for protozoan CT were interpreted from the research conducted by White,[1] Jarroll et al.,[2] and Rice et al.[3] It is important to note that from the CT relationships at pH 7 to 8, which are values common to most water systems, a rule-of-thumb can be applied. Systems using ground water as a source of supply should apply a minimum CT of 15 to 30 to provide for viral and bacterial inactivation, which are generally the contaminants of concern; i.e., the size of viral particles permits easier penetration of strata overlying aquifers while protozoan cysts are trapped and retained. (Protozoan cysts are about 1000 times larger than virus particles.) Systems using surface water, especially those located in areas where source waters attain low temperatures of 5°C, should apply a minimum CT of 100 to 150 for protozoan inactivation (at temperatures less than 5°C, the CT requirements will increase). Figure 1 graphically depicts the CT requirements included in Table 1.

With the CT value known, the actual contact time is determined through evaluation of the treatment facility, and the required chlorine concentration is determined mathematically. Determination of the actual contact time should be done carefully, for there are several factors that should be considered. A common mistake is using hydraulic displacement

Table 1
CHLORINE CONCENTRATION-CONTACT TIME (CT) RELATIONSHIPS FOR 99% INACTIVATION OF PATHOGENIC AGENTS

	Viral CT		Protozoan CT		
Water pH	0—5°C	10°C	5°C	15°C	25°C
6			80	25	15
7			100	35	15
7.0—7.5	12	8			
7.5—8.0	20	15			
8			150	50	15
8.0—8.5	30	20			
8.5—9.0	35	22			

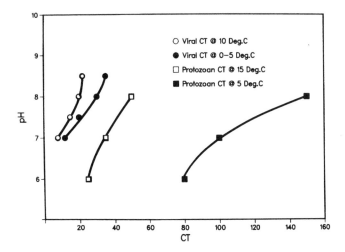

FIGURE 1. Chlorine concentration-contact time (CT) relationships for 99% inactivation of pathogenic agents.

calculations to compute contact time. For example, a 1 MG rectangular storage tank with inflow and outflow at opposite ends provides 2.4 hr of contact time when calculations consider the average daily plant output of 10 MGD, but the contact time actually provided may approach one fourth of this calculated value.[4] With an actual contact time of 36 min, the chlorine concentration based on 2.4 hr contact time would be insufficient. This difference in the calculated displacement or theoretical retention (contact) time and the actual retention (contact) time stems from short-circuiting and dead spaces that are common to both rectangular and circular storage tanks. The use of the average daily plant output also can lead to errors. Plant output varies with customer demands which change during the day; peaks occur in the morning and evening. Peak daily demands for some systems can be double the average daily plant output and if this is the case the contact time is underestimated by an additional 50%. If the chlorinator is not paced to respond to changes in plant output, fluctuation in demand must be taken into consideration when evaluating contact time and chlorine dosage requirements.

The preferred method of obtaining contact time is to conduct a tracer test. A simple and quick method is to increase the chlorine feed and determine chlorine residual concentrations of the basin effluent. The initial increase in concentration signals the flow-through time for a fraction of the plant output, and subsequent samples will reflect a steady rise in chlorine residual until the incremental dosage has been accounted for. The mixed flow and dead space components of basin hydraulics also can be estimated through application of the tracer methods discussed by Hudson.[5] Another simple and quick method for tracer tests is to cease fluoride feed and read the decrease in fluoride concentration over time. The initial decrease in fluoride concentration signals the flow-through time for a fraction of plant output and represents the contact time provided for a portion of the basin volume. The advantage of using chlorine and fluoride as tracers is that there is no need to introduce a dye, salt, or other chemical that may require special permission to introduce into a water system and may cause concern over its toxicity. Decreasing the concentration of fluoride or increasing the concentration of chlorine for a short time should not cause objections to conducting the test. The elapsed time between the increase or decrease of chemical dosage and the initial corresponding change in the basin effluent represents the contact time, and should be used in the CT relationship to calculate the concentration of chlorine which will be required to achieve a microbiocidal residual. CT relationships are based upon maintaining the chlorine concentration for the entire contact time so the dosage must be adjusted accordingly to compensate for chlorine demand and dissipation.

Tracer tests are normally not required where transmission or distribution mains are depended upon for contact time. The mains are pressurized, and under full-flow conditions the hydraulic characteristics approximate plug-flow. With plug-flow the theoretical and actual contact times are equivalent. However, they can be checked easily by increasing the chlorine dosage and measuring the increase in residual at an appropriate downstream location. Table 2 provides a ready reference for computing contact time in transmission and distribution mains. By dividing the plant output or flow into the pipe capacity adjusted for length available, the contact time can be determined. For example, at a plant output of 1 MGD and 1000 ft of 24-in. pipe available, the contact time would be 34 min.

B. pH and Temperature

The effectiveness of chlorine as a disinfectant is influenced by the pH and temperature of the water being dosed. The pH determines the amount of hypochlorous acid (HOCl) and hypochlorite ion (OCl^-) in solution. HOCl is the predominant form at lower pH values (pH 6 to 7.5) and OCl^- is the predominant form at higher pH values (pH 8 to 10). At pH 7 and 0°C, the dissociation of chlorine produces 83.3% HOCl and 16.7% OCl^-; at pH 9, 4.5% HOCl and 95.5% OCl^- are formed. It is important to know the dissociated forms of chlorine present at various pH values because HOCl is the form which possesses the destructive power as a microbiocide. According to White,[1] "the germicidal efficiency of HOCl is due to the relative ease with which it can penetrate cell walls. This penetration is comparable to that of water, and can be attributed to both its modest size (low molecular weight) and its electrical neutrality (absence of an electrical charge)." White[1] states that the hypochlorite ion "is a relatively poor disinfectant because of its inability to diffuse through the cell wall of microorganisms due to the negative electrical charge." Research[1] has shown that HOCl is 80 times more effective than OCl^- ion in the inactivation of *Escherichia coli* and about 150 times more effective for cysts of *Entamoeba histolytica*.

The primary impact on evaluation of chlorination is to note that the CT relationship increases with pH. Contact time, or the T portion of the CT relationship, is controlled by structural facilities (i.e., a contact basin, storage tank, or transmission main) which cannot be readily adjusted in size to compensate for a needed increase in CT. Therefore, the only immediate alternative is to increase chlorine concentration, or if pH is such that OCl^- is the predominant form, it may be necessary to adjust pH to a lower desirable level so that

Table 2
FLUID CAPACITY OF PIPES

Pipe diameter (d) (in.)	Capacity[a] (gal/100 ft)
3	37
4	65
6	147
8	261
10	408
12	588
14	800
16	1044
18	1322
20	1632
24	2350
30	3672
36	5288
42	7197
48	9400

[a] $4.08\ d^2$.

HOCl is the predominate form. Adjustment of pH should be considered when the chlorine concentration requirement becomes excessive and possibly objectionable.

Chlorine hydrolyzes very rapidly in water; hydrolysis occurs in tenths of a second at 65°F and a few seconds at 32°F. Although temperature has very little effect on the ability of chlorine to go into solution, the reaction of chlorine with microorganisms is affected. CT relationships are inversely related with water temperature, and as the temperature decreases the product of CT increases (Table 1). Therefore, water temperature is important and must be considered when selecting a bactericidal, viricidal, or cysticidal chlorine dosage.

C. Interferences

When chlorine is introduced into water, it not only reacts to form microbiocides, but also reacts with other substances to form organic and inorganic compounds that have little or no disinfection capability. The reaction with organic and inorganic substances constitutes chemical interference with disinfection through consuming a part of, or all of, the applied chlorine leaving little or no chlorine to react as a disinfectant. Organic and inorganic forms of nitrogen are among the compounds that exert a significant interfering effect with chlorination of surface water. Hydrogen sulfide, iron, and manganese are commonly found in ground waters and interfere with disinfection by combining with chlorine to form sulfates, ferric hydroxide, and manganese dioxide, respectively. The significance of this interference is exemplified by the requirement of 8.32 mg/ℓ of chlorine to oxidize 1 mg/ℓ of hydrogen sulfide.[1] In fact, the ability of chlorine to oxidize hydrogen sulfide, iron, and manganese is used to advantage as a treatment technique and is employed at some water plants for removal of these substances.

Microorganisms entrapped within the matrix of particulate matter are shielded from the effects of disinfection and this constitutes physical interference. Research[22-27] indicates that the protective effect of shielding by inorganic particulates is minimal and of little consequence as physical interfering substances; however, organic particulates such as those associated with sewage effluent solids, do afford protection and permit survival of microorganisms after long chlorine contact periods (60 min). A recent study of LeChevallier et al.[28] on the survival of coliform organisms in the presence of turbidity after 1 hr of contact time indicated there was 99.5% reduction at a concentration of 6 nephelometric turbidity units (NTUs) or less and approximately 20% reduction at a concentration of 8 to 13 NTUs. It was concluded that turbidity control to at least the 1 and 5 NTU concentrations as required by the EPA regulations is justified. While more research is needed to specifically define the degree of physical interference by turbidity and particulates, common sense dictates that where sewage discharges affect raw water quality, turbidity should be reduced by appropriate treatment in order for chlorination to be effective. Appropriate treatment to reduce turbidity normally requires structural measures (mixing, coagulation-flocculation, settling, filtration) which, if not already in place, require considerable time for planning, design, and construction. In an outbreak situation, emergency short-term measures must be implemented and if chlorination is the only treatment provided, it will have to be adjusted to compensate for the physical interference of turbidity.

An example is provided to illustrate the procedure that should be used in optimizing chlorination to cope with a pathogen challenge. An outbreak of illness caused by a protozoan parasite occurs in a community using an impoundment on a mountain stream as a source of supply. There are no sources of human contamination in the watershed. Chlorination is provided at the impoundment with contact time occurring in 5000 ft of 24-in. transmission main. The average daily demand is 2 MGD, which is increased by 50% during peak usage periods. Temperature of the water is 5°C, pH is 7, and turbidity is 1 NTU. Research data indicate that the CT is 100. The chlorine dosage requirement is calculated for peak output, and it has been assumed that the chlorinator feeds in proportion to flow.

$$\text{Peak output:} \quad 2 \text{ MGD} \times 150\% = 3 \text{ MGD}$$

$$\text{Contact time:} \quad \frac{2350 \text{ gal}^a}{100 \text{ ft}} \times \frac{5000 \text{ ft}}{3,000,000 \text{ gal/day}} = 0.039 \text{ days}$$

$$0.039 \text{ days} \times 1440 \text{ min/day} = 56 \text{ min}$$

Chlorine demand: Test with chlorine comparator kit yields a demand of 0.5 mg/ℓ at 56 min

$$\text{Chlorine dosage requirement:} \quad \text{Dosage} = \frac{CT}{T} + \text{Demand}$$

$$= \frac{100}{56} + 0.5$$

$$= 1.8 + 0.5$$

$$= 2.3 \text{ mg/ℓ}$$

[a] From Table 2.

The calculations account for all factors affecting chlorination except for extrapolating the CT relationship from laboratory research data to applied use. An increased dosage of 0.5 to 1.0 mg/ℓ would not be unreasonable, and for this example a dosage of 3 mg/ℓ might be chosen to account for unknown variables in extrapolation. Increasing the dosage to 3 mg/ℓ has the effect of adjusting the CT relationship by 40% (i.e., $1.8 + 0.7 \times 56 = 140$). The free chlorine residual that is produced, 2.5 mg/ℓ after 56 min of contact time, should be confirmed by monitoring.

The above example assumes a flow-proportioning chlorinator which automatically adjusts dosage in response to changes in plant output. Where a constant-feed chlorinator is used, several adjustments in dosage are required throughout the course of the day. For example, between the hours of 10 p.m. and 6 a.m., customer demand decreases and plant output responds accordingly. If demand decreases to 50% of the average, the chlorine residual is going to increase to objectionable concentrations unless the dosage is adjusted. With a 50% decrease in average output, or 1 MGD, and the chlorinator adjusted to maintain the 3 mg/ℓ dosage calculated for 3 MGD, the chlorine residual increases dramatically during low-demand hours to 8.5 mg/ℓ. In this example the resulting chlorine residual would likely cause widespread consumer complaints of tastes and odors.

Equally as important as providing the correct dosage is assuring that reliable chlorination is in operation 24 hr per day. Reliability demands dual-cylinder feed, auxiliary power, stand-by equipment and spare parts, and maintenance of records documenting operation. Dual-cylinder feed is necessary to ensure continuous dosage so if one cylinder becomes depleted there is an automatic switching to the fully charged one. A dual-cylinder hookup also is needed to ensure continuous feed during cylinder replacement. Changing an empty cylinder for a full cylinder requires about 30 min, and if a dual hookup is not provided, unchlorinated water will enter the distribution network. For a plant output of 1 MGD, 21,000 gal of unchlorinated water would be distributed during these 30 min. Auxiliary power is necessary to sustain operation during power outages. A stand-by chlorinator and spare parts are needed for uninterrupted operation during maintenance and repairs. Records to substantiate reliability in operation including pounds of chemical used, plant output on a daily basis, and chlorine residual measurements are also required. Current technology to ensure reliability in chlorination utilizes loop-controlled feed with remote sensing and automatic adjustment of dosage to a predetermined residual. The residual can be continuously analyzed and recorded on circular or strip chart recorders and an alarm system can be installed to warn of low residuals. The continuous record substantiates chlorination, or lack of it, and is a very important factor in verification of proper disinfection.

If treatment cannot be made capable of responding to various challenges within a short period of time, emergency measures such as a boil water order must be implemented to protect against possible waterborne transmission of disease. Boil water orders are not popular because they shift responsibility for providing safe water from the water utility to the consumer, they cause inconvenience for the consumer, and are often ignored. They are a nagging problem for regulatory agencies in assuring protection of public health because the agencies are unsure of the degree of compliance with the order and have no means of enforcing it. Other short-term measures which can provide safe water during an investigation or during the time required to implement treatment changes include use of bottled water, water from tank trucks, or emergency treatment by mobile units. In instances where bottled water or tank trucks are used, provisions should be made for evaluating the sources of water and ensuring that containers and tanks are free of contamination.

III. EVALUATION OF THE WATER SYSTEM
(Priority 2 if cases are not occurring)

An engineering evaluation of the water system including source, treatment, and distribution is required to identify sources of contamination, defects in facilities, and failures in operation that may have contributed to the cause(s) of the outbreak. The statistics presented in Chapter 5 should be used as a reminder of the historical causes of outbreaks, but it is important to keep an open mind in conducting a thorough, objective evaluation, as previously undetermined or unimportant causes of outbreaks may be responsible. The engineer should become familiar with the physical layout of the system, how it is operated, and any peculiarities that may be related to time or place with onset of illness. Findings from the evaluation

Table 3
SOURCES OF CONTAMINATION

Source	Contaminant
Point sources	
Waste water treatment plant (domestic and industrial)	C,B,V,P[a]
Water treatment plant (backwash, sludge, spills)	C,B,V,P
Septic tanks	B,V,P
Feed lots (with treatment/holding facility)	B,V,P
Sewerage systems (overflows at pumping stations and manholes, damaged pipes)	C,B,V,P
Highways, railroads (spills)	C
Beaver lodges	P
Cross-connections	C,B,V,P
Nonpoint sources	
Agricultural	
Livestock grazing	B,V,P
Application of chemicals (herbicides, pesticides, weedicides)	C
Application of fertilizer (manure, sludge, chemical)	C,B,V,P
Power line right-of-way (spraying)	C
Forests (spraying)	C

[a] C = chemical, B = bacteria, V = virus, P = parasite.

should be quickly communicated to others involved in the investigation, especially the epidemiologists, so that they are aware of the time of failures in operation, equipment breakdowns, or other events that may impact on the outbreak investigation.

A number of references[1,5-13] provide useful background information for evaluation of water systems; however, there is no substitute for experience gained through conducting on-site surveys of systems, preparing a written report, and defending the report and recommendations which frequently require capital improvements or an increase in operating budget.

A. Source of Supply

Two components are required to precipitate a waterborne disease outbreak. There must be a source of contamination and a route-of-entry for the contaminant to penetrate the water system. Statistics overwhelmingly show that outbreaks are caused when the source of water supply is affected by a contaminant and the treatment which is provided is inadequate or is overwhelmed, thus permitting entry of the contaminant into the distribution network (see Figure 48, Chapter 5). The source of contamination may be active on a continuous, intermittent, or sporadic basis. It may be a point source or a nonpoint source, and it may be visible or invisible. Table 3 lists the principal sources of contamination and the contaminants of concern.

When an actual or potential source of contamination has been identified, testing may be required to show that it is influencing water quality, or it may be so obvious[14] that the impact is apparent. Intermittent or sporadic contaminants and sources of contamination which are not readily apparent or invisible may require lengthy and more complex testing to determine impact on raw water quality. Dye testing has been used successfully in a number of outbreaks to document contamination of wells and springs by septic tanks and a waste water disposal pond. Rather dramatic results were obtained in a viral gastroenteritis outbreak in Colorado[15] when dye introduced into a septic tank through a cabin toilet appeared in a spring in less than 1 hr. The spring served as a source of water supply for a 3000-acre resort camp and also as a source of illness for 418 visitors. Similar, but not quite as dramatic results, were obtained in a shigellosis outbreak in Florida affecting 1200 persons.[16] Dye introduced into a septic tank through a church toilet appeared in a well serving the community in 9 hr.

Persistence and perseverance paid off in an outbreak of gastroenteritis in Missouri where 219 persons became ill after visiting a restaurant. The restaurant was served from a well that was contaminated by a waste water disposal pond. Dye appeared in the restaurant well 529 hr (22 days) after it was introduced into the pond.[17] In situations involving ground water sources, and especially in water systems served by a number of wells, it is helpful to collect samples of raw water for coliform analysis to determine if any are influenced by indicator organisms. This procedure aids in isolating a trouble spot for further extensive testing. In situations where animals served as a source or vehicle for contaminants, trapping and necropsy have been used to trace the source of contamination.[4,18]

It is to the investigator's advantage to identify and confirm sources of contamination so that epidemiologic investigations can be strengthened, sources removed, and required treatment provided. Devoting sufficient time to this part of the water system evaluation is especially rewarding if efforts are successful. Sources of contamination can either be halted or treated, use of the raw water that is affected can be terminated, or the water can be treated to offset the challenge. Finding the source of contamination presents the investigator with more alternatives in developing solutions. Where the source is not found, doubt lingers as to whether there will be recurrence even when precautionary and corrective measures supposedly address all conceivable challenge situations.

B. Treatment

Evaluation of treatment should include an inspection of facilities and a review of operating records to determine whether the plant is functioning properly. It is normally easier to understand treatment and operation if the inspection is conducted as water flows through the plant from raw water intake to finished water outlet. A schematic diagram should be drawn showing treatment units including mixing, coagulation-flocculation, settling, filtration, chlorination, points of chemical addition, and appurtenant units including intake, clearwell, pumping, etc. Information on unit size, volume, capacity, condition, age, and maintenance should be noted. Type, frequency, and records for operational monitoring and control should be obtained for review. Data on plant output, chemical dosage, operational down-time due to maintenance, failure, power outage, or other reasons should be collected. The epidemiological information should be used to identify a critical period of time when consumers were likely exposed to the contaminant, and treatment information for that period should be detailed and specific.

Chemical dosage for coagulation, pH control, and disinfection should be plotted vs. time to compare plant response with changes in water quality and plant output.[19] The comparisons should show that increases in raw water turbidity are accompanied by increases in the use of coagulants, and an increase in plant output is accompanied by an increase in the use of chlorine. The plotted data can reveal lapses in operation that may coincide with the critical period or it may simply provide an insight on how poorly or how well the plant is operated.

Table 4 lists information, guidelines, and simple tests that can be used in plant inspections to determine operational characteristics. Analysis of operating records and results of the plant inspection should reveal weaknesses that may permit a contaminant to penetrate the treatment train and enter the distribution network.

C. Distribution Network

The distribution network is the most difficult component of the water system to evaluate since nearly all of it is underground. Facilities that are visible, including chlorine booster stations, storage reservoirs, and pumping stations, should be inspected. Since these facilities are unmanned and are visited only periodically by operators and maintenance personnel, they are vulnerable to periods of inoperation. Records should be reviewed for down-time that may coincide temporarily with the outbreak. Unmanned facilities also are vulnerable to illegal entry which may be responsible for inoperative periods or contamination.

Table 4
GUIDES FOR PLANT INSPECTION[a]

Process	Guides	Remarks
Mixing		
Rapid mechanical	Detention: 30 sec	10—20 sec not uncommon $\left(\dfrac{\text{Basin volume}}{\text{Flow}} = \text{Detention}\right)$
	Drive motor: 1—2 HP/CFS	$G = \left[\dfrac{P}{\mu V}\right]^{1/2}$
	Temporal mean velocity gradient or "G" factor: 500—1000 sec^{-1}	where P = power to H_2O in ft lb/sec, μ = absolute viscosity[b] in lb sec/ft^2 and varies with H_2O temperature, and V = volume of basin in ft^3
Hydraulic	Efficiency varies with plant output	Turbulence decreases at low plant output during off-demand hours; operator has little control over the process
Coagulation-flocculation	Detention: 4 hr	$\left(\dfrac{\text{Wetted basin volume}}{\text{Flow}} = \text{Detention}\right)$ Check detention with tracer test[2]
	Flow-thru velocity: 0.5—1.5 ft/min	$\left(\dfrac{\text{Flow}}{\text{Wetted cross-section}} = \text{Velocity}\right)$
	PPS or peripheral paddle speed: 0.5—2.0 ft/sec	$\dfrac{\text{Perimeter of unit}}{\text{Time of 1 revolution}} = \text{PPS}$
Clarification		
Sedimentation	Detention: 4 hr	$\left(\dfrac{\text{Wetted basin volume}}{\text{Flow}} = \text{Detention}\right)$ Check detention with tracer test[2]
	Flow-thru velocity: ≤0.5 ft/min	$\left(\dfrac{\text{Flow}}{\text{Wetted cross-section}} = \text{Velocity}\right)$
	Overflow rate: ≯20,000 gal/day/ft of Weir length	$\left(\dfrac{\text{Weir length}}{\text{Daily flow}} = \text{Overflow rate}\right)$ Swirls, eddies, currents indicative of short-circuiting causing suspension of sediment and carryover to filters
Filtration	Control of spike	Turbidity measurement at 10-min intervals for 1 hr when backwashed filter is put on line
	Control of bumping	Monitor filter effluent during run for turbidity to detect sudden increases followed by decreases
	Uniformity of media	Conduct sieve analysis and determine uniformity coefficient
	Loss of media	Check measurements from top of filter box to top of media against dimensions of engineering plans or "as-built" drawings or check staining of walls in filter box

Table 4 (continued)
GUIDES FOR PLANT INSPECTION[a]

Process	Guides	Remarks
	Effectiveness of backwash	Visual inspection for even media surface; collapse of media away from walls and surface cracks; uniform dispersal and absence of "boil"; measurements for 25 to 30% bed expansion during backwash; analysis for mud ball concentration[4]
Disinfection	Concentration, contact time	See Section II.A
	pH, temperature	See Section II.B
	Interferences	See Section II.C

[a] For conventional plants, high-rate units may be different.
[b] μ @ 0°C = 3.74 × 10^{-5}; μ @ 10°C = 2.73 × 10^{-5}; μ @ 20°C = 2.11 × 10^{-5}.

Other information that may be related to the critical time period should be collected, including repair of main breaks, installation of new lines and services, recent construction work on the sewerage system, community events with a large number of visitors creating unusual demands on the system, occurrence of fires, and consumer complaints of low pressure, dirty, discolored, or strange-tasting water.

The engineer should gain an understanding of how water is distributed in the community, especially noting pressure zones, areas of the network that are isolated by valves, any sections that are served water from a different source, and any interconnections with other water supplies. Residences or establishments that maintain connections with the water system and have privately owned sources of supply, such as a well used by a homeowner for irrigation, should be noted and inspected. Areas of low pressure should also be examined for possible cross-connections. It is sometimes possible to recreate events to demonstrate the occurrence of backsiphonage under conditions of low or negative pressure.[20]

The information gained through familiarization with the distribution network will be useful in establishing an association by time or place with cases of illness. Spotting cases of illness on a map of the distribution network may show clustering that can be explained by a specific service area, equipment malfunction, main break, source of contamination, reservoir, peculiar event, or other happening that relates to a specific time or place of occurrence.

IV. SAMPLING FOR THE AGENT (Priority 3 if cases are not occurring)

If the outbreak is quickly recognized and initiation of an investigation is timely, the investigators may be fortunate in identifying the causative agent in samples collected from the water system; however, this kind of success has been rare in the past. In less than 1% of the outbreaks has the etiologic agent been identified in water samples. Investigators have a better chance of identifying the agent through analysis of clinical specimens from infected individuals. Microbiological agents multiply in the intestinal tract and are shed in large numbers which provide for better opportunities in isolating and identifying bacteria, viruses, or parasites in stool specimens. Conversely, the opposite occurs in water systems where microbiological agents die away in a hostile environment or are diluted in number to a concentration not detectable by current technology. Detecting the agent in the water system is a very convincing piece of evidence and contributes substantially to the success of the investigation. It not only strengthens the epidemiological evidence that an outbreak is waterborne, but also supports the engineer's evaluation of the weaknesses in the water system that require correction.

Collection of water samples for bacteria, virus, or protozoa should be guided by the epidemiological evidence or the appraisal of the health authority of the etiological agent. Large-volume sampling techniques[21] capable of isolating the agent from 100 to 500 gal of water have been developed for virus and parasites and should be used to increase the chances for capture. Specific sample volumes for bacteria are not determined but normally are limited by the suspended solids concentration of the water being sampled and clogging of the sampling device. Microbiological overgrowth of culturing media occurs and also interferes with analysis of large-volume samples for bacteria. The extent of sample collection is determined by the capability of the laboratory doing the analyses and the accompanying expense.

One of the drawbacks of identifying the causative agent for some pathogens is the length of time required for analysis. Viral assays are especially time-consuming, and in some cases several months are required for identification. While this should not detract from sample collection efforts, it does serve to dampen spirits for quickly determining the cause of illness in the community and does not help in allaying the immediate concern and fear of people who are stricken.

The engineer, through his evaluation of the water system and interpretation of information provided by the epidemiologist, can assist in selecting the sampling locations. If the source of water supply is suspected of being contaminated, sampling of surface or ground water sources should be conducted. Single or multiple sources of actual or potential contamination identified in Table 3 can be sampled to document active discharges. Bracketing a discharge location through upstream and downstream sampling provides a comparison of results to help identify a source of contamination. For example, a river used as a source of supply may have a creek tributary influenced by a waste discharge. Bracketing the tributary through river samples may show elevated coliform counts and indicate further sampling of the creek to identify the source of contamination. If a water system is served by a number of wells, raw water at each well should be sampled for indicator organisms to aid in isolating a trouble spot for follow-up and for large volume sampling for pathogens.

Through evaluation of treatment the engineer will be familiar with the weaknesses, and selection of sampling sites may be dependent upon bracketing the weakness to show that a process can be penetrated if challenged. Sampling filter effluent at the beginning of a run (immediately after backwash) for indicator organisms coupled with a breakdown or interruption in chlorination (changing cylinders) would show the effects of filter spike and the plausibility of pathogen challenge and penetration. If treatment is marginal or inefficient for removal and inactivation of indicator organisms, raw and finished water samples for coliform and pathogens should be collected to show challenge and possible penetration.

Collection of samples from the distribution network should be guided largely through the analysis of epidemiological data. If cases of illness are predominantly in one area of the distribution network and localized contamination is suspected, samples should be collected in hopes of identifying the source. An open reservoir or well serving that area of the community may be the trouble spot and collection of appropriate samples should be conducted for indicator organisms as well as the causative agent.

If the outbreak has subsided before the investigation begins and there have not been any recent cases of illness, there is always the possibility of stored water still available from the critical period. Comtaminated water still might be present if the water utility has valved off a reservoir or isolated a main for repairs. Swimming pools may have been filled and, if not chlorinated, may still harbor the agent. Pumper trucks from the local fire department may have filled tanks during the outbreak and could be a sampling source. Ice frozen during the outbreak should be considered as a source for identification of virus and parasites. Investigators should keep this in mind when initiating public health measures to prevent further illness from occurring. Ice from the critical period should not be used if a viral agent is suspected of causing the illness since they do survive freezing.

V. COMMUNICATION

During the investigation of the outbreak there is a need to exchange information among the participants and to communicate progress and results to the community. Frequent meetings of the participants are necessary so that all are informed of what others are doing to avoid duplication of effort and to prevent the "I thought *you* were going to do it" syndrome. It is important that the investigation proceed with a team effort with all members carrying out their assignments and reporting periodically to a team leader. The team leader should act as a spokesperson, communicating progress and results of the investigation to local authorities and the community. There should be early agreement between the team leader and local authorities on the methods of communicating with the community through the news media. Depending on the situation, it may be necessary for the team leader to communicate either directly, jointly, or indirectly through a local official. The situation can range from one of local officials not feeling comfortable with translation of scientific terminology and results and thus depending on the team spokesperson for periodic release of information through interviews, press conferences, etc. to one where a local governing body may have its own public affairs office which handles all communication with the press and community. It is especially important that investigators be aware of the local situation and act accordingly to ensure that communication does not by-pass or ignore traditional methods.

It is the engineer's responsibility to communicate all activities and findings to the team leader and local authorities. The engineer should be especially careful that findings are interpreted correctly and engineering terminology does not interfere with the understanding of the evaluation of the water system. Words and terms commonly used and understood by public health workers are uncommon and misunderstood by the public and the news media. Coliform, pathogen, etiological agent, epi-curve, contact time, and pH are examples which often require definition and explanation in order that findings are interpreted correctly. Without explanation, positive coliform results may appear in the press as "cauliflower" in the water. Persons interviewed by the press soon become known as "experts" and sometimes make the mistake of believing it. The belief leads to answering questions beyond the "expert's" capability rather than referring the question or responding, "I don't know but will find the answer for you". The temporary embarrassment of not knowing everything is much preferred to the problems encountered with explaining and correcting erroneous statements appearing in print.

The engineer's written report of evaluation and findings should be complete but not necessarily lengthy. It should include a description of the water system and how it is operated, a specific discussion of weaknesses or deficiencies that contributed to occurrence of the outbreak, identification of the route-of-entry of the contaminant, actions that were taken to abate the outbreak, description of tests or analyses that substantiate breakdowns in operation of facilities, a summary/conclusion section followed by recommendations for improvements that may address both short- and long-term needs. The form and content of the written report reflects individual professional style and may vary considerably for different situations. However, it should be written in a manner that it is understood by those who are expected to implement the recommendations and can be understood by the general public to aid in gaining community acceptance for the recommendations. Therefore, submitting a draft of the report for review prior to the final report may be helpful in avoiding misunderstanding.

REFERENCES

1. **White, G. C.**, *Handbook of Chlorination*, Van Nostrand Reinhold, New York, 1972, 302.
2. **Jarroll, E. L., Bingham, A. K., and Meyer, E. A.**, Effect of chlorine on *Giardia lamblia* cyst viability, *Appl. Environ. Microbiol.*, 41, 483, 1981.
3. **Rice, E. W., Hoff, J. C., and Schaefer, F. W.**, Inactivation of giardia cysts by chlorine, *Appl. Environ. Microbiol.*, 41, 250, 1982.
4. **Lippy, E. C.**, Tracing a giardiasis outbreak at Berlin, New Hampshire, *J. Am. Water Works Assoc.*, 70, 512, 1978.
5. **Hudson, H. E., Jr.**, *Water Clarification Processes: Practical Design and Evaluation*, Van Nostrand Reinhold, New York, 1981, 75.
6. The American Water Works Association, *Water Quality and Treatment*, 3rd ed., McGraw-Hill, New York, 1971.
7. **Ameen, J. S.**, *Community Water Systems Source Book*, 3rd ed., Technical Proceedings, High Point, N.C., 1965.
8. Water Supply Control — Bulletin 22, New York State Department of Health, Albany, N.Y.
9. Manual for Evaluating Public Drinking Water Supplies, U.S. Department of Health, Education and Welfare, Public Health Service Publ. No. 24, Washington, D.C., 1963.
10. Manual of Individual Water Supply Systems, U.S. Environmental Protection Agency, EPA-430/9-73-003, Washington, D.C., 1973.
11. **Bellack, E.**, Fluoridation Engineering Manual, U.S. Environmental Protection Agency, Washington, D.C., 1972.
12. Cross-Connection Control Manual, U.S. Environmental Protection Agency, EPA-430/9-73-002, Washington, D.C., 1973.
13. Great Lakes — Upper Mississippi River Board of State Sanitary Engineers, Recommended Standards for Water Works, Health Education Service, Albany, N.Y., 1976.
14. **Johnston, J. M., Martin, D. L., Perdue, J., McFarland, L. M., Caraway, C. T., Lippy, E. C., and Blake, P. A.**, Cholera on a Gulf Coast oil rig, *N. Engl. J. Med.*, 309, 523, 1983.
15. **Morens, D. M., Zweighaft, R. M., Vernon, T. M., Gary, G. W., Eslien, J. J., Wood, B. T., Holman, R. C., and Dolin, R.**, A waterborne outbreak of gastroenteritis with secondary person-to-person spread, *Lancet*, 1, 964, 1979.
16. **Weissman, J. B., Craun, G. F., Lawrence, D. N., Pollard, R. A., Saslaw, M. S., and Gangarosa, E. J.**, An epidemic of gastroenteritis traced to a contaminated public water supply, *Am. J. Epidemiol.*, 103, 391, 1976.
17. **Leyland, D.**, Personal communication, 1979.
18. **Lippy, E. C.**, Waterborne disease: occurrence is on the upswing, *J. Am. Water Works Assoc.*, 73, 57, 1981.
19. **Lippy, E. C. and Erb, J.**, Gastrointestinal illness at Sewickley, Pa., *J. Am. Water Works Assoc.*, 68, 606, 1976.
20. **Taylor, R. B.**, The Holy Cross episode, *J. Am. Water Works Assoc.*, 64, 230, 1972.
21. **Jakubowski, W., Chang, S. L., Ericksen, T. H., Akin, E. W., and Lippy, E. C.**, Large volume sampling of water supplies for microorganisms, *J. Am. Water Works Assoc.*, 70, 702, 1978.
22. **Boardman, G. D.** Protection of Waterborne Viruses by Virtue of their Affiliation with Particulate Matter, Ph. D. thesis, University of Maine, 1976.
23. **Hoff, J. C.** The relationship of turbidity to disinfection of potable water, in *Evaluation of the Microbiology Standards for Drinking Water*, Hendricks, C. W., Ed., EPA-570/9-78-006, 103, 1978.
24. **Boyce, D. S., Sproul, O. J., and Buck, C. E.**, The effect of bentonite clay on ozone disinfection of bacteria and viruses in water, *Water Res.*, 759, 1981.
25. **Walsh, D. S., Buck, C. E., and Sproul, O. J.**, Ozone inactivation of floc associated viruses and bacteria, *J. Environ. Eng. Div., Am. Soc. Civ. Eng.*, 106, 761, 1980.
26. **Stagg, C. H., Wallis, C., and Ward, C. J.**, Inactivation of clay-associated bacteriophage MS-2 by chlorine, *Appl. Environ. Microbiol.*, 33, 385, 1977.
27. **Scarpino, P. W.**, Effect of particulates on disinfection of enteroviruses in water by chlorine dioxide, EPA 600/2-79-054, 1979.
28. **LeChevallier, M. W., Evans, T. M., and Seidler, R. J.**, Effect of turbidity on chlorination efficiency and bacterial persistence in drinking water, *Appl. Environ. Microbiol.* 42, 159, 1981.

Section IV: Prevention of Waterborne Outbreaks

Chapter 10

REGULATIONS AND SURVEILLANCE

Peter C. Karalekas, Jr. and Floyd B. Taylor

TABLE OF CONTENTS

I. Regulations 234
    A. Introduction 234
    B. Modern Historical Development of Regulations 234
    C. Evaluation of Regulations on Bacteria, Turbidity, and Sanitary Surveys..236
        1. Bacteria 236
    D. Safe Drinking Water Act (SDWA) 239
    E. Current Regulations 240
        1. Coliform Organisms 240
        2. Turbidity..240
        3. Sanitary Surveys 240
        4. Public Notification 241
    F. Summary 242

II. Surveillance...242
    A. Responsibility of the Water Consumer 242
    B. Responsibility of the Supplier of Water 242
        1. Compliance with Regulations 243
        2. Record Keeping..243
        3. Public Notification 243
    C. Responsibilities of Local Health Officials 244
        1. Water Quality Monitoring..244
        2. Field Inspections and Sanitary Surveys 244
        3. Investigation of Outbreaks 249
        4. Noncommunity Water Systems 249
        5. Individual Water Systems..249
    D. Responsibilities of State Regulatory Agencies.........................250
        1. Administration and Program Development.....................250
        2. Enforcement 250
        3. Surveillance, Sanitary Surveys, and Technical Assistance 250
        4. Plan Review 251
        5. Laboratory Capability and Certification 251
        6. Disease Surveillance and Investigation..........................251
    E. Responsibilities of Federal Agencies 251
        1. U.S. Environmental Protection Agency.........................251
        2. U.S. Food and Drug Administration 252
        3. U.S. Public Health Service Centers for Disease Control 252

References..252

I. REGULATIONS

A. Introduction

Regulations governing or at least related to the biological quality of drinking water predate the modern era by several millenia. One must define what is meant by "regulations". In the modern sense, they are rules which are enforceable by municipal, state, or federal law. In ancient times, they were practices which were mandated by priestly, municipal, or military authority. Three cases in point will illustrate:

- Baker[1] cites a Sanskrit reference of about 2000 B.C. which reads "It is directed to heat foul water by boiling and exposing to sunlight and by dipping seven times into it a piece of hot copper, then to filter and cool in an earthen vessel." This is a curious mix of scientific fact and superstition. Certainly the boiling and exposure to sunlight would render the water sterile or at least disinfected. The efficacy of the seven dips of hot copper is less certain.
- The ancient Hebrews understood the value of pure water. They prohibited the laying of water conduits through cemeteries and also appointed special inspectors, who were officials of the community, whose assignment was the prevention of contamination of water supplies.[2]
- It is probable that the most rigidly enforced water supply regulations were Army regulations. Cyrus the Persian, according to Herodotus (484—425? B.C.E.)[3] was supplied in the field with boiled water, and the water was kept in silver jars. This is more evidence that the ancients understood that good health was related to good water. It is highly unlikely that Cyrus's medical department was aware of the bactericidal or bacteriostatic effect of silver. It just so happened that silver was used abundantly by the Persians.

Some further facts were recorded by Sextus Julius Frontinus, Curator Aquarum, in his great work, *The Two Books on the Water Supply of the City of Rome*. First he describes his duties as "contributing partly to the convenience, partly to the health even to the safety of the city."[4] Later he discusses the problem of turbidity. Describing the concern of the Emperor Trajan for the Rome water supply he writes, "What shall we say of the painstaking interest which our Emperor evinces for his subjects . . . that he thought he would be contributing too little to our needs . . . unless he should also increase its (Rome's water supply) purity and its palatableness? . . . For when has our city ever been without muddy or turbid water, even though there had only been moderate rainstorms?"[5]

From the days of Frontinus (97 A.D.) until the days of Snow (1813—1858), there is little evidence that man thought much about the link between good water and good health. Perhaps we should say that he wrote little or nothing about it. This brings us to the dawn of modern water supply regulations. Dr. Snow should be called the father of the sanitary survey. With painstaking care he traced sanitary drains and water lines in Albion Terrace, London, and linked cholera cases to foul-appearing and odorous drinking water.[6] Dr. Snow is, however, best remembered for controlling a terrible cholera outbreak in London's Golden Square area by having the handle removed from the Broad Street Pump on September 8, 1854.[7]

B. Modern Historical Development of Regulations

This discussion is confined to the constituents of bacteria, turbidity, and sanitary surveys and will trace the development of regulations governing these three. Before doing this, however, it is essential to cite an important aspect of sanitary science as it relates to water supply. We refer to the evolution of that remarkable series of documents now entitled *Standard Methods for the Examination of Water and Waste Water*. It is beyond the scope

Table 1 EDITIONS OF STANDARD METHODS			
Edition	Year	Edition	Year
First	1905	Eighth	1936
Second	1912	Ninth	1946
Third	1917	Tenth	1955
Fourth	1920	Eleventh	1960
Fifth	1923	Twelfth	1965
Sixth	1925	Thirteenth	1971
Seventh	1933	Fourteenth	1975
		Fifteenth	1980

Table 2 EDITIONS OF PUBLIC HEALTH SERVICE DRINKING WATER STANDARDS	
Edition	Year
First	1914
Second	1925
Third	1943
Fourth	1946
Fifth	1962

of this chapter to trace in detail the relationships between *Standard Methods* and the forerunner of the EPA drinking water regulations, namely the Public Health Service (PHS) Drinking Water Standards. It is, however, evident that the PHS standards grew as *Standard Methods* grew and that the latter preceded the former. Tables 1 and 2 illustrate this point. That this should be the case is natural, since before a substance can be regulated there must be a way to measure it.

The legislative path to the PHS Drinking Water Standards and then to the EPA Safe Drinking Water Act began in 1878 with passage of the first National Quarantine Act signed on April 29, 1878, entitled "An Act to Prevent the Introduction of Infectious or Contagious Diseases into the United States".

In 1890 an Act was passed (U.S. Statutes at Large, vol. 26, ch. 51, p. 31 Approved March 27, 1890) to prevent the introduction of contagious diseases from one State to another and for the punishment of certain offenses. This Act and the two which followed it dealt with the contagious diseases cholera, yellow fever, smallpox and plague of which the first is water borne. The two following acts were (U.S. Statutes at Large, vol. 27, ch. 114, p. 449 Approved February 15, 1893) and (U.S. Statutes at Large, vol. 28, ch. 300, p. 372, Approved August 18, 1894), the latter of which was an amendment to the former and granted additional quarantine powers and imposed additional duties upon the Marine-Hospital Service. Based on the Act of 1893 the first Interstate Quarantine Regulations were published on September 27, 1894 by Secretary of the Treasury J. G. Carlisle.

Finally, in 1901, an act (U.S. Statutes at Large, vol. 31, ch. 836, p. 1086 Approved March 3, 1901) was passed which granted still further quarantine powers and imposed additional duties on the Marine-Hospital Service largely with respect to vessels. This emphasis on vessels reflected the great importance of the maritime in those days and resulted in the issuance of quite specific instructions for the handling of drinking water on vessels.

On October 5, 1912 there was published Treasury Department Circular No. 49 whereby all commissioned officers of the Public Health Service traveling under official orders on trains and vessels in interstate traffic were directed to observe sanitary conditions and submit reports. One such report cited the occurrence of 122 cases of gastroenteritis, 42 of typhoid-like disease and 5 deaths among the 235 passengers on a Great Lakes steamer. On October 14, 1912 paragraph 13 was added to the interstate Quarantine Regulations thereby controlling the use of common drinking cups on interstate carriers. Paragraph 15 followed early in the next year and as this was the prelude to the drinking water standards of 1914 its short contents are quoted in full as follows:

'Water provided by common carriers on cars, vessels, or vehicles operated in interstate traffic for the use of passengers shall be under the following conditions:

(a) Water shall be certified by State or municipal health authorities, within whose jurisdiction it is obtained, as incapable of conveying disease. Provided: that water in regard to the safety of which a reasonable doubt exists may be used if the same has been treated in such a manner as to render it incapable of conveying disease and the fact of such treatment is certified by the aforesaid health officer.

(b) Provided: for ice coming in contact with water to be from a safe source certified by State or municipal health authorities, and,

(c) Provided: that the water containers shall be cleansed and thoroughly scalded with live steam at least once each week that they are in operation.'

Regulatory action in 1914 came in two parts, the latter of which would become known as the Treasury Standards for drinking water or as the first Public Health Service Drinking Water Standards. Paragraph 16 to the Interstate Quarantine Regulations was added June 4, 1914 and read as follows:

'No person undertaking to furnish water for drinking or culinary purposes to any vessel in any harbor of the United States intending to clear for some port within some other State or territory of the United States or the District of Columbia shall furnish for such purposes water taken from the water of such harbor or from any other place where it has been or may have been contaminated by sewer discharges. Any person violating this regulation will be liable to a penalty of not more that $500 or imprisonment of not more than one year or both at the discretion of the U.S. District Court.'

The impetus for this action was the occurrence during 1913 of 144 cases of typhoid fever among Great Lakes seamen.

Then on October 21, 1914, Secretary of the Treasury William Gibbs McAdoo promulgated the first modern drinking water standards which prescribed limits for bacteria. Before considering later editions of the standards themselves it is pointed out that between 1893 and 1974 there was no basic change in the legislation upon which standards were based. It is also interesting that when the 1914 drinking water standards were promulgated there were over 9000 supply sources to which they would apply. By 1974 these had dwindled to around 700.

Beginning in 1968 with HR 15899 a series of bills were introduced in both houses of Congress under the general heading of Safe Drinking Water Act or Bottled Water Control legislation. There were some fourteen in all and the last two (HR 13002 and S 433) became The Safe Drinking Water Act of 1974 which was signed by President Ford on December 14, 1974 as P.L. 93-523. This is a far sighted and far reaching piece of legislation which among other things directed the Administration of the Environmental Protection Agency to promulgate primary drinking water regulations which were to be applicable to all public water supplies in the United States and which were to be enforced by the United States Government but with provision for primary enforcement by the States. Before proceeding further to the subject at hand, I think it important to point out that, unlike all other environmental legislation under which the EPA protects the environment, P.L. 93-523 is an amendment of the Public Health Service Act, P.L. 410. It thus continues the long standing protection of the public health which has been the responsibility of the Public Health Service.

C. Evaluation of Regulations on Bacteria, Turbidity, and Sanitary Surveys

1. Bacteria

From the first PHS Drinking Water Standards to the most recently proposed EPA Primary Drinking Water Regulations, the coliform organism has been the surrogate for the pathogens. There has been, as we shall see, an interesting change in direction concerning the evaluation of the general biological condition of water supplies. This was the standard plate count. As the first PHS standards set the pattern for all future standards, it is of value to quote the short statement of the 1914 edition.[9]

THE BACTERIOLOGICAL STANDARD FOR WATER

1. The total number of bacteria developing on standard agar plates, incubated 24 hours at 37°C, shall not exceed 100 per cubic centimeter; provided that the estimate shall be made from not less than two plates, showing such numbers of colonies to indicate that the estimate is reliable and accurate.
2. Not more than one out of five 10 c.c. portions of any sample examined shall show the presence of organisms of the *(Bacillus coli)* group when tested as follows . . . (there then followed seven paragraphs dealing with the laboratory procedures of that day for determining members of the *(Bacillus coli)* group).
3. It is recommended as a routine procedure, that in addition to five 10 c.c. portions one 1 c.c. portion and one 0.1 c.c. portion of each sample examined be planted in a lactose peptone broth fermentation tube, in order to demonstrate more fully the extent of pollution in grossly polluted samples.
4. It is recommended that in the above designated tests the culture media and methods used shall be in accordance with the specifications of the committee on standard methods of water analysis of the American Public Health Association, as set forth in STANDARD METHODS OF WATER ANALYSIS (A.P.H.A., 1912).

The standard as recommended by the commission was submitted with the following report discussing the question of standards of purity for water in general: First Progress Report of Commission Appointed to Recommend Standards of Purity for Drinking Water Supplied to the Public by Common Carriers Engaged in Interstate Traffic.

The authors of this first standard and the accompanying discussion also noted the importance of the sanitary survey. They said:[9]

In view of the impossibility of accurately ascertaining the source and history of each supply examined reliance must be placed upon results of laboratory examination to a greater extent than is necessary or justified in estimating the quality of a supply from a known source with a known history.

It is requested that the recommendation of these hard-and-fast limits of bacteriological impurity be not interpreted as minimizing in any way the importance of field surveys in estimating the sanitary quality of water supplies in general. It is always desirable to obtain information from as many angles as possible, and this is, indeed, necessary in order to form an altogether fair estimate of an individual supply.

The 1925 PHS Standards placed greater emphasis on the sanitary survey by including the following[10] as part of the Standard:

I. AS TO SOURCE AND PROTECTION

(1) The water supply shall be —

(a) Obtained from a source free from pollution; or
(b) Obtained from a source adequately protected by natural agencies from the effects of pollution; or
(c) Adequately protected by artificial treatment.

(2) The water supply system, including reservoirs, pipe lines, wells, pumping equipment, purification works, distributing reservoirs, mains and service pipes, shall be free from sanitary defects.
NOTE. 1. Natural agencies affording more or less complete protection against the effects of pollution are, in surface waters: Dilution, storage, sedimentation, the effects of sunlight and the associated biological processes tending to natural purification; and, in the case of ground waters, percolation through the soil. Important items in the natural purification of ground water are the character and depth of the strata penetrated.
2. Adequate protection by artificial treatment implies that the method of treatment is appropriate to the source of supply; that the works are of sufficient capacity, well constructed, skillfully and carefully operated. The evidence that the protection thus afforded is adequate must be furnished by frequent bacteriological examinations and other appropriate analyses, showing that the purified water is of good and reasonably uniform quality, a recognized principle being that irregularity in quality is an indication of potential danger.
3. Sanitary defect means faulty condition, whether of location, design, or construction of works, which may regularly or occasionally cause the water supply to be polluted from an extraneous source, or fail to be satisfactorily purified.

The bacteriological standard was essentially the same as in 1914 except that there was no required standard plate count. The standard plate count has not been required in any U.S. Standard or Regulations since. However, the first reference to turbidity (in a general sense) appeared in a second PHS document.[10] It read as follows:

III. AS TO PHYSICAL AND CHEMICAL CHARACTERISTICS

The water should be clear, colorless, odorless, and pleasant to the taste, and should not contain an excessive amount of soluble mineral substances nor of any chemicals employed in treatment.

The 3rd edition of the PHS Standards was an abortive attempt to put standards and ways of meeting them in the same document. A general rebellion by the water works industry forced the PHS to separate the two so that the 1946 Standards were just that. *A Manual of Recommended Water Sanitation Practice* was published separately.

The 1943 and 1946 Standards were essentially the same regarding bacteria, turbidity, and sanitary surveys. There was, however, a marked difference over the 1925 edition. This was the inclusion of a requirement that a number of bacteriological samples must be collected (Table 3 and Figure 1). The theory behind the required number of samples was that population would be a rough but practical means of dealing with different-sized distribution systems. This theory is still used.

The 1943 to 1946 Standards also contained a table which defined a numerical number of coliforms which would govern the evaluation of a supply in the sense that coliform organisms had to be less than that number. This was <2.2 coliforms per 100 mℓ of sample if 10-mℓ

Table 3
MINIMUM NUMBER OF SAMPLES TO BE COLLECTED EACH MONTH FROM AN INTERSTATE CARRIER WATER SUPPLY

Population served	Minimum number of samples per month
≦ 2,500	1
10,000	7
25,000	25
100,000	100
1,000,000	300
2,000,000	390
5,000,000	500

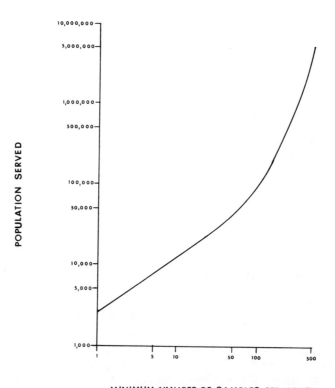

FIGURE 1. U.S. Public Health Service Drinking Water Standards, 1946.

portions were examined. There was also a mandatory turbidity limit of 10 ppm (silica scale). During the 16-year interval (the longest between any of the PHS Standards) between the 1946 and 1962 editions, many changes took place in the analytical procedures available to the public health professional and the water supply industry. Notable was the development of the membrane filter, a German invention during World War II, for measuring coliform organisms. Its great advantage was its speed in completing a coliform determination and a

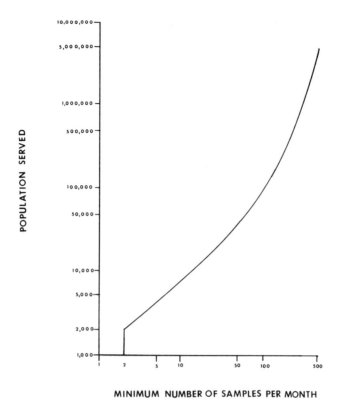

FIGURE 2. U.S. Public Health Service Drinking Water Standards, 1962.

facility for counting discreet organisms. When the membrane filter was used, the average monthly coliform content was not to exceed one with certain exceptions, and this is still the requirement today in the EPA Primary Drinking Water Regulations. There was an interesting change in the turbidity requirement in 1962. The level was dropped to five units, but it was no longer mandatory. The importance of the sanitary survey was emphasized by an expanded section on "Source and Protection of Supply". Also, the table of numbers of bacteriological samples required was dropped and a modified graph (Figure 2) was used instead.

Before proceeding further, it is important to again emphasize that the five editions of the Public Health Service Drinking Water Standards, as they evolved from a declaration in *Public Health* on bacteriological quality to a long list of biological, chemical, and physical constituents in 1962, are the foundation upon which rest all Western water supply standards and regulations.

D. Safe Drinking Water Act (SDWA)

We have already described the historical development of the Safe Drinking Water Act of 1974, PL 93-523. It might well be asked what led up to its passage.

A few reasons are

1. There was mounting concern over all facets of the environment during the 1960s, including water supply.
2. The PHS Drinking Water Standards applied to a small handful of U.S. water supplies and even then not directly. The line of enforcement was from the Surgeon General to the common carrier, not from the Surgeon General to the municipality.
3. There had been a reversal during the 1950s in the downward trend of waterborne outbreaks.[12]

4. The National Water Supply Survey conducted by the PHS in 1969 had shown an alarming number of supplies, especially in rural areas, which could not meet the 1962 PHS Standards.[13]
5. A report published by Dr. Robert Harris of the Environmental Defense Fund suggested that the high rate of bladder cancer in New Orleans was the result of high organic chemical content in the city water supply drawn from the Mississippi River.

The SDWA was noteworthy for several reasons. First, it was an amendment to the Public Health Service Act. Second, while it mandated federal control over nearly all U.S. public water supplies, it heavily emphasized that the states were to have primary responsibility for enforcement. Third, there would be a National Drinking Water Advisory Council to guide the EPA in its application of the Act. Fourth, the Act directed the National Academy of Sciences to make a study of the health aspects of water supply. Fifth, funds to finance state programs were authorized. Sixth, Emergency Powers were granted to the EPA Administrator, enabling him to take direct federal enforcement actions, and provision was made for fund authorization to encourage research and training. As the Act is extensive, the reader is urged to obtain a copy[14] for his or her own use. The immediate result of the Act was the development and publication by the EPA of National Interim Primary Drinking Water Regulations (IPDWR),[15]

E. Current Regulations
1. Coliform Organisms

Section 141.14 of the IPDWR details the microbiological contaminant levels permitted by the EPA. The reader is referred to this section but it may be summarized by saying that coliform organisms shall not exceed a monthly average of 1. The IPDWR also specified a minimum number of samples to be collected each month (Table 4).

2. Turbidity

Section 141.13 of the IPDWR mandates as follows:

S141.13 Maximum contaminant levels for turbidity.

The maximum contaminant levels for turbidity are applicable to both community water systems and non-community water systems using surface water sources in whole or in part. The maximum contaminant levels for turbidity in drinking water, measured at a representative entry point(s) to the distribution system, are:

(a) One turbidity unit (TU), as determined by a monthly average pursuant to S141.22, except that five or fewer turbidity units may be allowed if the supplier of water can demonstrate to the State that the higher turbidity does not do any of the following:

(1) Interfere with disinfection;
(2) Prevent maintenance of an effective disinfectant agent throughout the distribution system; or
(3) Interfere with microbiological determinations.

(b) Five turbidity units based on an average for two consecutive days pursuant to S141.22

The significance of this regulation is that it reinstates a mandatory requirement of the PHS Standards prior to 1962, but does so at the level of 1 turbidity unit (TU) and makes this a primary (health-related) maximum contaminant level (MCL).

3. Sanitary Surveys

While very little is said in the IPDWR about sanitary surveys, one short phrase in Section 141.14 establishes its importance. Section 141.14(b)(1)(1) describes the conditions under which a state had the authority to make certain modifications in the bacteriological require-

Table 4
MINIMUM NUMBER OF COLIFORM SAMPLES TO BE COLLECTED PER MONTH

Population served	Minimum number of samples per month	Population served	Minimum number of samples per month
25—1,000	1	90,001—96,000	95
1,001—2,500	2	96,001—111,000	100
2,501—3,300	3	111,001—130,000	110
3,301—4,100	4	130,001—160,000	120
4,101—4,900	5	160,001—190,000	130
4,901—5,800	6	190,001—220,000	140
5,801—6,700	7	220,001—250,000	150
6,701—7,600	8	250,001—290,000	160
7,601—8,500	9	290,001—320,000	170
8,501—9,400	10	320,001—360,000	180
9,401—10,300	11	360,001—410,000	190
10,301—11,100	12	410,001—450,000	200
11,101—12,000	13	450,001—500,000	210
12,001—12,900	14	500,001—550,000	220
12,901—13,700	15	550,001—600,000	230
13,701—14,600	16	600,001—660,000	240
14,601—15,500	17	660,001—720,000	250
15,501—16,300	18	720,001—780,000	260
16,301—17,200	19	780,001—840,000	270
17,201—18,100	20	840,001—910,000	280
18,101—18,900	21	910,001—970,000	290
18,901—19,800	22	970,001—1,050,000	300
19,801—20,700	23	1,050,001—1,140,000	310
20,701—21,500	24	1,140,001—1,230,000	320
21,501—22,300	25	1,230,001—1,320,000	330
22,301—23,200	26	1,320,001—1,420,000	340
23,201—24,000	27	1,420,001—1,520,000	350
24,001—24,900	28	1,520,001—1,630,000	360
24,901—25,000	29	1,630,001—1,730,000	370
25,001—28,000	30	1,730,001—1,850,000	380
28,001—33,000	35	1,850,001—1,970,000	390
33,001—37,000	40	1,970,001—2,060,000	400
37,001—41,000	45	2,060,001—2,270,000	410
41,001—46,000	50	2,270,001—2,510,000	420
46,001—50,000	55	2,510,001—2,750,000	430
50,001—54,000	60	2,750,001—3,020,000	440
54,001—59,000	65	3,020,001—3,320,000	450
59,001—64,000	70	3,320,001—3,620,000	460
64,001—70,000	75	3,620,001—3,960,000	470
70,001—76,000	80	3,960,001—4,310,000	480
76,001—83,000	85	4,310,001—4,690,000	490
83,001—90,000	90	$\geq$4,690,001	500

ments. The state is required to determine "the potential for contamination as indicated by a sanitary survey".

4. Public Notification

Highly controversial when first introduced, but nevertheless adopted, were requirements in the SDWA and the IPDWR that when a MCL for a primary constituent was exceeded, the public must be suitably notified. This notification could be in the form of newspaper, radio, or TV announcements, and inclusion of the information on water supply bills.

F. Summary

The SDWA has been in existence for nearly 10 years. It shows every evidence of being here to stay, and the water works industry as well as the environmentalists support its continuation. It is not perfect and continuing research will bring about changes. It is evident, for example, that a water which meets the bacteriological requirements can still be the source of waterborne outbreaks, e.g., giardiasis. However, much credit must be given to the succession of drinking water standards described in this section in contributing greatly to the high quality of drinking water in this country.

II. SURVEILLANCE

Protecting the health of water consumers and preventing the transmission of waterborne disease is a shared responsibility. The consumer himself, the supplier of water, and local, state, and federal health officials and regulatory agencies all have a role in the surveillance of water supply systems to ensure their continued safe operation and the protection of public health.

A. Responsibility of the Water Consumer

When a customer phones the water utility with a complaint about the quality of his water, this may be the first indication that something is amiss in the water system. The overwhelming percentage of customer complaints, of course, do not foretell an impending disease outbreak but are only symptomatic of the day-to-day variations in water quality that can be expected in a water system. For example, a nearby fire with heavy water usage can disturb sediments in water mains that can cause the customer's water to look rusty. This is usually a temporary problem that will quickly rectify itself with no further action required. However, it should be recognized that there may be circumstances during heavy demand and reduced water pressure during a fire that could result in backsiphonage[16] into a water system of a material that could present a health threat. Thus, customer complaints, rather than being considered a nuisance to water utilities, ought to be regarded as the first sentinel of danger and should be evaluated with the thought that there may be a problem in the water system that deserves immediate attention to prevent the transmission of waterborne disease.

In addition to notifying public officials of problems they encounter in water quality, pressure, or other related areas during normal events, the water consumer also plays a very important role in limiting waterborne outbreaks by following instructions during the outbreak itself. Boil water notices or advice to obtain water from other sources are only effective if the water consumer complies. While this may not be a great burden on individual homeowners, establishments such as restaurants, ice-making companies, and other facilities that rely on the water system may be reluctant to comply with such orders because they may lose customers. It is essential that they be made aware of the potential danger involved; there should also be a program on the part of health officials to check compliance with boil water orders to ensure the protection of the public health. Lastly, water consumers are likely to be the only source of a sample of water, several days old (i.e., from ice cubes), that may be needed in evaluating the water system for a disease outbreak.

B. Responsibility of the Supplier of Water

The foremost responsibility for supplying safe drinking water and protecting the health of water consumers rests with the operators of the water system. Craun[17] recently summarized how deficiencies in the facilities and operation of ground water systems are responsible for waterborne outbreaks. The literature is replete with many other examples of failures on the part of the system operators to provide adequate facilities to prevent the transmission of disease. Economic considerations are frequently given as reasons for failure to provide adequate facilities. Water rates may be inadequate to provide necessary funds, or elected officials may not provide the appropriate level of funding even though the water rates provide sufficient funds. A trend in recent years has been to enterprise accounting or independent water districts that can collect sufficient funds and expend funds to meet the number one priority of the system, i.e., compliance with health regulations in order to provide safe drinking water. There are a number of specific items which the operator must be aware of and responsive to.

1. Compliance with Regulations

The supplier or water has in recent years, as a result of the SDWA, played a significant role in the evolution and development of drinking water regulations. Larger water systems have been in the forefront of this effort and in achieving compliance with the regulations. Smaller water systems (i.e., those serving less than 10,000 people) have been much less successful in complying with current drinking water regulations.[17] The greatest incidence of waterborne disease outbreaks are found[15] in these smaller systems.

While close scrutiny of the coliform and turbidity regulations may find them inadequate in certain respects, one would certainly not advocate any lessening of the effort to bring water systems into compliance with the regulations. Therefore, it is incumbent upon the operators of the water system to fulfill their responsibilities of providing the necessary management and oversight of the water system to enable it to comply with drinking water regulations. The operators' responsibility extends beyond mere compliance to acting as the proponent in his or her community for water system improvements that will enable the water system to meet both current and future regulations. With regard to the future, progressively deteriorating sources[18] and existing treatment facilities may not even be adequate to meet current regulations in a few years, much less new regulations that could be more stringent. The system operator must anticipate these changes and plan for them.

2. Record Keeping

The SDWA and regulations[16] developed pursuant to the act mandate that certain records concerning water quality, monitoring, and sanitary surveys be kept by the supplier of water for periods ranging up to 10 years. This record keeping responsibility is merely the minimum record keeping necessary to meet legislative requirements.

In addition, there are many other records that the supplier of water should maintain for a variety of reasons.[19] These include providing an alert for changing raw water quality, aiding the operator in solving treatment problems, showing that the final product meets plant performance standards, determining equipment, plant, and unit process performance, aiding in answering complaints, anticipating routine maintenance, providing cost analysis data, providing future engineering design data, and finally, aiding in the investigation of waterborne outbreaks.

3. Public Notification

The SDWA also specifies in detail the manner and frequency in which the supplier of water is responsible for notifying the public in case of a failure to meet drinking water regulations. These notices are to be made through newspaper, radio, and television.

The purpose of these public notice provisions is based on several considerations, including the customers' right to know what the quality of his drinking water is. Further, they provide the framework by which customers can be made aware of circumstance in the system that threaten public health. In situations where a waterborne outbreak is occurring or is threatened, the supplier of water, in consultation with the state, may offer specific advice to consumers on how they may protect themselves. The obvious advice is, of course, to boil the water before drinking or disinfect it in some other appropriate manner.[20] While the legal requirements of public notice tend to focus on the negative, the intent is to promote some immediate positive action to protect health. Beyond that, however, the regulations are also designed to make consumers aware of problems that may require substantial water rate increases and investment to correct deficiencies. It is thought that the informed consumer will be more receptive and even supportive of expending funds to make the improvements.

The supplier of water should not limit notification to the public to problems or emergencies, but should take the initiative to provide information when the water system is in full compliance and beyond with regulations such those dealing with the aesthetics[21] of water. The positive informational approach can also be effective in dissuading consumers from seeking other sources of drinking water, such as unregulated and unsafe springs or wells. In summary, a consumer who is informed by the supplier of water as to both the problems when they occur and to the benefits of the water system is in the best position to protect his own health and prevent the transmission of disease.

C. Responsibilities of Local Health Officials

In many communities, local and county health departments play a very important role in the surveillance of public and private water supplies. Local health departments are closer to the action than either state or federal regulatory agencies, they have the specific mandate to protect public health in their community, and they are able to react very quickly when a disease outbreak occurs. Local health agencies can also participate in routine and special sampling and sanitary surveys as required by the SDWA, and may be the sole regulator of water systems serving the homeowner and other systems that are too small to be regulated by the SDWA.

1. Water Quality Monitoring

For public water systems that fall within the jurisdiction of the SDWA, the primary enforcement responsibility rests with either the state or the EPA, and the burden of water quality monitoring rests with the water system. In some areas, local health departments have assumed the responsibility for monitoring in some cases as a service to the water supply, in others as an additional check on the integrity of the water system and its operation. Furthermore, by conducting the required monitoring, particularly for coliforms, the local health department will receive immediate notification of unusual results. This will enable steps to be taken in both resampling and in checking with the system operator to determine if some deficiency exists that compromises the water system. If problems are identified, immediate action can then be taken to protect public health, if necessary. In summary, local health agencies should be encouraged to take an active role in the regulation and surveillance of water systems.

2. Field Inspections and Sanitary Surveys

The importance of field inspections of water systems was eloquently put forth by Moore,[22] who stated: "The fact that a water supply has been used for many years without causing recognized cases of disease can be relied on as evidence of present and future fitness only if conditions surrounding the supply remain unchanged. This is seldom true. Repeated field surveys and routine laboratory analyses of collected samples are the only real safeguard."

Moore[22] also stated that "The field survey is of surpassing importance in judging the fitness of water supply. It calls for a consideration of a great many items and should be entrusted only to persons who have by training and experience acquired a broad knowledge of those matters pertinent to the sanitation of water sources, to their physical development, and to the purification of water."[16] Table 5 provides a comprehensive form[23] for the field evaluation of water systems described by Moore, hereinafter termed "sanitary survey". A sanitary survey of a water system is defined as an on-site review of the water source,

Table 5
SANITARY SURVEY SAMPLE FORM[23]

Date of Survey _____
Name of Facility _____ System Identification _____
Owner _____ Telephone _____
Address _____
_____ County _____
Treatment Plant Telephone Number _____
Name of Operator _____ Certification _____
Water Purchased From _____ Water Sold To _____
(other than system)

SOURCE

1. What type of source? _____
2. What is the total design production capacity? _____ MGD
3. What is the present average daily production? _____ MGD
4. What is the maximum daily production? _____ MGD
5. Does system have an operational master meter? Yes _____ No _____
6. How many service connections are there? _____
7. Are service connections metered? Yes _____ No _____

WELLS   Yes   No

1. Is recharge area protected?
   Ownership _____ Fencing _____ Ordinances _____
2. What is nature of recharge zones?
   Agricultural _____ Industrial _____ Residential _____ Other _____
3. Is site subject to flooding?   ____   ____
4. Is well located in proximity of a potential source of pollution?   ____   ____
5. Depth of well _____ ft
6. Drawdown _____ ft
7. Depth of casing _____ ft
                                                                                           Yes   No
8. Depth of grout _____ ft
9. Does casing extend at least 12 in. above the floor or ground?   ____   ____
10. Is well properly sealed?   ____   ____
11. Does well vent terminate 18 in. above ground/floor level or above maximum flood   ____   ____
    level with return bend facing downward and screened?
12. Does well have suitable sampling cock?   ____   ____
13. Are check valves, blowoff valves, and water meters maintained and operating   ____   ____
    properly?
14. Is upper termination of well protected?   ____   ____
15. Is lightning protection provided?   ____   ____
16. Is intake located below the maximum drawdown?   ____   ____
17. Are foot valves and/or check valves accessible for cleaning?   ____   ____

                                                                                           Yes   No

SPRINGS AND INFILTRATION GALLERIES

1. Is the recharge area protected?
   Ownership _____ Fencing _____ Ordinances _____
2. What is the nature of the recharge area?
   Agricultural _____ Industrial _____ Residential _____ Other _____
3. Is site subject to flooding?   ____   ____
4. Is collection chamber properly constructed?   ____   ____
5. Is supply intake adequate?   ____   ____
6. Is site properly protected?   ____   ____
7. What conditions cause changes to quality of the water?
   _____
   _____

	Yes	No

SURFACE SOURCES

1. What is nature of watershed?
   Agricultural_____ Industrial_____ Forest_____ Residential_____
2. What is size of the owned/protected area of the watershed?
   _____
3. How is watershed controlled?
   Ownership_____ Ordinances_____ Zoning_____
4. Has management had a watershed survey performed? ____ ____
5. Is there an emergency spill response plan? ____ ____
6. Is the source adequate in quantity? ____ ____
7. Is the source adequate in quality? ____ ____
8. Is there any treatment provided in the reservoir? ____ ____
9. Is the area around the intake restricted for a radius of 200 ft? ____ ____
10. Are there any sources of pollution in the proximity of the intakes? ____ ____
11. Are multiple intakes, located at different levels, utilized? ____ ____
12. Is the highest quality water being drawn? ____ ____
13. How often are intakes inspected?_____
14. What conditions cause fluctuations in quality?
    _____
    _____
    _____

PUMPS

1. Number_____
   Type_____
   Location_____
2. Rated Capacity_____

	Yes	No

3. Are pumps operable? ____ ____
4. What is state of repair of pumps?_____
   _____
   _____
5. What type of lubricant is used?_____
6. Emergency power

   • What type_____
   • Frequency of testing_____
   • Record of primary power failures:_____ in last year.
   • Automatic_____ Manual_____ Switchover_____
   • Are backup pumps/motors provided? ____ ____

7. Is all electro/mechanical rotating equipment provided with guards? ____ ____
8. Are controls functioning properly and adequately protected? ____ ____
9. Are underground compartments and suction well waterproof? ____ ____
10. Are permanently mounted ladders for pumping stations sound and firmly anchored? ____ ____
11. Is facility properly protected against trespassing and vandalism? ____ ____

	Yes	No

TREATMENT UNITS (Note: Multiple units should have a separate information section completed for each unit.)

Prechlorination/Pretreatment Units

1. What chemical is used?_____
2. What amount is used?_____ lb/day
3. For prechlorination, has TTHM been evaluated? ____ ____
4. Where is point of application?_____
5. Is chemical storage adequate and safe? ____ ____
6. Are adequate safety devices available and precautions observed? ____ ____

	Yes	No

Mixing

1. Is mixing adequate based on visual observation? _____ _____
2. Is equipment operated properly and in good repair? _____ _____

Flocculation/Sedimentation

1. Is process adequate based on visual observation? _____ _____
2. Is equipment operated properly and in good repair? _____ _____

Filtration

1. Is process adequate based on visual observation? _____ _____
2. Are instrumentation and controls for the process adequate, operational, and being utilized? _____ _____
3. What type of filter is utilized?_____
4. Is equipment operated properly and in good repair? _____ _____

Postchlorination

1. Is adequate chlorine residual being monitored? _____ _____
2. Is the disinfection equipment being operated and maintained properly? _____ _____
3. Is there sufficient contact time (30 min minimum) between the chlorination point and the first point of use? _____ _____
4. Is operational standby equipment provided? If not, are critical spare parts on hand? _____ _____
5. Is a manifold provided to allow feeding gas from more than one cylinder? _____ _____
6. Are scales provided for weighing of containers? _____ _____
7. Are chlorine storage and use areas isolated from other work areas? _____ _____
8. Is room vented to the outdoors by exhaust grilles located not more than 6 in. above the floor level? _____ _____
9. Is a means of leak detection provided? _____ _____

	Yes	No

10. Is self-contained breathing apparatus available for use during repair of leaks? _____ _____
11. Are all doors hinged outward, equipped with panic bars, and at least one provided with a viewport? _____ _____
12. Are all gas cylinders restrained by chaining to wall or by other means? _____ _____
13. Have there been any interruptions in chlorination during the past year due to chlorinator failure or feed pump failure? _____ _____

	Yes	No

STORAGE

1. What type of water is stored?
   Raw_____ Treated_____
2. What type of storage is provided?
   Gravity_____ gal   Hydropneumatic_____ gal
3. Total number of days of supply?_____ days

Gravity Storage

1. Does surface runoff and underground drainage drain away? _____ _____
2. Is the site protected against flooding? _____ _____
3. Is storage tank structurally sound? _____ _____
4. Are overflow lines, air vents, drainage lines, or cleanout pipes turned downward or covered, screened, and terminated a minimum of 3 diameters above the ground or storage tank surface? _____ _____
5. Is site adequately protected against vandalism? _____ _____
6. Are surface coatings in contact with water approved? _____ _____
7. Is tank protected against icing and corrosion? _____ _____

8. Can tank be isolated from system? ____ ____
9. Is all treated water storage covered? ____ ____
10. What is cleaning frequency for tanks?_____
11. Are tanks disinfected after repairs are made? ____ ____

                                                                                                     Yes   No

Hydropneumatic

1. Does low pressure level provide adequate pressure? ____ ____
2. Are instruments and controls adequate, operational, and being utilized? ____ ____
3. Are the interior and exterior surfaces of the pressure tank in good physical condition? ____ ____
4. Are tank supports structurally sound? ____ ____
5. Is storage capacity adequate? ____ ____
6. What is cycle rate?_____

                                                                                                     Yes   No

DISTRIBUTION SYSTEM

1. Is proper pressure maintained throughout the system? ____ ____
2. What types of construction materials are used?
_____

3. Are plans of the water system available and current? ____ ____
4. Does the utility have an adequate maintenance program? ____ ____
5. Is the system interconnected with any other system? ____ ____

                                                                                                     Yes   No

CROSS-CONNECTIONS

1. Does the utility have a cross-connection prevention program? ____ ____
2. Are backflow prevention devices installed at all appropriate locations? ____ ____
3. Are cross-connections present at the treatment plant? ____ ____

                                                                                                     Yes   No

MONITORING

1. Is the operator competent in performing necessary tests? ____ ____
2. Are testing facilities and equipment adequate? ____ ____
3. Do reagents used have an unexpired shelf life? ____ ____
4. Are records of test results being maintained? ____ ____

                                                                                                     Yes   No

MANAGEMENT

1. Are personnel adequately trained? ____ ____
2. Are operators properly certified? ____ ____
3. Are there sufficient personnel? ____ ____
4. Are financing and budget satisfactory? ____ ____
5. Is an emergency plan available and workable? ____ ____
6. Is adequate safety and personal protective equipment provided? ____ ____
7. Are the facilities free of safety hazards? ____ ____

facilities, equipment, operation, and maintenance of a public water system for the purpose of evaluating the adequacy of such a source and its facilities, equipment, operation, and monitoring for producing and distributing safe drinking water.

Such a comprehensive survey may be beyond the capability of a local agency. In any case, state regulatory agencies should conduct these surveys as a major element in their regulatory program. Local agencies may wish to participate jointly with the state. They also assist both the state and the supplier of water in investigating nuisance complaints that may involve pollution of the source of supply (e.g., from septic tank violations, agricultural and farm runoff, erosion, illegal discharges, and any other activity that may threaten the source of supply). Customer complaints may also be subject to local health agency review in order to determine the acceptability of the water to consumers and as a tool to measure the performance of the supplier in providing potable water.

3. Investigation of Outbreaks

The local health agency should be cognizant of changes in the health status of persons within its jurisdiction to enable it to determine if a waterborne disease outbreak is underway. Contact through physicians and hospitals is essential in this area. Again, the local health agency may not possess the expertise to fully investigate an outbreak, but they should have the ability to determine if an outbreak of disease exists and to know whom to contact at the state or federal level for assistance.

4. Noncommunity Water Systems

Noncommunity water systems, such as health care facilities, restaurants, schools, motels, industries, gas stations, etc. are also regulated by provisions[24] of the SDWA as long as they provide water to 25 or more persons on a daily basis. In terms of preventing waterborne disease, current federal regulations require that standards be met for coliforms, turbidity, and nitrate. Sanitary surveys of these systems are encouraged, but as a practical matter and in view of limited resources these types of systems are not receiving the attention they deserve from some state regulatory agencies. In fact, local health agencies may be much more familiar with these systems, particularly restaurants, from their own surveillance activities. Once again, a cooperative effort between local and state officials may provide the best protection of the public health through a shared role in the surveillance of noncommunity water systems.

5. Individual Water Systems

Individual water systems for homeowners, etc. are not regulated by the SDWA. Although state and federal regulations on water quality are certainly useful in assessing water quality, they are not enforceable except by local ordinance or state law. Local health agencies in most cases must play the primary role in protecting the public from problems related to individual water supplies. While it should be recognized that the magnitude of the threat to public health is not as great from a private well as from a public water system, it is nevertheless a threat and should be treated accordingly.

Local health agencies can adopt regulations that pertain to both water quality and water system construction[25,26] of individual systems. Local health agencies or building inspectors should inspect new well installations[27] and require water samples to be taken and meet standards before an occupancy permit is given for any new housing. Before a house can be sold, similar requirements should be in effect in order to protect prospective new owners from an unsafe water supply. Nuisance complaint investigation can also play an important role in protecting sources of supply from contamination. Other local health activities may include licensing well drillers and equipment installers and issuing permits.[27] Finally, local health agencies should encourage the consolidation of small systems into larger systems to provide better management and operation.[28]

D. Responsibilities of State Regulatory Agencies

The framework for the surveillance of public water supply systems by state and federal agencies is determined by the provision of the SDWA and regulations[29] published pursuant to the Act. The major burden of responsibility is placed upon the state once they have received primary enforcement responsibility (primacy) from the EPA. In the few states that currently do not have primacy, the EPA, in conjunction with the state, operates a water supply surveillance program which must contain a number of specific program elements to ensure the protection of public health and the prevention of waterborne disease.[30,31]

1. Administration and Program Development

As in any organization, an overall management plan is essential for effectiveness in carrying out its mission. Major elements should include:

- Development and implementation of program policy
- Staffing and budgeting
- Program direction and supervision
- Development of legislative authority
- Planning
- Data management
- Public participation

2. Enforcement

This element will include the appropriate statutory authority at the state level similiar to the federal SDWA to establish, amend, and enforce all necessary regulations. Regulation must then be promulgated dealing with drinking water quality, operation and maintenance of water systems, and new design standards. After the adoption of regulations, it is necessary to develop procedures for administrative and judicial enforcement of regulations. From a practical standpoint, it is desirable to publish all of the state regulations in one document which should be placed in the hands of the operators of water supply systems so they are fully aware of the requirements that they are obliged to meet.

3. Surveillance, Sanitary Surveys, and Technical Assistance

A prompt review of microbiological data from water supply systems is of paramount importance in preventing a waterborne disease outbreak. A mechanism should be established to enable the state to receive, review, and respond to positive microbiological results within a very short time (i.e., within 24 hr). In addition to the requirement for taking additional samples, the response may range from at least a phone call to the supplier, a sanitary survey, to technical assistance, to a boil water notice, to administrative or judicial remedies depending on the circumstances involved.

As stated previously, the sanitary survey is of great importance in determining the capability of water systems to continually produce safe drinking water. A water sample is an infinitesimal portion of the total volume of water collected, processed, and distributed by a water system. Water samples alone cannot be relied upon as the sole determinant of the safety of all the water that is reaching the consumer. Therefore, sanitary surveys are essential in ensuring the continued capability of the water system to produce safe drinking water.[32] Table 5 is a comprehensive sanitary survey[23] form which can be used in the field to collect information.

An important adjunct to sanitary surveys is providing technical assistance to the operators of water systems. This includes supplying information on planning, design, operation, maintenance, treatment, quality control, cross-connection control, and any other aspect related to the production of safe drinking water. Formal training programs for and certification of water plant operators, supervisory personnel, and distribution system operators are also effective tools in raising the level of knowledge and competence of persons involved in providing water. Finally, a plan to provide drinking water in emergencies, like outbreaks, through field treatment, tank cars, and interconnections with other systems should be included in the technical assistance activity of each state.

4. Plan Review

All new water supply facilities should be subject to review and approval by the state regulatory agency. Using existing standards[33-35] which can be modified to suit local needs, items to be approved should include:

- New sources of supply
- Water treatment facilities
- Changes in water treatment
- Major water main extension
- New storage facilities
- Disinfection procedures for all system components

This process of review and approval should begin with the review of preliminary engineering feasibility studies and proceed to a review of construction plans, and finally to inspection of actual construction practices.

5. Laboratory Capability and Certification

The state should provide its own laboratory to perform the analyses required by their regulations and to perform special analyses such as the identification of pathogens in water. The burden of routine monitoring is usually placed on the supplier of water or may be assumed by a local health agency or private laboratory. In these cases, the state must ensure that other laboratories are performing analyses properly by adopting a formal laboratory certification and inspection program.

6. Disease Surveillance and Investigation

A coordinated activity between state and local agencies should be maintained to detect, investigate, and report suspected waterborne disease outbreaks. There are several ways in which the state can facilitate this activity, including

- Establishing an epidemiological team to investigate outbreaks
- Establishing cooperation between local and state health agencies to centralize reporting
- Seeking agreement from medical associations to report selected diseases related to water

E. Responsibilities of Federal Agencies

1. U.S. Environmental Protection Agency

Under the provision of the SDWA there is a joint responsibility between the EPA and the states to carry out the mandate of the law to protect public health through the provision of safe drinking water. The responsibilities of the state, described above, may also fall upon the EPA when a state has not received primacy. Under law, the EPA is also required to carry out the following responsibilities:

- Establishment of primary regulations for the protection of the public health
- Establishment of secondary regulations relating to the taste, odor, and appearance of drinking water

- Measures to protect underground drinking water sources
- Research and studies regarding health, economic, and technological problems of drinking water supplies
- A survey of the quality and availability of rural water supplies
- Aid to the states to improve drinking water programs through technical assistance, training of personnel, and grant support

In the areas of research and technical assistance, the EPA has maintained a program of investigation and reporting waterborne outbreaks in close cooperation with state and local agencies. At the request of states, field sanitary surveys are made of water supplies to detect deficiencies that may have caused a disease outbreak. Technical assistance is also offered to water systems on source protection and methods of improving treatment and operation to reduce the threat of disease.

Finally, the EPA, in cooperation with the Food and Drug Administration (FDA), maintains a surveillance activity over water supplies that provide drinking water to be used on carriers in interstate commerce. This activity involves approximately 600 of the larger water systems in the U.S. that provide water to buses, trains, planes, and ships that travel interstate. The EPA reviews bacteriological data and conducts joint sanitary surveys with the state to determine the capability of these systems to continually produce safe drinking water and prevent the interstate transmission of disease.

2. U.S. Food and Drug Administration

The SDWA requires that bottled water be regulated and this responsibility rests with the FDA. When the EPA issues primary drinking water regulations, the FDA is also required to issue regulations that ensure that standards for bottled drinking water conform to the primary regulations or publish reasons for not doing so. These regulations[36] have been promulgated and are essentially the same as the EPA primary regulations.

3. U.S. Public Health Service Centers for Disease Control

The Centers for Disease Control (CDC) have no direct regulatory involvement in public water supply systems. Their primary function in this area involves technical assistance to the states. It has been customary for states to report the unusual occurrence of disease outbreaks to the CDC. The CDC in turn has published those of wide interest in *Morbidity and Mortality Weekly Report* to alert the health community of new disease outbreaks and problems related to drinking water. The CDC has also provided epidemiologists to investigate specific outbreaks as well as assigning epidemic intelligence service officers to state health departments for extended periods to increase the ability of the states in the areas of detecting, responding to, and controlling disease outbreaks. Since 1971, the EPA has assisted the CDC in the investigation of waterborne outbreaks.

REFERENCES

1. **Baker, M. N.**, *The Quest for Pure Water*, American Water Works Association, Denver, Colo., 1948.
2. Public Health Practices Among the Ancient Hebrews, Muntner, Z., TAVRUAH, 1958, 9.
3. Herodotus, *History* (trans. by A. D. Godley), Loeb Classical Library, Harvard University, Cambridge, Mass., 1966.
4. **Frontinus, Sextus Julius,** *The Two Books on the Water Supply of Rome*, Trans. by C. Herschel, republished by New England Water Works Association, 1973, 27.
5. **Frontinus, S. J.,** *The Two Books on the Water Supply of Rome*, trans. by C. Herschel, republished by New England Water Works Association, 1973, 57.

6. **Snow, J.,** *Snow on Cholera being a Reprint of Two Papers by John Snow, M.D.,* The Commonwealth Fund, New York, 1936, 25.
7. **Snow, J.,** *Snow on Cholera being a Reprint of Two Papers by John Snow, M.D.,* The Commonwealth Fund, New York, 1936, 38.
8. **Taylor, F. B.,** Drinking water standards — principles and history, 1914—1976, *J. N. Engl. Water Works Assoc.,* 239, 1977.
9. Bacteriological Standard for Drinking Water, Public Health Report, Reprint 232, 2966, 1914.
10. Drinking Water Standards, Public Health Report, Reprint 1029, 3, 1925.
11. Drinking Water Standards, Public Health Report, Reprint 1029, 4, 1925.
12. **Craun, G. F. and McCave, L. J.,** Review of the causes of water-borne disease outbreaks, *J. Am. Water Works Assoc.,* 65, 74, 1973.
13. Community Water Supply Study, Analysis of National Survey Findings, U.S. Department of Health, Education and Welfare, Public Health Service, Washington, D.C., 1970.
14. Public Law 93-523, 93rd Congress, S433, December 16, 1974, Office of Public Affairs (A-107), U.S. Environmental Protection Agency, Washington, D.C., 1974.
15. National Interim Primary Drinking Water Regulations, *Fed. Regist.,* Wednesday, December 24, 1975.
16. Cross Connection Control Manual, U.S. Environmental Protection Agency, Washington, D.C., 1973.
17. **Craun, G. F.,** *Current Trends in Waterborne Disease Transmission Through Contaminated Groundwater 1971—1982,* National Association of Environmental Health, Grand Rapids, Mich., 1984.
18. **Karalekas, P. C., Jr.,** Watershed management and water quality, *J. N. Engl. Water Works Assoc.,* 91, 1, 1977.
19. Manual of Instruction for Water Treatment Plant Operators, New York State Department of Health, Albany.
20. Safe Drinking Water in Emergencies, U.S. Department of Health, Education and Welfare, Public Health Service, Publ. No. 287, Washington, D.C., 1964.
21. National Secondary Drinking Water Regulations, *Fed. Regist.,* 42, 17143, 1977.
22. **Moore, E. W.,** Sanitary analysis of water, in *Preventive Medicine and Public Health,* 10th ed., Sartwell, P. E., Ed., Appleton-Century-Crofts, New York, 1973, 1077.
23. Sanitary Survey Training Student's Text, U.S. Environmental Protection Agency, Washington, D.C., 1983.
24. Implementation of drinking water standards, *Fed. Regist.,* 44, 40557, 1979.
25. Manual of Individual Water Supply Systems, U.S. Environmental Protection Agency, EPA-430/9-74-007, Washington, D.C., 1982.
26. Manual of Water Well Construction Practice, U.S. Environmental Protection Agency, EPA III A-350-76, Washington, D.C., 1976.
27. Environmental Health Planning, U.S. Department of Health, Education and Welfare, Public Health Service, Publ. No. 2120, Washington, D.C., 1971.
28. Environmental Health Planning Guide, U.S. Department of Health, Education and Welfare, Public Health Service Publ. No. 823, Washington D.C., 1967.
29. State Public Water System Supervision Program Grants and National Interim Primary Drinking Water Regulations Implementation, *Fed. Regist.,* 41, 2912, 1976.
30. A Manual for the Evaluation of a State Drinking Water Supply Program, U.S. Environmental Protection Agency, Washington, D.C., 1974.
31. **Taylor, D. H. and Hutchinson, G. D.,** Evaluation of state drinking water quality surveillance programs, *J. Am. Water Works Assoc.,* 67, 428, 1975.
32. Manual for Evaluation Public Drinking Water Supplies, U.S. Environmental Protection Agency, Washington, D.C., 1971.
33. *Water Treatment Plant Design,* American Water Works Association, Denver, Colo., 1968.
34. Recommended Standards for Water Works, Report of the Committee of the Great Lakes-Upper Mississippi Board of State Sanitary Engineers, Health Education Service, Albany, 1982.
35. *AWWA Standards,* American Water Works Association, Denver, Colo., 1984.
36. Quality Standards for Foods With No Identity Standards Bottled Water, *Fed. Regist.,* 44, 12169, 1979.

Chapter 11

BARRIERS TO THE TRANSMISSION OF WATERBORNE DISEASE

Gary S. Logsdon and John C. Hoff

TABLE OF CONTENTS

I.	Introduction	256
II.	Filtration	256
	A. Process Descriptions	256
	B. Microorganism Removal Capabilities	257
	1. Slow Sand Filters	257
	2. Rapid Rate Filters	260
	3. DE Filters	263
III.	Disinfection	268
	A. General Considerations	270
	B. Microorganism Inactivation Capabilities	270
	1. Chlorine Species	270
	2. Chlorine Dioxide	271
	3. Ozone	271
	C. Relevance to Disinfection Practice	271
IV.	Summary	272
References		273

I. INTRODUCTION

As water supply practice has evolved in the U.S. in the past 100 years, efforts to provide safe drinking water and prevent outbreaks of waterborne disease have focused on protection of water quality and treatment to improve water quality. The use of both water quality protection and water treatment is known as the multiple barrier concept.

The multiple barrier concept provides for measures to reduce contamination of the water source and to eliminate pathogens should they be present. Some water systems rely on barriers to protect raw water quality, but those using major river systems must rely more heavily on treatment of the water than on protection of the source. Some systems using ground water rely only on sanitary well construction and the natural protection given to the ground water, and do not disinfect. Such systems are vulnerable if the aquifer or well becomes contaminated.

The need and the opportunity for water quality protection measures can be established by conducting sanitary surveys. Water quality protection can be obtained by a great variety of techniques, including sanitary construction of wells, restriction of human activity on watersheds, treatment of waste water, location of water works intakes upstream from pollution discharges, and cross-connection control programs.

Water treatment techniques include clarification by slow sand filters, diatomaceous earth (DE) filters, or rapid rate granular (sand, dual media, or mixed media) filters, with the latter preceded by coagulation and (usually) by flocculation and sedimentation. Disinfection in the U.S. is generally accomplished by the use of chlorine or chloramine. Some waters are disinfected with ozone or chlorine dioxide, or occasionally by iodine or by ultraviolet light.

This chapter discusses water treatment techniques and disinfection practices to prevent transmission of disease. Water quality surveillance and source protection are discussed in Chapter 9.

II. FILTRATION

Water filtration can remove a variety of particulate substances, including soil particles, algae, and microorganisms. Water filtration improves the aesthetic quality of drinking water. Properly operated water filtration plants can clarify water, making disinfection more effective, and can remove microorganisms from water, leaving fewer in the water for the disinfection barrier to inactivate. Three water filtration processes generally used are slow sand filtration, rapid sand or rapid rate granular media filtration, and DE filtration.

A. Process Descriptions

Slow sand filtration originated in Great Britain in the 1800s. In this process, uncoagulated water is applied to a sand bed about 1 m deep at a filtration rate (approach velocity) of about 0.1 m/hr. Removal of microorganisms is aided by biological processes that occur on and in the filter bed, where an ecosystem becomes established with extended use. On the top of the media, a biologically active scum layer (schmutzdecke) builds up which assists in filtration.

As the water enters the schmutzdecke, biological action breaks down some organic matter, and inert suspended particles may be physically strained out of the water. The water then enters the top layer of sand where more physical straining and biological action occur and attachment of particles onto the sand grain surfaces takes place. Also, some sedimentation may occur in the pores between the media grains where velocity is sufficiently slow.

The filter must be drained and the top layer of sand removed when the depth of the schmutzdecke layer increases, clogging the top sand layer and causing head loss through the filter to reach a predetermined level. This period of scraping is the only time during normal operations in which additional manpower may be required to operate the filter. The amount of time between scrapings will depend on how much material is filtered out of the water. Thus, high-turbidity levels would shorten the cycle. Slow sand filter runs can be as short as 1 or 2 weeks when turbid or algae-laden waters are treated. The runs can last for several months when very clear waters are filtered.

Slow sand filters are uncomplicated and easy to operate, but they can be used successfully only with a very good quality of raw water (turbidity usually less than 10 nephelometric turbidity units (NTUs) and no undesirable inorganic or organic chemicals present). They also require large land areas per volume of water treated, and thus would probably be limited to use for small water systems in the U.S.

About the year 1900, engineers in the U.S. began to recommend the use of rapid sand filters for treatment of muddy surface waters like those found in the Ohio and Mississippi River valleys. These filters treat water that has been conditioned by coagulation, flocculation, and sedimentation to remove most of the particulate matter, and this results in longer filter runs. Rapid sand filters typically were operated at rates of 5 m/hr in the first half of this century. Since 1950, use of dual media (coal and sand) or mixed media (coal, sand, and garnet or ilmenite) has permitted operation of rapid rate filters at rates as high as 10 to 24 m/hr. For clear waters (10 NTUs or lower) the sedimentation and sometimes flocculation processes may be omitted; this process is known as direct filtration. In direct filtration, all removal of particulate matter occurs within the filter. Effective coagulation is essential for successful operation of rapid rate filters, with or without sedimentation.

During WW II, an intense effort was made to develop a filtration process that would give practically complete removal of *Endamoeba histolytica* cysts. The result was the DE filtration process. In DE filtration, a thin coating (3 to 5 mm) of DE is placed on a filter septum by recirculating filtered water that contains diatomite. After the filter cake is established in the precoating process, raw water that has been dosed with a small amount of DE is passed through the filter. Particles are removed, and they accumulate on the filter cake. Because DE (the body feed) is added to the raw water, the good hydraulic characteristics of the filter cake are maintained, and long filter runs can be attained. When a run is terminated, the filter cake is removed, disposed of, and fresh diatomite is used to recoat the clean septum.

Each of the filtration processes described can provide a barrier to the transmission of waterborne pathogens. The remainder of this section of the chapter describes the factors that influence the pathogen removal efficacy of the filtration processes described previously.

B. Microorganism Removal Capabilities
1. Slow Sand Filters

The ability of slow sand filters to remove bacteria from drinking water was shown first by the decline in disease and death rates for waterborne diseases such as cholera and typhoid fever in areas served by water systems employing this treatment. After appropriate microbiological methods were developed, the efficacy of slow sand filters for removal of bacteria, virus, and *Giardia* cysts was evaluated.

A number of factors influence the efficiency of microorganism removal by slow sand filters. Results of a number of investigations have been tabulated for purposes of comparison.

Early work by Hazen[1] and recent results from Colorado State University (CSU)[2] show that finer sand is more effective, although the use of smaller-sized media results in greater head loss and therefore shorter filter runs (Table 1).

Because slow sand filtration is influenced by biological processes, the condition of the filter bed has a strong influence on removal of microorganisms, especially coliform and plate count bacteria[3-5] and polio virus. Virus removal by sterilized sand was negligible,[6] and virus removal by clean sand was variable, but often poor.[7] Filter scraping and schmutzdecke removal, which are done when terminal head loss is reached, have in some instances

Table 1
EFFECT OF MEDIA SIZE ON REMOVAL OF MICROORGANISMS BY SLOW SAND FILTRATION

Effective size (mm)	Reported removal percentages		
	Serratia marcesens[1]		Total coliform[2]
0.09	99.98	99.98	—
0.14	99.96	99.97	—
0.30	99.87	99.99	—
0.26	—	99.90	—
0.29	—	99.84	—
0.3	—	—	87
0.38	—	99.84	—
0.5	—	—	83

Table 2
EFFECT OF FILTER BED CONDITION ON MICROORGANISM REMOVAL BY SLOW SAND FILTRATION

Location	Organism	Condition of filter bed	Removal percentage or concentraion in effluent	Ref.
Pittsburgh	Plate count bacteria	First 2 days after scraping	139—459/mℓ	3
	Plate count bacteria	Established biological population after first 2 days since scraping	59—112/mℓ	3
Iowa State University	Total coliform	First 2 days after scraping	83.0—99.8%	4
	Total coliform	Remainder of runs, excluding first 2 days after scraping	96.7—100%	4
Public Health Service	Poliovirus type I	Clean sand	22—96%	7
London	Poliovirus type I	Sterilized sand	Little or no removal	6
London	Poliovirus type I	Established biological population	≧99.9%	6
Colorado State University (CSU)	Total coliform	New filter sand	93%	5
	Total coliform	Filter bed scraped and disturbed	95.3%	5
	Total coliform	Established schmutzdecke	99.89%	5
	Giardia cysts	New filter sand	≧99.0%	5
	Giardia cysts	Bed scraped	≧98.7%	5
	Giardia cysts	Established schmutzdecke	≧99.7%	5

decreased bacteria removal efficiency. Research with *Giardia* cysts indicates that after a biological population has become established in a new sand bed, the removal of the schmutzdecke has little or no deleterious effect on removal of the cysts[5] (Table 2). This may indicate that the cysts are more readily removed than bacteria. The need for an initial ripening period should be considered when new slow sand filters are brought on line, as water quality will gradually improve over a period of weeks of operation.

Water temperature has an important effect on slow sand filter effluent quality (Table 3). For the full-sized filters or very large pilot filters studied by Hazen,[3] Jordan,[8] and Logsdon et al.,[9] data are summarized by season rather than for operation at a specific temperature. Work by Poynter and Slade[10] and studies performed at CSU[2,5] were done with pilot filter columns about 1 ft in diameter, and specific temperature data are available. Removal of all organisms has been shown to decline as temperatures decrease and biological processes slow

Table 3
EFFECT OF TEMPERATURE OR SEASON ON MICROORGANISM REMOVAL BY SLOW SAND FILTRATION

Location	Organism	Temperature or season	Removal (%)	Ref.
Pittsburgh	Plate count bacteria	February to November	Generally ≧99	3
	Plate count bacteria	December and January	97.2—98.9	3
Indianapolis	*Bacillus coli*	March to November	99.2—99.9	8
	B. coli	December to February	94.1—97.2	8
Vermont	Plate count bacteria	June to September	98.8	9
	Plate count bacteria	October to January	98.1	9
	Total coliform	June to September	99.4	9
	Total coliform	October to January	97.9	9
Colorado State University (CSU)	Total coliform	15°C	97	2
	Total coliform	5°C	87	2
London	Poliovirus I	16—18°C	≦99.997	10
	Poliovirus I	5—8°C	99.68	10
Vermont	*Giardia lamblia* cyst	21°C	99.99	9
	G. lamblia cyst	0.5°C	93.7	9
CSU	*Giardia* cyst	15°C	≧99.9	5
	Giardia cyst	5—15°C	≧99.9	5

Table 4
EFFECT OF FILTRATION RATE ON MICROORGANISM REMOVAL BY SLOW SAND FILTRATION

	Filtration rate (m/hr)						
Organism	0.02	0.04	0.08	0.12	0.20	0.40	Ref.
	Removal Percentage Observed						
Bacillus prodigiosus or *Serratia marsescens*	99.98	99.86	99.89	99.69	—	—	1
Total coliform	—	99.96	—	99.67	—	98.98	5
Fecal coliform	—	99.84	—	98.45	—	98.65	5
Plate count	—	91.40	—	89.47	—	87.99	5
Giardia cyst	—	99.991	—	99.994	—	99.98	5
Poliovirus I	—	—	—	—	99.93	99.78	10
	—	—	—	—	99.997	99.865	10

down. If water demand permits, slow sand filters should be operated at lower rates in winter than in summer.

Although some exceptions exist, higher filtration rates generally result in a slight decrease in efficiency for removal of microorganisms (Table 4). Higher filtration rates cause higher shear stresses within the pore spaces of the filter bed and reduce the time required for water to pass through the bed, thus reducing time of exposure to the ecosystem in the bed.

Evaluation of the effect of filter bed depth on microorganism removal by Hazen,[1] and more recently at CSU,[2] showed that shallow filter beds are less effective than deeper beds. According to Huisman and Wood,[11] a typical bed depth for a new slow sand filter would be 1.2 to 1.4 m, and clean sand should be added to a slow sand filter when repeated scrapings have reduced the sand bed depth to 0.5 to 0.8 m.

The removal of microorganisms by slow sand filtration is less effective when the sand is clean (as in a new filter), when water temperature is close to 0°C, when filtration rates are in the higher portion of the usual design rates, and when scraping to remove sand has decreased the depth of the filter bed. Of the factors that tend to degrade filter performance, water temperature cannot be controlled by the design engineer. In the northern U.S., and in mountain communities where raw water is very cold for long periods of time, a conservative filter design should be used to compensate for the reduction in efficiency that is caused by the cold temperature.

2. Rapid Rate Filters

Rapid filtration processes and processes used to condition water for filtration (coagulation, flocculation, and sedimentation) can bring about reduction in the numbers of microorganisms in water. Several investigations on the microorganism removal capability of clarification processes are presented in this part of the chapter.

Studies on removal of viruses by coagulation and sedimentation were reviewed by Sproul,[12] who concluded that removal of enteric viruses from water would range from 90 to over 99.99%. This conclusion was based on studies of poliovirus and coxsackie A2 in which metallic coagulants (ferric or aluminum salts) were used. Sproul also reported that maximum virus removal occurred at or near the metallic coagulant dose that resulted in maximum turbidity removal.

Engelbrecht et al.[13] reported that when ferric chloride or aluminum sulfate was used for coagulation, and when turbidity removal was optimum, typical removals for total coliforms by coagulation-sedimentation ranged from 90 to 99%.

Cyst removal by coagulation and sedimentation has also been evaluated. The U.S. Army tested its Mobile Water Purification Unit, Model 1940, and found that sedimentation removed 98.5 and 99.8% of applied *E. histolytica* cysts in two tests in which alum and soda ash were used for coagulation.[14] Arozarena[15] performed jar tests for removal of *G. muris* cysts from a low turbidity (1 to 2 NTUs) gravel pit water and found that cyst removal by alum coagulation was about 90% or higher in the pH range of 6 to 9, with maximum removals exceeding 99% (Table 5). Cyst removal by alum coagulation could be improved by use of cationic polymer to supplement alum and by use of nonionic polymer to strengthen floc. Arozarena found that cyst removal by sedimentation usually was similar to turbidity removal when aluminum sulfate was used as the coagulant.

Microorganism removal by coagulation and rapid filtration has been evaluated by numerous investigators.[3,4,7,9,14,16-18] Water filtration studies have repeatedly shown that effective coagulation is needed to attain high percentages of removal of microorganisms by rapid filtration on a consistent basis. Microorganisms, like other small particles, must be destabilized to be removed efficiently in the rapid filtration process. This has been demonstrated when conventional treatment processes, including sedimentation, have been used, and when direct filtration without sedimentation has been employed (Table 6). Rapid rate filtration of uncoagulated water cannot be depended upon to give consistently high removals by microorganisms. Good coagulation was considered to have been attained when filtered water turbidity was well below 1.0 NTUs. Inadequate coagulation is that which resulted in filtered water turbidity close to or greater than 1.0 NTUs. In some cases, comments on the quality of coagulation were provided by the researchers.[3,14,18]

A number of aspects of filter operation can adversely affect microorganism removal by rapid rate filtration. Among these are interruption of chemical feed, poor filter efficiency at the beginning of a run (before ripening has occurred), sudden increases in filtration rate, and turbidity breakthrough that sometimes occurs with higher head loss at the end of a run.

Interruption of coagulant feed can cause rapid degradation of water quality. This was shown by Hazen's[3] experiments with the Warren Filter at Pittsburgh in 1898. Raw water and coagulant were introduced at one end of the elliptical settling basin, which had a flow-through time of 40 min when the filter operated at 2 gpm/sf (4.9 m/hr). During a filter run that began on May 1 at 8:36 a.m., in which coagulant feed was shut off at 9:02 a.m., the bacteria count was 28 per milliliter at 9:30 a.m. that morning. It increased to over 1000 per

Table 5
SUMMARY OF JAR TEST
***GIARDIA MURIS* CYST**
REMOVAL EFFICIENCIES
WITH ALUM COAGULATION[15]

25 mg/ℓ alum, raw water 2.3 NTU

pH	*Giardia* cyst (% removal)	Settled turbidity (NTU)
5.9	58.4	1.5
6.6	>89.6	0.20
7.5	>88.8	0.20
7.9	>89.9	0.28
8.6	>90.5	0.35

pH 6.4—6.7, raw water 2.1 NTU

Alum (mg/ℓ)	*Giardia* cyst (% removal)	Settled turbidity (NTU)
5	95.7	3.7
10	>99.4	0.82
25	>99.4	0.29
50	>99.4	0.29

pH 7.2, 10 mg/ℓ alum, raw water 1.4 NTU

Nonionic polymer (mg/ℓ)	*Giardia* cyst (% removal)	Settled turbidity (NTU)
0	65.4	0.24
0.01	95.0	0.25
0.05	96.2	0.18
0.1	>96.2	0.17
0.5	>96.2	0.18
1.0	>92.3	0.19

pH 6.5—6.7, 5 mg/ℓ alum, raw water 1.6 NTU

Cationic polymer (mg/ℓ)	*Giardia* cyst (% removal)	Settled turbidity (NTU)
0	96.0	0.40
0.1	98.8	0.48
0.2	99.1	0.52
0.4	99.3	0.58
0.8	>99.0	0.78
1.6	96.3	1.0

Table 6
EFFECT OF PRETREATMENT ON MICROORGANISM REMOVAL BY RAPID FILTRATION

Processes used					Organism	Percentage removal reported			Ref.
Coagulation	Flocculation	Sedimentation	Sand media	Dual or mixed media		Good coagulation	Inadequate coagulation	No coagulation	
×		×	×		Bacteria	98.8%	—	35—86%	3
×		×	×		Bacteria	98.5%	—	38—70%	3
				×	Coliform			1—99.9%	9
×				×	Coliform	86—96.5%	—		4
×				×	Poliovirus 1	90—99%	—	1—50%	7
×	×	×		×	Poliovirus 1	>99.7%	—		7
×				×	Giardia muris	59—99.98%	23—94%	59—94%	16
×	×			×	G. lamblia	95—99.97%	64—92%	48%	17
×				×	Giardia cyst	83% of samples have removal ≥ 99.0%	—	Removal <80% for half of samples	9
×			×		Endamoeba histolytica cyst	99.6—99.9%	96%	80—88%	14
×	×	×	×		E. histolytica cyst	99.94—99.97%	—	98.3%	14
×	×				E. histolytica cyst	99.99%	—	—	18

milliliter at 11:00 a.m. and remained at that level for the next 12 hr. Several days later, when coagulation was resumed, and the filter was backwashed, the bacteria count decreased from over 1000 to under 10 per milliliter in 3 hr. Arozarena[15] demonstrated that interruption of coagulant feed caused an increased concentration of *G. muris* cysts in filtered water. In a run with an alum dose of 22 mg/ℓ, cyst concentration rose from less than 80 cysts per liter with coagulation to 790 cysts per liter and later 1760 cysts per liter after the chemical feed was shut off. The large increase in cysts occurred even though filtered water turbidity remained at less than 1 NTU; turbidity increased from 0.27 NTUs with coagulation to only 0.37 NTUs and later 0.57 NTUs without coagulation. In a second experiment of this nature, water was coagulated with 5 mg/ℓ alum and 0.9 mg/ℓ cationic polymer. Filtered water turbidity was 0.28 NTUs, and cyst concentration was 70 per liter. After chemical feed was shut off, turbidity increased to only 0.38 NTUs, but the cyst concentration increased to 3310 cysts per liter.

Filtered water quality can be poor at the beginning of a run, during the initial period after backwashing before a filter has ripened. In experiments with the Warren Filter at Pittsburgh, Hazen[3] showed that bacteria counts in filtered water were as much as 5 to 10 times higher during the first 20 min after backwashing, as compared to counts at the end of the prior run or 50 to 60 min after the new run had begun. Logsdon et al.[16] reported that *G. muris* cyst concentrations observed during the ripening phases (the first 30 min) of three filter runs were from 10 to 25 times higher than cyst concentrations measured later in the runs, but in three other runs, this phenomenon was not observed. These results suggest that the practice of filtering to waste for perhaps 10 to 30 min immediately after backwash may have merit.

The detrimental effect of rate changes on filtered water quality has been demonstrated by Cleasby et al.[19] and by DiBernardo and Cleasby[20] for filters removing iron and for filters operated for turbidity removal. Similar results were observed in a filter operated for *Giardia* cyst removal.[16]

Turbidity breakthrough at the end of a filter run can also result in large numbers of microorganisms passing into filtered water. Robeck[7] showed that an increase in the poliovirus concentration in filtered water was associated with a turbidity increase at the end of a filter run. Logsdon et al.[16] showed that the passage of large numbers of *G. muris* cysts was also associated with turbidity breakthrough. The increase in cyst concentration was not in direct proportion to the increase in turbidity, but was much greater than the turbidity increase. In one run in which cysts were added to the raw water during the first 31 hr of operation, a turbidity increase from 0.4 to 0.7 NTUs resulted in an increase in *Giardia* cyst concentration from 740 to 2900 cysts per liter. The latter concentration was observed 2 hr after the supply of cysts had been exhausted. Turbidity breakthrough can cause previously stored contaminants to be dislodged from the filter bed and discharged into the filtered water. Under such circumstances, effluent concentration from a filter can exceed the influent concentration for the contaminant.

These results of rapid rate filtration experiments show the need for careful control of coagulation chemistry and for continuous or frequent monitoring of the effluent turbidity at each filter in a treatment plant. Rapid rate filters are most effective for removal of microorganisms when coagulation pH is controlled at the optimum range, when the dose of coagulant is adequate, when floc is strong enough to resist breakthrough, when abrupt filtration rate increases are avoided, when filtered water turbidity is maintained well below 1.0 NTU, and when filtered water turbidity increases are not allowed to occur during a filter run.

3. DE Filters

DE filtration can remove a variety of microorganisms. Performance is related to the size of the microorganism, the grade of DE used, the thickness of precoat, and the presence or absence of special coatings on the DE.

Two investigations of virus removal have been conducted[21,22] with T-2 bacteriophage and poliovirus Mahoney Type I inoculated in the water. Malina et al.[21] showed that bacteriophage removal was lower when a more porous grade of DE was used and higher when a finer grade of DE was used. The best removal was achieved when DE was coated with hydrous iron oxides or when cationic polymer was added to the raw water. One 12-hr run was conducted in which cationic polymer (Dow® C-31, 0.07 mg/ℓ) was added to raw water, and the feed water contained 1700 pfu/ℓ (plaque forming units per liter) of T-2 bacteriophage. No bacteriophages were detected in 11 of the 12 samples of the filtered water, but in one sample, 2 pfu/ℓ were found. This filtration technique thus attained a reduction of T-2 virus of 99.88% or better in these experiments.

Malina et al.[22] enhanced poliovirus removal with the use of coated DE filter aid or the use of cationic polymer in the raw water compared to uncoated diatomite, and two 12-hr runs were performed to evaluate these techniques for poliovirus removal. In one run, Hyflo® filter aid was coated with 1 mg of Dow® C-31 cationic polymer per gram of filter aid, and 2350 pfu/ℓ of polio virus was added to the feed water. No viruses were recovered in 11 of the 12 filtered water samples, indicating a virus removal exceeding 99.95%. Removal was greater than 99% in the single sample that did contain 15 pfu/ℓ of virus. In another 12-hr run, in which uncoated DE was used and 0.14 mg/ℓ of Dow® C-31 cationic polymer was added to the raw water as a supplemental treatment, no viruses were recovered from any of the 12 filtered water samples. Because the raw water was spiked with 2950 pfu/ℓ, the recovery of only 1 virus pfu/ℓ would have yielded a 99.96% removal rate; thus, the performance exceeded this percentage.

Hunter et al.[23] studied coliform removal by DE filtration and concluded that penetration of the filter by coliforms was greater when the filter cake was more permeable. The most permeable grade tested (Hyflo®) had the lowest coliform removal (about 90% when raw water coliform count exceeded 1000 per 100 mℓ). Removal for Celite 512® exceeded 99.0%. The finest grade of DE tested, Standard Super-Cel®, gave coliform removals of 99.86% or better when the raw water coliform count was 19,000 per 100 mℓ. When Standard Super-Cel® was used, no coliforms were detected when raw water coliforms were in the range of 6 per 100 mℓ to 7000 per 100 mℓ.

When coliforms did pass through the filter, the level of coliforms in filtered water was higher when influent coliform count was higher. In actual practice, however, disinfection would be carried out, so DE filtration would not be relied upon as the sole barrier to passage of bacteria in drinking water. For better control of coliforms, the benefit of using more permeable filter aid resulting in longer runs should be weighed against using less permeable filter aid, thus getting shorter runs but better coliform removal.

Research performed at CSU[24] with the DE test filter has confirmed the earlier observation of Hunter et al.[23] that fewer total coliforms pass through the DE filter as finer grades of diatomite are used and has shown that removals of standard plate count organisms, total coliforms, and turbidity all improve as finer grades of DE are used (Table 7). Because of the size differences of bacteria vs. the fine clays in the source water, the removal of turbidity-causing particles was less than the removal of microorganisms for most grades of DE. The finest grade tested, FilterCel®, was very effective for turbidity removal, but this grade usually is not used in potable water treatment because of the high head loss caused by small particle size.

Turbidity removal by DE filtration and membrane filtration was compared to learn what sizes of particles were causing the turbidity in the Horsetooth Reservoir water used in the CSU research.[24] A 5-μm filter permitted passage of about 98% of the turbidity, more than 88% of turbidity passed by Celite 503®; a 1.2 μm filter passed 64%, similar to the 72% passed by Celite 512®; the 9% passed by a 0.2-μm membrane filter was similar to the 2% passed by FilterCel®. These results indicate that particle sizes smaller than 1 μm were causing a substantial amount of the turbidity in this water.

Table 7
AVERAGE REMOVAL PERCENTAGES OF TOTAL COLIFORM BACTERIA, STANDARD PLATE COUNT BACTERIA, TURBIDITY, AND *GIARDIA* CYSTS WITH VARIOUS GRADES OF DIATOMACEOUS EARTH[9]

Grade of diatomaceous earth	Celite 545®	Celite 535®	Celite 503®	Hyflo Super-Cel®	Celite 512®	Standard Super-Cel®	Filter Cel®
Number of tests	15	2	10	4	2	3	2
Median particle size (μm)	26.0	25.0	23.0	18.0	15.0	14.0	7.5
Total coliform removal (%)	49.46	85.45	69.41	92.60	97.95	99.90	99.86
Standard plate count removal (%)	58.42	80.51	69.10	75.29	78.63	99.11	99.85
Turbidity removal (%)	13.36	12.50	11.97	23.56	31.14	49.77	97.64
Giardia cyst removal (%)	>99.7	>99.5	>99.7	>99.4	—	—	—

Table 8
REMOVAL PERCENTAGES OF TOTAL COLIFORM BACTERIA, STANDARD PLATE COUNT BACTERIA, AND TURBIDITY FOR DIATOMACEOUS EARTH GRADES C-545 AND C-503 COATED AT VARIOUS ALUM CONCENTRATIONS[24]

		Average removals		
Grade	Alum coat (% by wt)	Total coliform (%)	Standard plate count (%)	Turbidity (%)
C-545	2	—	—	66.11
	4	99.02	95.02	86.41
	5	99.86	98.56	98.38
C-503	5[a]	98.01	79.31	79.06
	5	99.56	93.25	94.41
	5[b]	96.33	99.57	98.61
	8	99.83	99.52	98.80

[a] Bodyfeed concentration was increased to 50 ppm after 3 hr of testing at 25 ppm.
[b] Bodyfeed concentration 50 ppm; all other bodyfeed rates were 25 ppm.

Research with alum-coated DE showed that diatomite onto which aluminum hydroxide had been precipitated was more effective for removal of standard plate count organisms, total coliforms, and turbidity than the uncoated DE provided by the manufacturer and generally used for precoat and body feed. The use of alum coating resulted in a dramatic improvement in filtered water quality (Table 8).

Research on DE filtration during World War II led the U.S. Army to state, "A filtration unit properly designed and operated would be expected to remove completely the cysts of *Endamoeba histolytica*."[14] The success of the U.S. Army in using DE filtration for *E. histolytica* cyst removal led to later research on DE filtration for removal of *Giardia* cysts.

Research on DE filtration, conducted with a 1-ft^2 filter, for removal of 9-μm radioactive beads (*Giardia* cyst models), *G. muris* cysts, and *G. lamblia* cysts has shown that DE filtration is an effective treatment method for cyst-contaminated waters. Data obtained with radioactive beads indicated that a precoat thickness of 1.0 kg/m^2 (0.2 lb/ft^2) should be used.[16] EPA research showed that a DE filter with adequate precoat and a clean septum could consistently remove more than 99% of the *G. muris* cysts applied to the filter.[16] Research with *G. lamblia* cysts at the University of Washington, also with the 1-ft^2 filter, confirmed these results. Cyst removals for four filter runs ranged from 99.02% to above 99.87% in 12 separate determinations.[17] Similar results were obtained at McIndoe Falls, Vt., with a 10-ft^2 filter. In a test in which 8×10^6 *G. lamblia* cysts were added to the raw water, removal was 99.97%.[9]

The most thorough study of *Giardia* cyst removal conducted to date is the work done at CSU.[9,24] Filtration rates of 1 to 4 gpm/sf (2.4 to 9.8 m/hr) were evaluated, and a variety of DE grades were studied using the 1-ft^2 filter (Table 9).

Of the 26 runs, cysts were detected in only 1 run, and that had a very high influent cyst concentration of 34,000 cysts per liter. When influent cyst concentrations ranged from 100 to 10,000 cysts per liter, no cysts were detected in the filtered water. These experiments were performed with Horsetooth Reservoir water, which had very fine turbidity-causing particles. With this water source, cyst removal efficiency was not related to turbidity removal by DE filters. Results indicate that DE filtration is effective for cyst removal at rates of 2.4 to 9.8 m/hr.

Field tests for removal of *Giardia* cysts were performed at the Cache la Poudre River and

Table 9
GIARDIA CYST TURBIDITY DATA FOR DIATOMACEOUS EARTH FILTRATION TESTS[9]

Run no.	Celite® grade	Filtration rate (m/hr)	Temp. (°C)	Length of test (min)	Influent *Giardia* cyst concentration Added to feed water (cysts/ℓ)	Influent *Giardia* cyst concentration Detected in feed water (cysts/ℓ)	No. of cysts detected in effluent	Effluent *Giardia* concentration (cysts/ℓ)	*Giardia* cyst percent removal %	Turbidity Influent NTU	Turbidity Effluent NTU
13	545	2.44	5	90	100	27.7	0	< 0.701	> 99.299	3.5	3.3
18	545	2.44	13	90	100	50.5	0	< 0.461	> 99.539	4.2	3.6
20	545	2.44	5	90	500	229.0	0	< 0.465	> 99.907	4.4	3.5
28	545	2.44	5	340	770	75.0	0	< 0.326	> 99.958	4.6	3.8
42	545	2.44	13	145	10,000	—	0	< 0.112	> 99.998	9.1	6.9
43	545	2.44	15	150	5,460	—	0	< 0.108	> 99.998	—	—
45	545	2.44	14	260	8,850	—	0	< 0.063	> 99.999	—	—
41	545	2.44	12	55	3.36 × 10⁴	—	1,700	25.148	99.925	—	—
49	545	2.44	10.5	980	2,467	—	0	< 0.0004	> 99.999	9.9	8.4
14	545	4.88	5	90	100	44.4	0	< 0.425	> 99.575	3.5	3.4
19	545	4.88	13	90	100	65.9	0	< 0.323	> 99.323	4.4	3.6
21	545	4.88	5	90	500	330.0	0	< 0.326	> 99.326	4.4	3.6
17	545	9.76	5	90	100	47.0	0	< 0.443	> 99.557	4.2	3.7
26	545	9.76	5	90	500	31.5	0	< 3.342	> 99.332	4.2	3.8
15	535	2.44	5	90	100	45.5	0	< 0.423	> 99.577	3.6	3.2
16	535	4.88	5	90	100	45.5	0	< 0.453	> 99.547	3.6	3.1
11	503	2.44	5	90	100	76.4	0	< 0.257	> 99.743	3.5	3.3
25	503	2.44	5	90	500	127.0	0	< 0.691	> 99.862	4.4	3.7
44	503	2.44	15	275	5,460	—	0	< 0.058	> 99.998	—	—
12	503	4.88	5	90	100	40.0	0	< 0.481	> 99.519	3.6	3.3
22	503	4.88	13	90	100	59.0	0	< 0.357	> 99.643	4.2	3.6
24	503	4.88	5	90	500	127.0	0	< 0.895	> 99.821	4.3	3.5
27	503	9.76	13	90	100	33.0	0	< 0.532	> 99.468	4.2	3.7
9	Hyflo	2.44	5	90	100	41.4	0	< 0.478	> 99.522	3.7	3.0
10	Hyflo	4.88	5	90	100	27.2	0	< 0.694	> 99.306	3.5	3.1
45	545	4.88	14	40	8,850	—	0	< 0.408	> 99.995	6.6	—

at Dillon, Colo., at filtration rates of 2.4 to 9.8 m/hr. Celite 545® was used in the six filter runs in which cyst removal was studied. In five of the six runs, no cysts were detected, and removals exceeded 99.93 to 99.99%, depending on influent cyst concentration and the volume of filtered water sampled for cysts. Filtered water turbidity in these runs ranged from 0.3 to 1.5 NTUs. During one of the six filter runs, cysts passed through the filter, and when the filter was subjected to leak testing, the existence of a leak was confirmed. The turbidity of the filtered water in the run in which the leak occurred was 7.3 NTUs, and cyst removal was 99.4%. These results suggest that when DE filtration equipment is properly maintained, this process is capable of removing over 99.9% of *Giardia* cysts from raw water, and that process efficiency is related to proper operation and maintenance of equipment rather than to effluent turbidity.

Studies cited above indicate that as particle size decreases from cysts to bacteria to virus, the straining mechanism for particle removal by DE filtration becomes less effective. Research has shown, however, that the use of cationic polymer or aluminum or iron salts to form a coating on the DE can substantially enhance the removal of very small particles by DE filtration so that this process can also become a barrier to the passage of bacteria and viruses. Use of DE has been shown repeatedly to be effective for the removal of cysts.

III. DISINFECTION

Disinfection is usually the final step in drinking water treatment and the primary means to prevent waterborne outbreaks. In some instances disinfection is the only treatment provided. Although disinfection may be conducted continuously throughout the treatment process to prevent growth of nuisance microorganisms, the major goal of the process is to ensure inactivation of waterborne pathogens. Disinfection is not intended to sterilize water, i.e., to kill all living organisms in the water.

Selection of the disinfection process to be using in drinking water treatment depends on a number of factors[25] including

1. Efficacy against waterborne pathogens (bacteria, viruses, protozoans, etc.)
2. Accuracy with which process can be monitored and controlled
3. Ability to maintain effective residual to counteract possible contamination after treatment
4. Effects on aesthetic quality
5. Availability of the technology for application on the scale needed, including cost considerations

The disinfecting capabilities of a wide variety of chemical and physical agents have been investigated, but relatively few meet the above criteria and can be realistically considered for current use in drinking water disinfection. Given the heavy reliance on disinfection in preventing waterborne disease transmission, those responsible for water treatment are understandably reluctant to use innovative disinfection techniques in the absence of overwhelming favorable evidence because chlorine has been studied extensively and has a long history of use in the U.S. The increasing concern about the possibility of adverse health effects of disinfectants and their by-products has caused a reassessment of various disinfectants to find a suitable disinfectant that will not produce undesirable by-products. Chlorine has been found to react with certain precursors to form chlorinated by-products. Other disinfectants, such as chloramine, may not produce these same by-products but may not be as effective as chlorine in inactivating pathogens.

The suitability of a large number of possible drinking water disinfection methods was evaluated recently by the Safe Drinking Water Committee of the National Research Council[26] (Table 10). Although several of the agents provided adequate inactivation, the Committee

Table 10
STATUS OF POSSIBLE METHODS FOR DRINKING WATER DISINFECTION

Disinfection agent	Suitability as inactivating agent	Limitations	Suitability for drinking water disinfection[a]
Chlorine	Yes	Efficacy decreases with increasing pH; affected by ammonia or organic nitrogen	Yes
Ozone	Yes	On-site generation required; no residual; other disinfectant needed for residual	Yes
Chlorine dioxide	Yes	On-site generation required; interim MCL 1.0 mg/ℓ	Yes
Iodine	Yes	Biocidal activity sensitive to pH	No
Bromine	Yes	Lack of technological experience; activity may be pH sensitive	No
Chloramines	No	Mediocre bactericide, poor virucide	No[b]
Ferrate	Yes	Moderate bactericide; good virucide; residual unstable; lack of technological experience	No
High pH conditions	No	Poor biocide	No
Hydrogen peroxide	No	Poor biocide	No
Ionizing radiation	Yes	Lack of technological experience	No
Potassium permanganate	No	Poor biocide	No
Silver	No	Poor biocide; MCL 0.05 gm/ℓ	No
UV light	Yes	Adequate biocide; no residual; use limited by equipment maintenance considerations	No

[a] This evaluation relates solely to the suitability for controlling infectious disease transmission.
[b] Chloramines may have use as a secondary disinfectant in the distribution system in view of their persistence.

From Reference 26. With permission.

regarded only free chlorine, ozone, and chlorine dioxide as suitable for actual use in drinking water disinfection. Other disinfectants were not recommended for a variety of reasons, including poor biocidal activity, lack of technological experience, and adverse health effects to some population groups. For many utilities, conversion of systems from free chlorine to chloramines is the simplest and most economical method for reducing formation of trihalomethanes in finished drinking water, and these utilities have increased their reliance on chloramines in spite of the poor biocidal capability of chloramines.

The most important single factor considered in choosing a particular disinfectant is its effectiveness against various types of pathogenic microorganisms. A large body of literature on the biocidal characteristics of drinking water disinfectants has been established. The information is based primarily on laboratory studies conducted under closely controlled conditions. For comparison of biocidal efficiency, Baumann and Ludwig[27] suggested the use of an equation based on log-log plots of time vs. disinfectant concentration needed to inactivate a certain percentage (e.g., 99%) of a given microorganism under defined pH and temperature conditions. The general equation is $C \cdot t' = K$ in which C is the disinfectant concentration, t' is the contact time, and K is the constant for a particular microorganism. The concept was to provide a method for determining contact time and disinfectant concentration needed to inactivate a particular pathogen of concern under defined conditions, and this formula has been used for the disinfection data in this chapter.

In actual water treatment practice the disinfectant concentration is generally known to a greater degree of accuracy than the contact time. Often, contact time is estimated on the basis of theoretical flow through or displacement time (water volume divided by flow rate). Few chlorine contact basins have conditions of plug flow, however, so the actual retention

Table 11
EFFECT OF TEMPERATURE ON RATE OF INACTIVATION OF *E. COLI* BY CHLORINE DIOXIDE[29]

Temp. (°C)	Chlorine dioxide conc. (mg/ℓ)	Contact time (min)	$C \cdot t'^{a}$
5	0.30	1.8	0.54
15	0.30	1.3	0.39
25	0.30	0.98	0.29

[a] Chlorine dioxide concentration times contact time for 99% inactivation.

Table 12
INACTIVATION OF MICROORGANISMS BY FREE CHLORINE SPECIES

Test microorganism	Free chlorine species	pH	Temp. (°C)	Conc. (mg/ℓ)	Contact time (min)	$C \cdot t'^{a}$	Ref.
Escherichia coli	HOCl	6.0	5	0.1	0.4	0.04	31
	OCl⁻	10.0	5	1.0	0.92	0.92	31
Poliovirus I	HOCl	6.0	5	0.5	2.1	1.05	32
	OCl⁻	10.0	5	0.5	21	10.5	32
Endamoeba histolytica cysts	HOCl	6.0	5	5.0	18	90	33
G. lamblia cysts	HOCl	6.0	5	2.0	40	80	34

[a] Concentration times contact time for 99% inactivation.

time for a portion of the water flowing through the contact basin is only a fraction of the theoretical time because of short circuiting. Additional information on this is presented in Chapter 9.

A. General Considerations

Disinfectant effectiveness is influenced by several physical and chemical factors. One of the most important of these is water temperature. For all chemical disinfectants, inactivation rates decrease as water temperatures decrease. Because of this effect, cold water temperature represents a "worst case" condition for disinfection effectiveness. Most of the information presented below is based on experiments conducted at 5°C or less. Water pH also can affect inactivation rates. In the case of free chlorine, water pH determines the proportions of the active agents, hypochlorous acid (HOCl) and hypochlorite ion (OCl⁻) present. Since the HOCl species is a much more effective disinfectant, lower pH values (pH 6 to 7), which result in the formation of HOCl, are favorable for rapid inactivation while higher pH values (pH 8 to 10) at which OCl⁻ predominates result in slower rates of inactivation. With chlorine dioxide (ClO_2), which does not dissociate into different chemical species, inactivation is more rapid at higher pH values (pH 9) than at lower pH value (pH 7).[28,29] The disinfection capability of ozone appears to be unaffected by pH[30] (Table 11).

B. Microorganism Inactivation Capabilities
1. Chlorine Species

Results of a number of studies of the disinfection efficiency of free chlorine species for inactivation of representative enteric bacteria, enteroviruses, and protozoan pathogens are shown in Table 12. The polioviruses are much more resistant than *Escherichia coli* to both

Table 13
INACTIVATION OF MICROORGANISMS BY INORGANIC CHLORAMINES

Test microorganism	Combined chlorine species	pH	Temp. (°C)	Conc. (mg/ℓ)	Contact time (min)	C · t'ᵃ	Ref.
E. coli	NH$_2$Cl	9.0	5	1.0	175	175	35
	NH$_2$Cl	9.0	15	1.0	64	64	35
	NHCl$_2$	4.5	15	1.0	5.5	5.5	36
Poliovirus I	NH$_2$Cl	9.0	15	10	90	900	35
	NHCl$_2$	4.5	5	100	140	14,000	36
	NHCl$_2$	4.5	15	100	50	5,000	36
G. lamblia cysts	NH$_2$Cl/NHCl$_2$	7.5	3	2.4	220	528	37

ᵃ Concentration times contact time for 99% inactivation.

species of free chlorine, and cysts of *G. lamblia* and *Endamoeba histolytica* are in turn nearly 100-fold more resistant than poliovirus. The much greater bactericidal and virucidal efficiency of HOCl as compared with OCl⁻ also is evident.

Results of similar studies using inorganic chloramine disinfectants are shown in Table 13. In general, for all types of microorganisms, C · t' values for chloramines are higher than C · t' values for free chlorine species. C · t' values for the enteroviruses are extremely high while values for *G. lamblia* cysts are lower, in contrast to the findings for free chlorine species. Although *E. histolytica* cyst inactivation data suitable for C · t' comparison are not available, studies[38] that have been done indicate that *E. histolytica* cyst resistance to inorganic chloramines is similar to that of *G. lamblia* cysts. Again, increased inactivation rates at higher temperatures are seen.

2. Chlorine Dioxide

C · t' data for ClO$_2$ are shown in Table 14. The data indicate that at pH 7.0, ClO$_2$ is a somewhat weaker bactericide and virucide than HOCl. In contrast to free chlorine, the efficiency of ClO$_2$ increases as pH increases, as the poliovirus results indicate that ClO$_2$ is a more efficient virucide than HOCl. Protozoan cyst data suitable for C · t' comparisons are not available.

3. Ozone

C · t' values for ozone are shown in Table 15. *G. muris* cysts are on the order of tenfold more resistant than poliovirus 1, which in turn is about tenfold more resistant than *Escherichia coli*. Overall comparison of the ozone C · t' values with those of ClO$_2$ and chlorine species show that ozone is a much more potent biocide than the other disinfectants.

C. Relevance to Disinfection Practice

The laboratory studies clearly indicate that there are differences in relative disinfectant efficiency related to the chemical nature of the disinfectant, the type of microorganism, and physical factors such as pH and temperature. In general, enteric bacteria are the most sensitive to all disinfectants discussed; the enteroviruses are intermediate in resistance, followed by protozoan cysts. An anomaly may exist for inorganic chloramine in that *Giardia* cysts appear to be more sensitive to this disinfectant than enteroviruses. The reason for this is not presently understood. Although less quantifiable, there are similar indications from studies of inactivation of *Endamoeba histolytica* cysts. When the disinfectants considered are compared on a weight basis, ozone causes the most rapid inactivation of all types of microorganisms. Chlorine dioxide and free chlorine in the form of HOCl are about equivalent. Chlorine

Table 14
INACTIVATION OF MICROORGANISMS BY CHLORINE DIOXIDE[29]

Test microorganisms	pH	Temp. (°C)	Conc. (mg/ℓ)	Contact time (min)	C · t'[a]
E. coli	7.0	5	0.30	1.8	0.54
Poliovirus I	7.0	5	0.5	12.0	6.0
	7.0	21	0.3	5.0	1.5
	9.0	21	0.4	1.0	0.4

[a] Concentration times contact time for 99% inactivation.

Table 15
INACTIVATION OF MICROORGANISMS BY OZONE

Test microorganisms	pH	Temp. (°C)	Conc. (mg/ℓ)	Contact time (min)	C · t'[a]	Ref.
E. coli	7.2	1	0.07	0.083	0.006	39
	7.2	1	0.065	0.33	0.022	39
Poliovirus I	7.2	5	0.15	1.47	0.22	40
G. muris cysts	7.0	5	0.15	12.9	1.94	41

[a] Concentration times contact time for 99% inactivation.

dioxide has an advantage over free chlorine in that it is less likely to be depleted through reactions with ammonia. Both free chlorine in the form of OCl^- and chloramines are much weaker biocides than chlorine dioxide.

The potential for interference with disinfection efficiency by protection of microorganisms associated with various types of particles has been demonstrated for chlorine,[42,43] chlorine dioxide,[44] and ozone.[45,46] Association with organic particles such as waste water effluent solids and cell debris offers a much higher degree of protection than association with inorganic particles such as clays and alum floc. None of the disinfectants appears to possess any particular ability to penetrate such particles. The results of these studies illustrate the need for physical removal of particles in preparing water for the disinfection step in order to ensure protection against waterborne disease transmission.

IV. SUMMARY

Water clarification processes that are properly designed and operated can be an effective barrier to the passage of microorganisms into finished drinking water. At conventional treatment plants that employ coagulation, flocculation, sedimentation, and filtration, the sedimentation process constitutes one barrier to passage of pathogens, and a second is provided by filtration. At rapid rate direct filtration plants, slow sand filtration plants, and DE filtration plants, sedimentation is not used, so the only physical removal step (barrier) is the filtration process. When these latter processes are considered, engineers should be careful to verify that the raw water quality is appropriate for filtration without prior sedimentation.

Water filtration plants must be operated and maintained properly in order to achieve their potential for microorganism removal. At plants without sedimentation, extra care must be taken to ensure that the filtration process is an effective barrier. This chapter discussed facets of plant operation that need attention.

Disinfection is a major barrier to the passage of pathogens into finished drinking water, and the one upon which the most reliance has been placed in this century. Although numerous chemical and physical agents might inactivate microorganisms, in actual practice only chlorine, chlorine dioxide, and ozone find wide acceptance for drinking water treatment. The vast majority of water utilities in the U.S. use chlorination, either free chlorine or chloramine, or both for disinfection.

Engineers and treatment plant operators need to be aware that disinfection efficacy can be influenced by the type and concentration of disinfectant used, contact time, water temperature, water turbidity, type of microorganism, and in some instances, pH. Some pathogens are more difficult to inactivate than indicator organisms, so caution must be used when assessing the efficacy of disinfection in terms of inactivation of indicator organisms. In spite of these caveats, disinfection remains the most effective and most relied upon barrier in water treatment. For many ground waters, disinfection constitutes the only treatment barrier, with major reliance being placed on source protection barriers. In these situations, consistent attainment of effective disinfection is particularly important.

REFERENCES

1. **Hazen, A.**, *The Filtration of Public Water Supplies*, 3rd ed., John Wiley & Sons, New York, 1913, 32.
2. **Brink, D. R., McElroy, J. M., Al-Ani, M., Bellamy, W. D., Howell, D., Rau, D. M., and Hendricks, D. W.**, Removal of *Giardia lamblia* from Water Supplies; Appropriate Treatment Technologies for Small Systems, Environmental Engineering Technical Report 5847-83-3, Colorado State University, Fort Collins, September, 1983, 1.
3. Report of the Filtration Commission of the City of Pittsburgh, Pennsylvania, January, 1899, pp. 124, 125, 138, 139, 152—154, 167—168, 170—171.
4. **Cleasby, J. L.**, Slow sand filtration and direct in-line filtration of a surface water, in *Proc. AWWA Seminar Innovative Filtration Techniques*, American Water Works Association Publ. No. 20176, Denver, Colo., 1983, 10.
5. **Bellamy, W. D. et al.**, Removing *Giardia* cysts with slow sand filtration, *J. Am. Water Works Assoc.*, 77(2), 52, 1984.
6. **Windle Taylor, E.**, Forty-Fourth Report on the Results of the Bacteriological, Chemical, and Biological Examination of the London Waters for the Years 1969—1970, Metropolitan Water Board, London, 52.
7. **Robeck, G. G., Clarke, N. A., and Dostal, K. A.**, Effectiveness of water treatment processes in virus removal, *J. Am. Water Works Assoc.*, 54, 1275, 1962.
8. **Jordan, H. E.**, Fifteen years filtration practice in Indianapolis, in *Proc. XIII Annu. Convention Indiana Sanitary and Water Supply Assoc.*, 1920.
9. **Logsdon, G. S., Hendricks, D. W., Pyper, G. R., Hibler, C. P., and Sjogren, R.**, Control of *Giardia* cysts by filtration: the laboratory's role, in *Proc. 11th Annu. AWWA Water Quality Technol. Conf.*, American Water Works Association, Denver, Colo., 1984, 265.
10. **Poynter, S. F. B. and Slade, J. S.**, The removal of viruses by slow sand filtration, *Prog. Water Technol.*, 9, 75, 1977.
11. **Huisman, L. and Wood, W. E.**, *Slow Sand Filtration*, World Health Organization, Geneva, 1974, chap. 4.
12. **Sproul, O. J.**, Critical Review of Virus Removal by Coagulation Processes and pH Modifications, EPA-600/2-80-004, U.S. Environmental Protection Agency, Cincinnati, Ohio, 1980, 2.
13. **Engelbrecht, R. S. et al.**, Acid-Fast Bacteria and Yeasts as Indicators of Disinfection Efficiency, EPA-600/12-79-091, U.S. Environmental Protection Agency, Cincinnati, Ohio, 1979, 7.
14. Efficiency of Standard Army Water Purification Equipment and of Diatomite Filters in Removing Cysts of *Endamoeba histolytica* from Water, War Department Report 834, July 3, 1944, pp. 27, 63—82.
15. **Arozarena, M. M.**, Removal of *Giardia muris* Cysts by Granular Media Filtration, M.Sc. thesis, University of Cincinnati, Ohio, 1979, 34.
16. **Logsdon, G. S. et al.**, Alternative filtration methods for removal of *Giardia* cysts and cyst models, *J. Am. Water Works Assoc.*, 73, 111, 1981.
17. **DeWalle, F. B., Engeset, J., and Lawrence, W.**, Removal of *Giardia lamblia* Cysts by Drinking Water Treatment Plants, EPA-600/2-84-069, U.S. Environmental Protection Agency, Cincinnati, Ohio, 1984.
18. **Baylis, J. R., Gullans, O., and Spector, B. K.**, The efficiency of rapid sand filters in removing the cysts of amoebic dysentery organisms from water, *Public Health Rep.*, 51, 1567, 1936.

19. **Cleasby, J. L., Williamson, M. M., and Baumann, E. R.**, Effect of filtration rate changes on quality, *J. Am. Water Works Assoc.*, 55, 869, 1963.
20. **DiBernardo, L. and Cleasby, J. L.**, Hydraulic considerations in declining-rate filtration, *J. Environ. Eng. Div. Am. Soc. Civil Eng.*, 106, 1023, 1980.
21. **Malina, J. F., Jr., Brown, T. S., and Moore, B. D.**, T-2 Bacteriophage Removal by Diatomaceous Earth Filtration, University of Texas, Austin, 1971.
22. **Malina, J. F., Jr., Moore, B. D., and Marshall, J. L.**, Poliovirus Removal by Diatomaceous Earth Filtration, University of Texas, Austin, 1972.
23. **Hunter, J. V., Bell, G. R., and Henderson, C. N.**, Coliform organism removals by diatomite filtration, *J. Am. Water Works Assoc.*, 58, 1160, 1966.
24. **Lange, K. P., Bellamy, W. D., and Hendricks, D. W.**, Filtration of *Giardia* Cysts and Other Substances, Volume 1, Diatomaceous Earth Filtration, EPA-600/2-84-114, U.S. Environmental Protection Agency, Cincinnati, 1984.
25. **Symons, J. M., Carswell, J. K., Clark, R. M., Dorsey, P., Geldreich, E. E., Heffernan, W. P., Hoff, J. C., Love, O. T., McCabe, L. J., and Stevens, A. A.**, Ozone, chlorine dioxide, and chloramines as alternatives to chlorine for disinfection of drinking water: state of the art, in *Ozone and Chlorine Dioxide Technology for Disinfection of Drinking Water*, Katz, J., Ed., Noyes Data Corp., Park Ridge, N.J., 1980, 2.
26. National Research Council, Safe Drinking Water Committee, The disinfection of drinking water, in *Drinking Water and Health*, Vol 2, National Academy Press, Washington, D.C., 1980, 5.
27. **Baumann, E. R. and Ludwig, D. D.**, Free available chlorine residuals for small nonpublic water supplies, *J. Am. Water Works Assoc.*, 54, 1379, 1962.
28. **Benarde, M. A., Israel, B. M., Olivieri, V. P.,and Granstrom, M. L.**, Efficiency of chlorine dioxide as a bactericide, *Appl. Microbiol.*, 13, 776, 1965.
29. **Cronier, S. D.**, Destruction by Chlorine Dioxide of Viruses and Bacteria in Water, M.Sc. thesis, University of Cincinnati, Ohio, 1977, 85.
30. **Morris, J. C.**, The role of ozone in water treatment, in *Proc. 96th Annu. Conf. Am. Water Works Assoc.*, Vol. 2, American Water Works Association, Denver, Colo., 1976, 26.
31. **Scarpino, P. V., Lucas, M., Dahling, D. R., Berg, G., and Chang, S. L.**, Effectiveness of hypochlorous acid and hypochlorite ion in destruction of viruses and bacteria, in *Chemistry of Water Supply Treatment and Distribution*, Rubin, A. J., Ed., Ann Arbor Science, Ann Arbor, Mich., 1974, 359.
32. **Engelbrecht, R. S., Weber, M. J., Salter, B. L., and Schmidt, C. A.**, Comparative inactivation of viruses by chlorine, *Appl. Environ. Microbiol.*, 40, 249, 1980.
33. **Snow, W. B.**, Recommended residuals for military water supplies, *J. Am. Water Works Assoc.*, 48, 1510, 1956.
34. **Jarroll, E. L., Bingham, A. K., and Meyer, E. A.**, Effect of chlorine on *Giardia lamblia* viability, *Appl. Environ. Microbiol.*, 41, 483, 1981.
35. **Siders, D. L., Scarpino, P. V., Lucas, M., Berg, G., and Chang, S. L.**, Destruction of viruses and bacteria in water by monochloramine, *Abstr. Annu. Meet. Am. Soc. Microbiol.*, Washington, D.C., E27 (Abstr.), 1973.
36. **Esposito, P., Scarpino, P. V., Chang, S. L., and Berg, G.**, Destruction by dichloramine of viruses and bacteria in water, *Abstr. Annu. Meet. Am. Soc. Microbiol.*, Washington, D.C., G99 (Abstr.), 1974.
37. **Bingham, A. K. and Meyer, E. A.**, Unpublished data, 1984.
38. **Stringer, R. and Kruse, C. W.**, Amoebic cysticidal properties of halogens in water, in *Proc. Natl. Specialty Conf. Disinfection*, American Society of Civil Engineers, Amherst, Mass., 1970, 319.
39. **Katznelson, E., Kletter, B., and Shuval, H. I.**, Inactivation kinetics of viruses and bacteria in water by use of ozone, *J. Am. Water Works Assoc.*, 66, 725, 1974.
40. **Roy, D., Engelbrecht, R. S., and Chian, E. S. K.**, Comparative inactivation of six enteroviruses by ozone, *J. Am. Water Works Assoc.*, 74, 660, 1982.
41. **Wickramanayake, G. B., Rubin, A. J., and Sproul, O. J.**, Inactivation of *Naegleria* and *Giardia* cysts in water by ozonation, *J. Water Pollut. Control Fed.*, 56(8), 983, 1984.
42. **Hoff, J. C.**, The relationship of turbidity to disinfection of potable water, in Evaluation of the Microbiology Standards for Drinking Water, Hendricks, C. H., Ed., EPA-570/9-78-OOC, U.S. Environmental Protection Agency, Washington, D.C., 1978, 103.
43. **Hoff, J. C. and Geldreich, E. E.**, Comparison of the biocidal efficiency of alternative disinfectants, *J. Am. Water Works Assoc.*, 73, 40, 1981.
44. **Brigano, F. A. O., Scarpino, P. V., Cronier, S., and Zink, M. L.**, Effect of particulates on inactivation of enteroviruses by chlorine dioxide, in Progress in Wastewater Disinfection Technology, Section 42, Venosa, A. D., Ed., EPA-600/9-79-018, 1979, 86.
45. **Walsh, D. S., Buck, C. E., and Sproul, O. J.**, Ozone inactivation of floc associated viruses and bacteria, *J. Environ. Eng. Div. Am. Soc. Civil Eng.*, 106, 711, 1980.
46. **Foster, D. M., Emerson, M. A., Buck, C. E., Walsh, D. S., and Sproul, O. J.**, Ozone inactivation of cell- and fecal-associated viruses and bacteria, *J. Water Pollut. Control Fed.*, 52, 2174, 1980.

EPILOGUE

Gunther F. Craun*

The incidence of infectious waterborne diseases has declined dramatically in this country over the past 100 years. This has been achieved largely through the development of community water systems, construction of filtration plants and disinfection facilities, and protection of water sources from human waste discharges. Waterborne outbreaks, however, continue to occur. While waterborne outbreaks are no longer a major cause of illness in this country and few deaths have been associated with recent outbreaks, we should nevertheless strive to eliminate this residual occurrence of outbreaks.

Recommendations to reduce the occurrence of waterborne disease outbreaks have been published[1] by the American Water Works Association's Committee on the Status of Waterborne Diseases in the United States and Canada for which I served as chairman. The reader is referred to the complete report,[1] which strongly urges that water supply surveillance and regulatory programs be based on waterborne outbreak data. The report[1] also notes the limitations of coliform surveillance of water systems, and although the Safe Drinking Water Act prescribes coliform surveillance for finished water, the Committee feels that it is "more important to know the quality of the source water and potential sources of contamination so that source protection and treatment can be provided. A water supply surveillance program should emphasize frequent engineering evaluation and sanitary surveys to identify and correct potential deficiencies. Microbiological resources are better applied to assessing raw water qualtiy, identifying sources of contamination, and evaluating the efficiency of treatment . . . ''

Some 14 years ago, Lee McCabe and I recommended the disinfection of untreated water as a means of reducing waterborne disease outbreaks. I am still of the opinion that this recommendation is valid today, especially for smaller ground water systems subject to intermittent or continuous contamination which cannot be prevented. This is not to suggest decreased emphasis on the protection of ground water sources from contamination, but to provide an increased level of protection. Wells in limestone areas are subject to contamination from distant sources which may not be identifiable, and it often requires much bacteriological sampling to determine when a well has become contaminated. In one instance, several outbreaks occurred over the summer at a recreational camp, but it was not until after the third outbreak that daily sampling of the well revealed an intermittent contamination. Previous sampling of the well was conducted daily, but only every second week, and this was not sufficient to detect the contamination. Because many outbreaks occur due to interrupted or inadequate disinfection, it is also important to provide proper operation and surveillance to ensure continuous, effective disinfection. To protect against the waterborne transmission of giardiasis in untreated or disinfected only surface water supplies, it may be necessary in some instances to pretreat and filter water in addition to disinfecting, and this is currently being considered by the EPA.

Recent concerns that certain by-products of chlorination may be carcinogenic must be tempered with considerations of the benefits provided by chlorination. As Dr. Wolman so aptly stated, the chlorination of water supplies has saved many lives. Reducing the levels of chlorine added to water or using alternate disinfectants may be appropriate to reduce or prevent the formation of by-products, but the effects of these actions on the occurrence of waterborne disease must be carefully weighed. Emphasis should be placed instead on reducing the levels of precursor compounds in water sources.

---

* This Epilogue was written by Gunther F. Craun in his private capacity. No official support or endorsement by the Environmental Protection Agency or any other agency of the Federal Government is intended or should be inferred.

Although this volume provides only a brief introduction on the association of water quality and noninfectious disease, this is likely to be the major concern in the future, as more organic contaminants are being found in water supplies. Public health officials cannot rest on past accomplishments but must meet these new challenges to continue to assure the public that drinking water is safe. Epidemiologic methods which have proved successful for infectious diseases cannot be strictly applied to the study of the effects of chronic low-level exposures to various water contaminants. These may contribute to diseases having latency periods of many years. Case-comparison and cohort epidemiologic studies can contribute to our understanding of these associations but must be carefully designed and conducted so they can be properly interpreted. Although risk estimates provided from toxicologic data are useful, uncertainties still exist, and these uncertainties can be minimized only when data are available from human populations.

The contributions of dedicated epidemiologists, engineers, sanitarians, microbiologists, and other health officials who investigate and report outbreaks are gratefully acknowledged. This volume would not be possible without their contributions. It is hoped that this volume will contribute to a better understanding of the causes of waterborne disease and their prevention and will stimulate epidemiologic studies of the association between disease and chronic exposures to chemical contaminants in drinking water.

REFERENCE

1. Committee Report, Waterborne disease in the United States and Canada, *J. Am. Water Works Assoc.*, 73, 528, 1981.

Index

INDEX

A

Acidity, water supply, 50
Acrylamide, 47
Acute chemical poisonings
  cases, number of, 75, 94—95, 107—111, 124—125, 132
  clinical and epidemiological features, 13—14, 19
  deaths from, 87, 89—91
  engineering aspects of investigating outbreaks of, 217—218, 224
  etiologic agents, 93—96, 124—125, 148—149, 153—155, 163—166
  outbreaks, number of, 124—125, 132
  physical causes, 4—5, 132, 135, 137, 148—149
  seasonal occurrence, 99, 101
Acute gastroenteritis, 75, 91—93, 96—97, 108—114
Adenoviruses, 27
AD-EVA broth, 199
Adipates, 47
AD MPN multitube method, 198—199
Aerobic colony test, see also Standard plate count test, 46, 172, 180
Aeromonas
  hydrophila, 31—32, 34
  sobria, 34
  sp. (aeromonads), 32, 34, 38, 204—205
Aeromonas membrane agar, 206—207
Aeromonas septicemia, 32—33
Agar test methods, 196—203, 205—207
Age group-determined incidence rate, 185—186
Agents, see Etiologic agents; specific agents by name
Airborne causes, water-related illnesses, 7—8
Alachlor, 47
Aldicarb, 47
Algae, 7—9
Alkalinity, water, 51
Alpha-particle activity, 8, 46—47
Alum coagulation, 260—261, 263
Alum-coated diatomaceous earth, 266, 268
Aluminum, 47
Aluminum hydroxide, 266
Aluminum salts, 260, 268
Aluminum sulfate, 260
Amebiasis
  cases, number of, 94—96, 105, 107, 109—110, 122
  clinical and epidemiological features, 14, 18
  deaths from, 87—88, 90—91, 105
  etiologic agents, 94—96, 105
Amebic dysentery, 105
Amebic meningoencephalitis, 36—37
Amides, 48
Amines, 48—49
Amphibole asbestos, 58—59
Anabaena flos-aquae, 9

Antibiotic inhibitors in culture media, 204—207
Antimony, 14, 19
APC test, see Aerobic colony test
Arsenic, 45, 47, 56—57, 124—125, 148
  intake, excessive, results of, 57
  safe levels, 57
Asbestos, 47, 58—60
Associations, epidemiologic, see Epidemiologic associations
Atrazine, 47
Attack rates, see Incidence rates
Average daily dietary intake, inorganic constituents, 48—50, 54—55
Average daily plant output, 219
Azide dextrose MPN multitube method, 198—199
Azide-sorbitol agar, 199

B

Bacillary dysentery, 75
Bacillus
  coli, 259
  prodigiosus, 259
Backsiphonage
  engineering evaluation of, 218, 227
  epidemiologic investigations, 180—181, 191
  outbreak cause data, 131—132, 134—135, 141, 147, 149, 151, 166
  outbreak statistics, 112, 125, 127—128, 131—132, 134—135, 139—143, 147, 149, 151, 155, 166
  regulations and surveillance, 242
  seasonal occurrence data, 135, 141
Bacterial diseases and agents, see also specific diseases and agents by name
  bacterial inactivation by disinfection, 268—272
  bacterial removal by filtration, 257—260, 262—265, 268
  clinical and epidemiological features, 12—17
  engineering aspects of investigating, 216, 218, 220—221, 224—225, 227—229
  epidemiologic investigations, 172—173, 175—176, 181—183
  general discussion, 5, 7
  identification of pathogens and indicator organisms, 196—210
  incubation period, 216
  outbreak statistics, 92, 96, 112—113, 142
  regulations on bacteria in drinking water, 44, 236—242, 244, 249, 252
  transmission barriers to, 257—260, 262—265, 268—272
  water-contact diseases, 24, 38
Bacterial gastroenteritis, 28, 202
Bacterial indicators, see Indicator organisms
Bacteriophage T-2, 264
Balantidium coli, isolation of, 209

BAN medium, 206—207
Barium, 45, 47
Barriers, to disease transmission, see Transmission barriers
Bennett's agar, 206—207
Benzene, 8, 48, 64, 163, 166
Beryllium, 47
Beta particle radioactivity, 8, 46—47
Beverage/food-specific incidence rate, 186—187
BHI broth, 199
Bile broth medium, 199
Biocidal characteristics, disinfectants, 269—272
Biocidal efficiency equation, 269
Bismuth sulfite agar, 200
Blaser's Campy-BAP medium, 203
Blood pressure, high, see Hypertension
Body defense mechanisms, compromised, 24, 38
Boil water orders, 192, 223, 242
Bopp's medium, 206
Bradley's classification, water-related diseases, 4—6
Brilliant green agar, 200
Brilliant green bile broth, 197
Bromide, 61
Bromine, disinfection with, 269
Bromodichloromethane, 61
Bromoform, 61
Bronchopneumonia, 31
Butyle's medium, 203
Byproducts of chlorination, contamination by, 44—45, 61—63, 268, 275

C

Cadmium, 14, 19, 45, 47, 52—54, 177
　intake, excessive, results of, 54
　maximum contaminant level (MCL) for, 53—54
Calcium, 50—53
Campgrounds, outbreak data for, 137—139, 142, 144
Campylobacter
　fetus, 96, 114
　　ssp. *jejuni*, see *Campylobacter jejuni*
　jejuni, 12—13, 106, 115—117, 167, 176
　　isolation of, 202—203
　sp., 163, 202—203
Cancer, see also Carcinogens
　chemical contaminant-associated, 49—50, 57—63
　chlorinated water-cancer relationship, 62—63
　mortality, 8, 49—50, 58—59, 62
　radon-associated, 8
　risk of, chemical contaminants and, 59—60, 62—63
Carbofuran, 47
Carbonate, 52
Carbonate water hardness, 50
Carbon dioxide, 166
Carbon tetrachloride, 8, 48, 64
Carcinogens, see also Cancer; specific types by name, 44, 48—50, 54, 57, 59, 61, 64—65, 275

deaths from, 50—53
Cardiovascular disease, chemical contaminants causing, 50—53, 57
Case histories, obtaining, 174—175
Cases
　defining, 175
　epidemiologic investigations, 184—185
　number of, see also as subheading beneath specific illnesses by name
　　1920—1980, 74—76, 85—89, 94—95, 105—111, 113—116, 125, 127, 129—133, 141—144, 147—155
　　1971—1980, 111, 142
　　1981—1983, 162—168
　community water system-caused, 125, 127, 129—133, 141—142, 145—148, 151, 162, 166
　individual water system-caused, 129—133, 141—142, 148, 152, 162, 166
　noncommunity water system-caused, 125, 129—133, 141—142, 148, 152, 162, 166
　water-contact diseases, 24—26, 30—31, 33—36
　searching for, 174, 179
Cationic polymers, 260—261, 263—264, 268
Causal associations, criteria for, 178
Causes, see Etiologic agents; Physical causes
CDC, see Centers for Disease Control
Celite 503® diatomaceous earth, 264—267
Celite 512® diatomaceous earth, 264—265
Celite 535® diatomaceous earth, 265, 267
Celite 545® diatomaceous earth, 265—268
Centers for Disease Control, 252
Centrifugation methods, cyst recovery, 208
Cercarial dermatitis, 37
Chemical agents and diseases, see also Acute chemical poisonings; Etiologic agents; specific agents and diseases by name, 44—65
　deaths from, 49—53, 58—59, 62
　drinking water standards, 44—48
　engineering aspects of investigative outbreaks caused by, 217—218, 224
　epidemiologic investigations, 47, 49—60, 62—63, 172, 176—177, 181—183, 191—192
　incubation periods, 12—14, 217
　inorganic contaminants, 13—14, 19, 44—60
　organic contaminants, 44—45, 47—48, 60—65
　outbreak statistics, 92, 124—125, 128—130, 137, 141—143, 148—149, 163—166
Chemical analysis tests, 180
Chemical precipitation water softening method, 50
Chi-square (χ^2) statistical method, 187—191
Chlamydia sp., 173
Chloramine disinfection process, 61, 123, 256, 268—269, 271—272
　inorganic, microorganisms inactivated by, 271
Chlordane, 47, 124—125, 148
Chloride, 46, 50, 52
Chlorinated hydrocarbons, 45, 47
Chlorination, water system
　byproducts of, contamination by, 44—45, 61—63, 268, 275

cancer-chlorinated water relationship, 62—63
chlorine as tracer, 220
concentration and contact time, 218—223, 227, 269—273
during outbreaks, 192
epidemiologic investigations, 180—181, 192
evaluation of, 216, 218—223
inadequate, 113, 115, 143, 166—167, 192
interferences, 221—223, 272
interrupted, 117, 123, 143, 166—167, 180—181, 223
microorganism inactivation capabilities, 270—272
optimizing, 222
outbreak statistics
1920—1980, 81, 113, 115, 117, 121—123, 143—144
1981—1983, 166—167
pH and temperature effects, 220—221, 269—273
reliability of, 223
serum cholesterol and, 53
swimming pools and recreational water, 25, 27, 180—181
systems, sanitary survey form for, 246—247
transmission barrier, function as, 256, 268—273
Chlorinators, 222—223
Chlorine demand test, 180
Chlorine dioxide disinfection process, 256, 269, 271—273
  E. coli inactivation by, temperature effects, 270
  microorganism removal capabilities, 271—272
Chlorine residual test, 180
Chlorobenzene, 8
Chloroform, 8, 61—62
Chlorophenoxies, 45, 47
Cholera, 5, 12—14, 102, 166, 235
Cholesterol, serum, chlorination and, 53
Chromates, 125, 148
Chromium, 45, 47, 53
Chromobacterium
  lividum, 207
  sp., 38, 206—207
  violaceum, 32—33, 206—207
Chrysotile asbestos, 58—59
Ciliates, 209
CIN medium, 202
Cisterns, see Storage facilities
Citrobacter sp., 197—198
Clam digger's itch, 37
Clarification process inspection, 226
Classification, water-related diseases, 4—6
Clinical syndromes, diseases and poisonings
  characteristic, 12—19
  defining, 180—182
  duration of signs and symptoms, 182—183
  epidemiologic investigations, 175—177, 181—183
  percentage of clinical manifestations of affected persons, 181—182
  summary of, 12—14
CMM, 202
Coagulant feed interruption, effects of, 260—263

Coagulation process, 226, 260—263, 272
Coated diatomaceous earth, 266, 268
Coincidental associations, 178
Coliforms
  densities, high, in water-contact disease outbreaks, 25—26
  engineering aspects of investigating, 221—222, 225, 228—229
  epidemiologic investigations, 172, 179
  isolation and identification of, 196—198, 201—202, 204
    fecal coliforms, 197—198
    total coliforms, 196—197
  outbreak statistics, 112—114, 121—123, 162, 167
  regulations on coliforms in drinking water, 46, 236—241, 243—244, 249, 275
  survival, in presence of turbidity, 221—222
  total, removal by filtration, 257—258, 260, 262, 264—266
  verification tests for, 197—198
Coliform test, 172
Color, water, 44, 46, 217
Common place contingency table, 188
Common-source outbreaks, single- and multiple-event, 183
Communication, engineer's report, 217, 224, 229
Community water systems
  cadmium and lead levels in, 53—54
  cases caused by, number of, 125, 127, 129—133, 141—142, 145—148, 151, 162, 166
  chemical contaminants in, 44—65
  deaths associated with, 87, 89, 91—92
  defined, 74
  disinfection and filtration of, see Chlorination; Disinfection; Filtration
  distribution system contamination data, 131—132, 134—135, 141
  drinking water standards, 44—48
  gastroenteritis caused by, 113—114, 116
  giardiasis caused by, 122—124
  growth of (1800—1958), 79
  hepatitis A caused by, 119
  largest outbreaks in, causes of, 125, 127
  occurrence and distribution of outbreaks in
    1920—1980, 74—92, 141—142, 146—147
    1971—1980, 141, 146
    1981—1983, 162—166
  outbreak reporting, 78—85
  outbreaks caused by, number of, 124—127, 129—132, 141—142, 145—148, 151, 162, 166
  physical causes of problems in, 124—138, 141, 143, 145—147, 151
  seasonal occurrence of problems in, 126—129, 133, 135, 142, 147, 162
  sodium concentrations, 54—56
  storage facility contamination data, 133, 136, 141
  surveillance, importance of, 168
  water quality monitoring, local health officials, 244

Complaints, water customer, 242
Complement fixation test, 175
Concentration effects, disinfectant (chlorine), 218—223, 227, 269—273
Confirmed case, definition and guidelines for, 175—177, 183
Conjunctivitis, 173
Constant-feed chlorinators, 222—223
Consumer surveillance responsibilities, drinking water standards, 242
Contact dermatitis, 9
Contact time, disinfectant (chlorine), 218—222, 227, 269—273
  computing, in transmission and distribution mains, 220—221
  method of obtaining, 219—220
Contamination mode, determining, see also Physical causes, 173—174, 179—181
Contamination sources, see also Physical causes
  collection of samples from, 228
  engineering aspects of investigating, 223—225, 228
  specific, classification of, 125—155
2 × 2 Contingency table, 188—191
Control, outbreaks, 192
Cooked Meat Media, 202
Copper contamination
  nature and characteristics of, 14, 19, 46—47, 52—53
  outbreak statistics, 124—125, 148, 163—166, 177
Corrosivity and corrosive effects, water, 44, 46, 52—54, 60, 148, 166—167
Coxsackie A and B viruses, 26—27
Coxsackie A2 virus, 218, 260
Cross-connections
  engineering evaluation of, 216, 218, 224, 227
  epidemiologic investigations, 180—181, 191
  outbreak cause data, 131—132, 134—135, 141, 146, 149, 155, 166—167
  outbreak statistics, 105, 112, 114, 121, 124—125, 127—128, 131—132, 134—135, 139—143, 146, 149, 155, 166—167
  sanitary survey form for, 248
  seasonal occurrence data, 135, 141
Crustaceans, 6
CT, see Contact time, disinfectant
Customer surveillance responsibilities, drinking water standards, 242
Cutaneous dermatitis, 37
Cutting oil, 148
Cyanide, 47
Cyanobacteria, 8—9
Cyclops sp., 6
Cysts, see also *Entamoeba histolytica* cysts; *Giardia* cysts; *Giardia lamblia* cysts
  concentration of, effects on filtration, 266—268
  inactivation by disinfection, 270—272
  removal by filtration, 257—263, 265—268
Cytopathogenic viruses, 114

D

2,4-D, 45, 47
Daily dietary intake, see Average daily dietary intake
Dalapon, 47
Data analysis, epidemiological investigations, 180—191
  statistical methods, 187—191
Deaths, see also as subheading beneath specific illnesses by name
  1920—1980, 75, 78, 86—92, 105, 120, 142
  cancer, 8, 49—50, 58—59, 62
  cardiovascular disease-caused, 50—53
  chemical contaminant-caused, 49—53, 58—59, 62
  community water system-associated, 87, 89, 91—92
  individual water system-associated, 88—89, 91—92
  mean number, 89, 92
  noncommunity water system-associated, 88, 91—92
  water-contact diseases, 31, 33, 36—37
Defense mechanisms, body, compromised, 24, 38
DE filtration, see Diatomaceous earth filtration
Dental fluorosis, 58
Developer fluid, 148
Diagnosis, verification of, 174—175
Diarrhea, see also specific types by name
  clinical syndrome, 13—14, 175—177, 181—183
  defined, 175
  epidemiologic investigations, 173, 175—177, 181—182
  etiologic agent and indicator organism isolation, 201, 208
  general discussion, 5
  outbreak statistics, 75, 90—91, 102, 105—106, 111—112, 114—115, 153—154, 166
Diatomaceous earth filters and filtration, 200, 257, 263—268, 272
  coated DE, 266, 268
  grades, effects of, 264—265
  microorganism removal capabilities, 263—268
  process description, 257
Dibromochloromethane, 61
Dibromochloropropane, 47
Dibromomethane, 47
Dichlorobenzene, 48, 64—65
Dichloroethane, 48, 64
Dichloroethylene, 48, 64—65
Dichloropropane, 47
Dietary intake, daily, see Average daily dietary intake
Diethylamine, 49
Dimethylamine, 48—49
Dinoseb, 47
Dioxin, 47
Diquat, 47
Direct associations with water supply, 183, 186—187

Discharge location, bracketing, 228
Disease surveillance and investigation, state regulatory agencies, 251
Disinfection, water system, see also specific techniques by name
  barrier to disease transmission, function as, 256, 268—273
  biocidal characteristics of disinfection agents, 269—272
  biocidal efficiency equation, 269
  byproducts, contamination by, 44—46, 61—63, 268, 275
  concentration, 218—223, 227, 269—273
  contact time, 218—222, 227, 269—273
  continuous, effective programs, 275
  epidemiologic investigations, 180—181, 192
  evaluation of, 216, 218—224, 227, 268—269
  general effectiveness, factors influencing, 270—273
  inadequate, 113, 115, 119—120, 124—128, 131—132, 139—140, 143—152, 154, 166—167, 180—181, 192, 221
  interferences, 221—223, 227, 272
  interruption of, 117, 119, 123—128, 131—132, 139—140, 143—145, 148—152, 154, 166—167, 180—181, 223
  methods, suitability of, Safe Drinking Water Committee evaluation, 268—269
  microorganism inactivation capabilities, 270—272
  normal, overwhelming of, 130, 132, 143, 149, 224
  outbreak statistics
    1920—1980, 78—81, 105, 113, 115, 117, 119—128, 130—132, 139—140, 143—152, 154
    1981—1983, 162—168
  pH and temperature, 220—221, 227
  plant process inspection, 227
  process selection, factors influencing, 268, 273
  recreational water, 25, 27, 38, 180—181
  systems, sanitary survey forms for, 246—247
  temperature affecting, 269—272
Distribution, outbreaks, see Occurrrence and distribution, outbreaks
Distribution and transmission mains, computing contact time in, 220—221
Distribution systems (water)
  barriers to transmission of disease, 256—273
  collection of samples, 228
  contamination data for, 131—132, 134—135, 141—152, 154, 166—168
  engineering evaluation of, 223, 225—228
  plant inspection, guides for, 225—227
  sanitary survey form for, 248
Distributor responsibilities, drinking water standards, 242—244, 249
Dow® C-31 cationic polymer, 264
Dracontiasis, 5—6, 100—101
Dracunculus medinensis, 5
Drake's medium, 203—204
Drinking water-caused diseases, see also specific types by name
  barriers to disease transmission, 256—273
  chemical agents and diseases, 44—65
  clinical and epidemiological characteristics, 12—19
  clinical syndromes and incubation periods, 12—14
  engineering aspects of outbreak investigation, 216—229
  epidemiologic investigations, 172—192
  etiologic agents, see Etiologic agents; specific agents by name
  general discussion, 4—5, 9, 275—276
  outbreak statistics
    1920—1980, 74—155
    1971—1980, 93—97, 107, 111, 142—155
    1981—1983, 162—168
  pathogen and indicator organism isolation and identification, 196—210
  preventive regulations and surveillance, see also Regulations; Surveillance, 234—252
  Safe Drinking Water Act, see Safe Drinking Water Act
  types, see also specific types by name, 12—19
Drinking water standards, see Standards, drinking water
Drinking water systems, see Community water systems; Individual water systems; Noncommunity water systems
DSF medium, 204
Dual-cylinder feed, chlorination systems, 223
Dulcitol Selenite broth, 200
Duration, signs and symptoms, 182—183
Dye tests, 179—180, 224—225
Dysentery, 75, 105

E

Ear infections, 27—29, 204
  seasonal occurrence, 29
EC broth, 197
Echoviruses, 27
Edelstein's medium, 206
Endothall, 47
Endotoxins, 7—9
Endrin, 45, 47
Engineering aspects, outbreak investigation, 216—229
  communication, 217, 224, 229
  disinfection, evaluation of, 216, 218—224, 227
  etiologic agent, sampling for, 217, 227—228
  general discussion, 216—218
  water system, evaluation of, 216—217, 223—227
Enrichment media and procedures, 196—203
Entamoeba histolytica, 13—14, 18, 105, 122, 177
  isolation of, 209
Entamoeba histolytica cysts, 220, 257, 260, 262, 266, 270—271
Enteric diseases and pathogens, see also specific types by name, 5, 24, 37—38, 172, 196, 198, 260, 270—271

Enteritis, see also Gastroenteritis, 27
Enterobacter sp., 197—198
Enterococci, isolation of, 199—200
Enteropathogenic *E. coli* strains, 114—115
Enterotoxigenic *E. coli*, see also Toxigenic *E. coli* diseases, 13—15, 114—115, 182
  isolation of, 208
Enterotoxigenic *E. coli* gastroenteritis, 13—15
Enteroviruses, 26—27, 118, 177, 260, 270—271
Enumeration, pathogens and indicator organisms, 196—210
Environmental Protection Agency, drinking water regulation by, 44—48, 235—236, 239—241, 244, 250—252
EPA, see Environmental Protection Agency
Epichlorohydrin, 47
Epidemic curve, construction of, 183—184
Epidemic hepatitis non-A, non-B, 119
Epidemiological characteristics, diseases and poisonings, 12—19, 175—177
Epidemiologic associations
  direct associations with water supply, 183, 186—187
  making, 173, 175, 183—191
  person associations, 178, 185
  place associations, 175—178, 183—185, 188
  statistical significance, evaluating, 187—191
  time associations, 175, 178, 183—184
Epidemiologic investigations, procedures for, 172—192
  chemical contaminants, 47, 49—60, 62—63, 172, 176—177, 181—183, 191—192
  clinical syndromes, 175—177, 181—183
  data analysis, 180—191
    statistical methods, 187—191
  decision to make, criteria for, 173
  engineering investigations and, 216—217, 224—225, 227—228
  general discussion, 172, 192, 276
  hypotheses
    formulating and modifying, 173, 178, 180—181
    testing, 180—191
  illness reports, obtaining, 172—173
  indicators of disease, detecting, 172—173
  post-investigation action, 191—192
  pre-investigation action, 173—174
  protocols and methodology, 174—181
  summarizing, 191
  tests and testing, 172—173, 175, 179—180, 187, 191
  water-contact diseases, 24, 26—29
Epoxy-glass fiber filters, 202
Escherichia coli
  epidemiologic investigations, 172, 175, 182
  general discussion, 7
  inactivation of, 220, 270—272
  isolation of, 197—198, 208
  pathogenic, types, see also specific types by name, 114—115

waterborne outbreak statistical studies, 90—91, 95—96, 110—111, 114—115, 119, 121, 153
water-contact disease studies, 27
Escherichia coli test, 172
Esculin-azide agar, 199
ETEC, see Enterotoxigenic *E. coli*
Ethyl acrylate, 148
Ethylene dibromide, 47
Etiologic agents, see also specific types by name; also as subheading beneath specific illnesses by name
  clinical and epidemiological characteristics of, 12—19
  engineering aspects of investigating, 216—219, 228—229
  epidemiologic investigations, 173, 175—177, 180—183, 191
  general discussion, 4—9, 275—276
  identification of, 180—183, 191, 200—210
  inactivation of, 218—219
  incubation periods, 12—14
  isolation and enumeration of, 200—210
  outbreak statistics
    1920—1980, 91—125, 148—149, 153—155
    1971—1980, 148—149, 154—155
    1981—1983, 162—167
  sampling for, 217, 227—228
  water-contact diseases, 24—38
Expected rate table, 188—190
Eye infections, 173

F

FDA, see Food and Drug Administration
Fecal coliforms, isolation of, 197—198
Fecal coliform test, 172
Fecal indicator bacteria, 196—200, 209
Fecal streptococci, isolation of, 198—200
Fecal streptococci test, 173
Federal agencies, see also specific agencies by name responsibilities, drinking water standards, 251—252
Feeley's medium, 206
Ferrate, disinfection with, 269
Ferric chloride, 260
Ferric hydroxide, 221
Ferric salts, 260, 268
Fiberglass epoxy depth filters, 201
Field inspections, see Sanitary surveys
Filter aids, 264
Filter bed condition, effect on slow sand filtration, 257—258
Filter bed depth, effects on microorganism removal capabilities, 259—260
FilterCel® diatomaceous earth, 264—265
Filters and filtration, pathogen isolation with, 196—209, 238—239
Filter scraping, 257—258
Filtration, water system

barrier to disease transmission, function as, 256—268, 272
engineering evaluation of, 216, 225—227
epidemiological investigations, 180—181
inadequate, 123, 125, 127, 131—133, 139, 145—147, 150, 162—167, 180—181
interruption of, 162
microorganism removal capabilities, 257—268
outbreak statistics
  1920—1980, 78—81, 105, 123, 131—133, 139, 145—147, 150
  1981—1983, 162—167
process, inspection of, 225—227
process descriptions, 256—257
systems, seasonal occurrence of outbreaks in, 133, 139
temperature affecting, 258—260, 267
units, sanitary survey form for, 247
Filtration rate, effects on microorganism removal capabilities, 259, 263, 266—268
Fisher's exact statistical method, 187—191
Flatworms, 37
Fletcher's semisolid medium, 207
Flocculation process, 226, 260, 262, 272
Flow-proportioning chlorinators, 222
Fluid capacity, pipes, 220—221
Fluoridation, water systems, 148—149
  deficiencies in, 124—125
  fluoride as tracer, 220
Fluoride contamination
  concentrations, excessive, results of, 58
  epidemiologic investigations, 176
  nature and characteristics of, 13—14, 19, 45, 47, 52, 58
  optimum fluoride level, 58
  outbreak statistics, 124—125, 148—149, 164, 166
Fluorosis, dental, 58
Foaming agents, 46
Food/beverage-specific incidence rate, 186—187
Food and Drug Administration, 252
Frequency distribution, incubation periods, 182
Fuel oil, 148

G

Gastroenteritis
  acute, see Acute gastroenteritis
  cases, number of, 26, 94—95, 106—111, 113—116, 118—121, 132, 184—185
  classification, 75
  clinical and epidemiological features, 13—15, 17—18
  community and noncommunity water system involvement, 112—114, 116
  deaths from, 90—91
  dye testing for, 224—225
  early cases of, reporting, 235
  epidemic curve, 184
  epidemiologic investigations, 173, 175, 182—185
  etiologic agents
    general discussion, 6—7, 9
    isolation and enumeration of, 200, 202
    outbreak statistics, 91—97, 103, 105—106, 108—117, 149, 153—155, 163—167
    water-contact forms, 26—28, 32, 200
  hepatitis A associated with, 119
  incidence rates, 184
  outbreaks, number of, 26, 96, 112—121, 132
  physical causes, 6—7, 26—28, 132, 135, 137, 149
  seasonal occurrence, 96—100
  shellfish-associated, 6—7
  viral (nonbacterial), see Viral (nonbacterial) gastroenteritis
  water-contact caused, 26—28, 32, 200
Giardia
  lamblia, 13—14, 18—19, 46, 95, 105, 116, 121—122, 148, 162, 177, 186, 262
    cysts, 259, 266, 270—271
  muris, 262
    cysts, 260—263, 266, 272
  sp., 162—166
  isolation, 208—209
Giardia cysts, 122—124, 162, 208—209
  transmission barriers to, 257—263, 265—268, 270—271
  turbidity data, 266—267
Giardiasis
  cases, number of, 94—95, 105, 107, 110—111, 121—124, 132
  clinical and epidemiological features, 14, 18—19
  community, noncommunity, and individual water system involvement, 122—124
  epidemiologic investigations, 173, 186
  etiologic agents, 93—97, 102, 121—124, 149, 153—155, 162—166
  incidence rate, 186
  outbreaks, number of, 81, 84, 96—97, 122—124, 132
  physical causes, 132, 168
  prevention of, 275
  seasonal occurrence, 98—101
  waterborne outbreaks and all causes compared, 105
Glasses of water/beverage ingested per day, incidence rate based on, 187
Glyphosphate, 47
GN broth, 201
Gram-negative broth, 201
Gram-negative and Gram-positive bacteria, 28
Gravity storage unit, sanitary survey form for, 247—248
Groundwater contamination, see Surface and groundwater facility contamination
Group D streptococci, 198—199
Guinea worm disease, see Dracontiasis

H

7H10 medium, 205
7H11 medium, 205
Halophilic *Vibrio* sp., 28—29, 35, 38
Hardness, water, 50—53
Health officials, local, responsibilities of, drinking water standards, 244—249
Heavy metals, see also specific types by name, 13—14, 19, 45—47, 52—54, 177
Helminths, 5
Hemagglutination inhibition assay, 175
Hemorrhagic jaundice, see also Heptospirosis, 35
Hepatitis A
   cases, number of, 26, 94—95, 107—111, 118, 132
   clinical and epidemiological features, 13, 17
   community, noncommunity, and individual water system involvement, 118—119
   deaths from, 90—91
   epidemiologic investigations, 173, 177
   etiologic agents, 6—7, 26, 93—97, 118—119, 149, 153—155, 163—164, 166
   gastroenteritis associated with, 119
   incubation period, 216
   isolation of virus, 209
   outbreaks, number of, 26, 96—97, 118—119, 132
   physical causes, 26, 132, 135, 149
   seasonal occurrence, 97, 99, 129
   shellfish-associated, 6—7
   waterborne outbreaks and all causes compared, 103—104
   water-contact caused, 26
Hepatitis B, 118
Hepatitis non-A, non-B, 118—119
Herbicides, 125, 148, 224
Hexachlorocyclopentadiene, 47
High blood pressure, see Hypertension
Hotels, outbreak data for, 138—139, 142, 144
Humic substances, 61
Hydraulic displacement calculations, 218—219
Hydrocarbons, chlorinated, 45, 47
Hydrogen ion, 46
Hydrogen peroxide, disinfection with, 269
Hydrogen sulfide, 221
Hydropneumatic storage units, sanitary survey form for, 248
Hydroquinone, 148
Hyflo® diatomaceous earth, 264—265, 267
Hypersensitivity pneumonitis, 7
Hypertension, chemical agents causing, 52—53, 55—56
Hypochlorite ion, 220, 270—272
Hypochlorous acid, 220—221, 270—271

I

Identification, pathogens and indicator organisms, 196—210
Illness reports, obtaining, 172—173
IM-MF method, 197—198
Inactivation, microorganisms, 218—220, 268—273
Inactivation capabilities, microorganism, disinfection systems, 270—272
Inadequate water treatment, see Chlorination, inadequate; Disinfection, inadequate; Filtration, inadequate
Incidence rates, 184—191
   by age group, 185—186
   defined, 184
   expected rate table, 188—190
   food/beverage-specific, 186—187
   by month, 184—185
   number of glasses of water/beverage ingested per day, 187
   by residence, 184—185
   by sex, 185—186
   tables, usefulness of, 186—187
   vehicle-specific, 186
Incubation periods (latent periods)
   bacterial, viral, and parasitic diseases, 12—14, 176—177, 182—183, 216
   calculating, 182—183
   chemical agents, 12—14, 217
   clinical and epidemiological features, 12—14
   defined, 182
   engineering aspects of outbreak investigation, 216—217
   epidemiologic investigations, 176—177, 182—183
   frequency distribution, 182
   summary of, 12—14
Indicator organisms, 191, 228, 273
   detecting, 172—173
   isolation, identification, and enumeration of, 196—200, 209
   theory of, 196
Individual water systems
   arsenic concentration, 57
   cases caused by, number of, 129—133, 141—142, 148, 152, 162, 166
   chemical contaminants in, 48—50, 52, 56—57, 60—61, 63—64
   deaths associated with, 88—89, 91—92
   defined, 74
   disinfection of, see Chlorination; Disinfection
   distribution system contamination data for, 131—132, 134—135, 141
   giardiasis caused by, 124
   hepatitis A caused by, 118—119
   monitoring of, local health officials, 244
   occurrence and distribution of outbreaks in 1920—1980, 74—76, 78, 86—87, 89, 91, 93, 141—142, 146—147
   1971—1980, 141, 146

1981—1983, 163—166
outbreaks caused by, number of, 129—132, 141—142, 146, 148, 152, 162, 166
physical causes of problems in, 125, 127—137, 141, 143, 148, 152
seasonal occurrence of problems in, 127, 129, 134, 136, 142, 147, 162
storage facility contamination data, 133, 136, 141
typhoid fever caused by, 117
Industrial facilities, outbreak data for, 138—140, 143, 145
Infantile diarrhea, 110—111
Infectious agents, see Etiologic agents; specific agents by name
Infectious conjunctivitis, 5
Infectious diseases, see also specific types by name
clinical and epidemiological characteristics of, 12—19
clinical syndromes and incubation periods, 12—14
Infectious hepatitis, see Hepatitis A
Inhalation exposures, 8
Inorganic chemical agents, 13—14, 19, 44—60
maximum contamination levels (MLC) for, 44—45
Inorganic chemical contaminants, see also specific types by name, 13—14, 19, 44—60
Inorganic chloramines, microorganisms inactivated by, 271
Inorganic nitrogen, 221
Interferences, disinfection systems, 221—223, 227, 272
Interrupted water treatment, see Chlorination, interrupted; Disinfection, interrupted
Invasive *E. coli* strains, 114
Investigation of outbreaks, see Epidemiologic investigation
Iodine disinfection process, 145, 256, 269
Ion exchange water softening method, 50, 52—53
Ionizing radiation, disinfection with, 269
Iron, 45—46, 221, 263
Iron salts, 260, 268
Isolation, identification, and enumeration, etiologic agents and indicator organisms, 196—210
etiologic agents, methods, 200—210
general discussion, 196, 209—210
indicator organisms, methods, 196—200, 209

J

Jar tests, 180, 260—261

K

Kerosene, 148
KF agar, 198—200
Klebsiella sp., 197—198

L

Laboratory practices
epidemiologic investigations, 172, 174—175, 182—183
etiologic agent and indicator organism identification, 196—210
laboratory capability and certification, 251
Lactose positive *Vibrio* sp., 33
Lancefield's Group D streptococci, 198—199
Largest outbreaks, causes of, 125, 127—128, 148—149, 162, 166
Latent period, see Incubation period
Lauryl sulfate tryptose broth, 196—197
Leaching, of metals from plumbing, 52—54, 124—125, 148, 166—167, 218
Lead contamination
intake, excessive, results of, 54
maximum contaminant level (MCL), 54—55
nature and characteristics of, 14, 19, 45, 47, 51—54
outbreak statistics, 148, 164, 166
Leaded gasoline, 148
Legionella
bozemanii, 31
pneumonia, 206
pneumophila, 7, 31
sp., 32, 46, 206
Legionellosis, 7, 31
Legionnaire's Disease Bacillus, 31
Leptospira
autumnallis, 35
caneiola, 35
canicola, 35
icterohemorrhagia, 35
interrogans, 35—36
pomona, 35—36
sp., 35—36, 38, 207—208
Leptospirosis
epidemiologic investigations, 173
waterborne, 94, 96, 102, 108
water-contact caused, 24, 35—36, 38
Local health officials, responsibilities of, drinking water standards, 244—249
Loop-controlled feed, chlorination systems, 223
LTB broth, 196—197
Lowenstein-Jensen medium, 205
Lubricating oils, 148
Lindane, 45, 47
Lypopolysaccharide endotoxin, 9

M

Magnesium, 50, 52—53
mA medium, 204
Management, sanitary survey form for, 248
Manganese, 46, 52, 221
Manganese dioxide, 221
Maximum contaminant levels (MCL), see also Recommended maximum contaminant levels, 44—46, 48, 53—54, 56, 64

mC filter method, 196—197
MCL, see Maximum contaminant level
mE agar, 199—200
Mean number of deaths, 89, 92
Median, calculating, 182—183
Media size, effect on slow sand filtration, 257—258
mE-EIA medium, 199—200
Membrane filtration techniques, 196—209, 238—239, 264
m-Endo method, 196—197
Meningoencephalitis, amebic, 36—37
mEnt agar, 199—200
M-enterococcus agar, 199—200
Mercury, 45, 47
Metallic coagulants, 260—261, 263
Metals, see also Heavy metals; specific metals by name
  leaching from plumbing, see Leaching
Methemoglobinemia, 5, 45—48, 124
  number of cases, 47—48
Methoxychlor, 45, 47
mFC membrane filter method, 197—198
Microbial tests, 172—173
Microbiocides, 220—221
Microbiological agents, engineering aspects of investigating outbreaks caused by, 216—229
Microbiological methods, etiologic agent and indicator organism identification, 196—210
Microorganisms, see also specific types by name, 24, 37—38, 172, 187, 192, 196, 204, 221, 256—273
  inactivation by disinfection, 268—273
  removal by filtration, 256—268
Microporous filters, 203
Middlebrook 7H10 and 7H11 media, 205
Mixing process inspection, 226
Molluscicides, 37
Molybdenum, 47
Monitoring, sanitary survey form for, 248
Monthly incidence rate, 184—185
Mortality, see Deaths
Most probable number methods, 196—199, 201, 203—204, 206
Motels, outbreak data for, 138—139, 142, 144
MPN methods, see Most probable number methods
mSD agar, 199
mTEC method, 198
m-T7 medium, 197—198
Multiple barrier concept, see also Transmission barriers, 256
Multiple-event common-source outbreaks, 183
Municipal water systems, see Community water systems
Mycobacteria, 7
Mycobacteriosis, 29—30
Mycobacterium
  balni, 29
  intracellulare, 204—205
  kansasii, 29—30, 204—205
  marinum, 29—30, 204
  platy, 29
  scrofulaceum, 205
  sp., 38, 204—205
  szulgai, 29—30
  tuberculosis, 29
  xenopi, 204

N

Naegleria
  flowleri, 36
  sp., 36—37
National Interim Primary Drinking Water Regulations, 44, 240—241
National Primary Drinking Water Regulations, 44—47, 239—241, 251—252
National Quarantine Act, 235—236
National Secondary Drinking Water Regulations, 44, 46
Nematodes, 5
Nephelometric turbidity units, 222, 257, 260—261, 263, 267—268
Nickel, 47
Nitrates, 44—50, 52, 124—125, 148, 164, 166, 249
  average daily dietary intake, 48—50
  maximum contamination level (MLC) for, 45
Nitrites, 45, 48—50, 173
  average daily dietary intake, 48—49
Nitrogen, inteference by, 221
Nitrosamines, 48—49
Nitrosation, 48
N-Nitroso compounds, 48—49
27-nm particles, virus-like, 112—113
Nonbacterial (viral) gastroenteritis, see Viral (nonbacterial) gastroenteritis
Noncarbonate water hardness, 50
Noncommunity water systems
  cases caused by, number of, 125, 129—133, 141—142, 148, 152, 162, 166
  deaths associated with, 88, 91—92
  defined, 74
  disinfection of, see Chlorination; Disinfection; Filtration
  distribution system contamination data, 131—132, 134—135, 141
  gastroenteritis casued by, 112—113
  giardiasis caused by, 123—124
  hepatitis A caused by, 119
  largest outbreaks, causes of, 125, 128
  monitoring of, local health officials, 249
  occurrence and distribution of outbreaks in
    1920—1980, 74—78, 84—91, 93, 141—142, 146
    1971—1980, 141, 146
    1981—1983, 162—166
  outbreak reporting, 84—85
  outbreaks caused by, number of, 129—132, 141—142, 146, 148, 152, 162, 166
  physical causes of problems in, 125, 127—138, 141, 143, 148, 152

seasonal occurrence of problems in, 127, 129, 133—134, 136, 142, 146, 162
storage facility contamination data, 133, 136, 141
surveillance, importance of, 168
Nonionic polymers, 260—261
Nontyphoidal *Salmonella* sp., 13
Norwalk agent, 28, 162—164, 166, 177, 182—183, 196
Norwalk gastroenteritis, 13, 17
Norwalk-like agents, 112—114, 177
Norwalk virus, 6—7, 13, 17, 26, 112—114, 209
Notification, see Public notification; Reporting
Number of samples per month, drinking water-regulations, 237—241

O

Occurrence and distribution, see also Seasonal occurrence and variation
  1920—1980, 75—93, 142, 146—147
  1971—1980, 141, 146—147
  1981—1983, 162—166
  community water systems, see Community water systems, occurrence and distribution of outbreaks in
  determining, 174, 178
  individual water systems, see Individual water systems, occurrence and distribution of outbreaks in
  noncommunity water systems, see Noncommunity water systems, occurrence and distribution of outbreaks in
  reduction of, recommendations for, 275
Odor, water, 44, 46, 217, 223
One degree of freedom, chi-square table for, 189—190
Opportunistic pathogens, isolation and enumeration of, 200—210
Organic chemical agents, see also specific types by name, 8, 47—48, 60—65
  maximum contaminant levels (MLC) for, 44—45
Organic compound volatilization, 8
Organic nitrogen, 221
Otitis externa, 28—29, 204
Otitis media, 29
Outbreaks
  common-source, 183
  confirmed, 175—177, 183
  controlling, 192
  definitions, 74—75
  epidemiologic investigation of, see also Epidemiologic investigations, 172—192
  etiology, see Etiologic agents; Physical causes
  importance of, 96—101
  incidence rates, see Incidence rates
  investigations of
    agent and indicator organism identification methods, 196—210
    difficulty of conducting, 216, 228
    engineering aspects, 216—229
  epidemiologic procedures, 172—192
  local health officials, 249
  isolation of etiologic agents in, 200—202, 208—209
  largest, causes of, 125, 127—128, 148—149, 162, 166
  magnitude of, 86—87
  number of, see also as subheading beneath specific illnesses by name
    1920—1980, 75—78, 81—97, 102—103, 106—107, 112—119, 124—127, 129—133, 141—155
    1971—1980, 82, 142
    1981—1983, 162, 165—168
    community water system-caused, 124—127, 129—132, 141—142, 145—148, 151, 162, 166
    individual water system-caused, 129—132, 141—142, 146, 148, 152, 162, 166
    noncommunity water system-caused, 129—132, 141—142, 146, 148, 152, 162, 166
    reported and expected, compared, 83—84
    reporting, 78—85, 162—165
    undetected or unreported, 85
    water-contact diseases, 24—26, 35—37, 112, 162
  occurrence and distribution, see Occurrence and distribution; Seasonal occurrence and variation
  physical causes, see Physical causes
  prevention of
    barriers to transmission, 256—273
    epidemiologic investigations, 192
    regulations and surveillance, 234—252
  propagated, 183
  reporting, 4, 78—85, 162—165, 192
  seasonal occurrence of, see Seasonal occurrence and variation
  statistics
    1920—1980, 74—155
    1971—1980, 93—97, 107, 111, 142—155
    1981—1983, 162—168
    cases, number of, see Cases, number of
    deaths, see Deaths
    definitions, 74—75
    general discussion, 74
    occurrence and distribution, see Occurrence and distribution; Seasonal occurrence and variation
    reporting, 78—85, 162, 165
Overwhelming, disinfection system, 130, 132, 143, 149, 224
Oxidase test, 204—205
Ozone disinfection process, 256, 269—273
  microorganism removal capabilities, 271—272

P

PAHs, 47
PAM, see Primary amoebic meningoencephalitis

Paramethylamino phenol, 148
Parasites, see also specific types by name, 13—14, 18—19, 37, 92, 181—182, 208—209, 222, 227
Parasitic diseases, see also specific types by name
  clinical and epidemiological features, 14, 18—19
  engineering aspects of investigating, 216, 222, 224, 227
  epidemiologic investigations, 173, 175, 177
  incubation period, 216
Paratyphoid fever, 25, 94, 96, 102—103, 108
Parks, outbreak data for, 137—139, 142, 144
Particles, 27-nm, virus-like, 112—113
Particulate matter, interference by, 221—222, 272
Parvollke virus, 137
Pathogenic *Escherichia coli,* types, see also specific types by name, 114—115
Pathogens, see Etiologic agents; specific types by name
PCBs, 47, 148
Pentachlorophenol, 47
Peptone-Sorbitol-Bile Salts media, 202
Percentage of clinical manifestations, affected persons, 181—182
Person associations, 178, 183, 185
Pesticides, see also specific types by name, 14, 19, 44—45, 47—49, 56, 125, 148, 181, 191, 224
Petroleum products, 14, 19, 125, 148
Pfizer Selective Enterococcus (PSE) agar pourplate procedure, 198—199
pH
  adjustment, water treatment by, 54
  disinfection system effects, 220—221, 227, 269—273
  high, disinfection method utilizing, 269
  rapid sand filtration affected by, 260—261, 263
  tests, 180—181
Phenol, 148
Phenol red lactose broth, 197
Photon radioactivity, 46—47
PHS, see Public Health Service
Phthalates, 47
Physical causes, outbreaks, see also Contamination sources; specific causes by name; also as subheading beneath specific illnesses by name
  1920—1980, 124—155
  1971—1980, 142—155
  1951—1983, 167—168
  community water system problems, 124—138, 141, 143, 145—147, 151
  contamination, sources of, specific, classification of, 125—155
  contamination mode, determining 173—174, 179—181
  contributory factors, table, 181
  distribution systems, data for, 131—132, 134—135, 141—152, 154, 166—168
  epidemiologic investigations, 173—175, 179—181, 183, 191

  hotels, motels, and rooming houses, data for, 138—139, 142, 144
  individual water system problems, 125, 127—137, 141, 143, 148, 152
  industrial facilities, data for, 138—140, 143, 145
  largest outbreaks, 125, 127—128, 148—149, 162, 166
  miscellaneous reasons, data for, 6—9, 136—137, 141—143, 148, 151, 166—167
  noncommunity water-system problems, 125, 127—138, 141, 143, 148, 152
  parks, campgrounds, and recreational areas, data for, 137—139, 142, 144
  schools, data for, 138—140, 143, 145, 149
  springs, data for, 128—129, 134—135, 140, 142—143, 149—150, 166—167
  storage facilities, data for, 133, 136, 141, 147, 150
  treatment deficiencies, 133, 138, 162—168
  water-contact diseases, 24—28
  wells, data for, 128—133, 135—137, 139—140, 142—144, 149—150, 154, 162, 166, 168
Pichloram, 47
Picornavirus, 118
Pipes, fluid capacity of, 220—221
Place associations, 175—178, 183—185, 188
Plant inspection, guides for, 225—227
Plate count bacteria, removal of, see also Standard plate count test, 257—259, 264—266
Plug-flow, 220, 269—270
Plumbing, leaching of metals from, see Leaching
Pneumonia, 30—32
Poisonings, chemical, see Acute chemical poisonings; chemical agents; specific agents by name
Poliomyelitis, 94, 96, 105, 109
Polioviruses, 218, 257—260, 262—264, 270—272
Polychlorinated biphenyls, 47, 148
Postchlorination units, sanitary survey form for, 247
Post-investigative action, epidemiological investigations, 191—192
Potassium, 51—52
Potassium permanganate, disinfection with, 269
Prechlorination/pretreatment units, sanitary survey form for, 246
Pre-investigative action, epidemiological investigations, 173—174
Preston enrichment broth, 203
Preston medium, 203
Presumptive case, defined, 175
Pretreatment effects, on rapid sand filtration, 260, 262
Primary amoebic meningoencephalitis, 36—37
Primary Drinking Water Regulations, National, 44—47, 239—241, 251—252
Probability determinations, in statistical analysis, 189—191
Propagated outbreaks, 183
Protozoa, 5, 208—209, 218—219, 222, 228, 268, 270
Protozoan cysts, 218, 271

PSB media, 202
PSE agar pour plate procedure, 198—199
Pseudomonas
   aeruginosa, 28, 173, 203—204
   putrefaciens, 31—32
   sp. (pseudomonads), 28—29
Public Health Service Centers for Disease Control, 252
Public Health Service Drinking Water Standards, 235—240
Public notification (reporting), results of regulation and surveillance programs, 241, 243—244, 252
Pumps, sanitary survey form for, 246
Purification, see Filtration
Pyrrolidone, 48

R

Radionuclides, 8, 192
   maximum contaminant levels (MLC) for, 44, 46
   regulation of, 44, 46—47
Radium 226 and 228, 46—47
Radon, 8, 47
Rapid sand filtration, 79—80, 272
   microorganism removal capabilities, 260—263
   process description, 257
Recommended maximum contaminant levels (RMCL), see also Maximum contaminant levels, 44—45, 48, 64—65
Record keeping, suppliers of drinking water, 243
Recreational areas, outbreak data for, 137—139, 142, 144
Recreational water-contact diseases, see Water-contact diseases
Regulations, drinking water, 44—48, 234—244
   current, 240—241
   general discussion, 234, 242
   historical development of, 234—240
   number of samples per month, 237—241
   surveillance and enforcement, see also Surveillance, 243—244, 249—252
Removal, microorganisms, by filtration, 256—268, 272
Removal capabilities, microorganism, filtration systems, 257—268
Reports
   of illness, obtaining, 172—173
   of outbreaks, 4, 78—85, 162—165
      submitting, 192
   public notification, results of regulation and surveillance programs, 241, 243—244, 252
Reservoirs, see Storage facilities
Residential incidence rate, 184—185
Respiratory illnesses, 7—8, 27, 30—32, 206
RMCL, see Recommended Maximum Contaminant Levels
Rooming houses, outbreak data for, 138—139, 142, 144
Rosolic acid, 197

Rotaviruses, 13, 17—18, 28, 111—112, 114, 166, 177
Rotavirus gastroenteritis, 13, 17
RS medium, 204
Ryall and Moss selective media, 206—207

S

Safe Drinking Water Act, 44, 235—236, 239—244, 249—252, 275
Safe Drinking Water Committee (National Research Council) evaluation of disinfection method suitability, 268—269
Safe water programs, 192
Salmonella
   cholerae-suis, 15
   dublin, 15
   flexneri, 16
   paratyphi, 15, 25
   sp., 13, 15, 25, 114—115, 176
      isolation of, 200—201
   typhi, 13, 15—16, 25, 93, 117, 129, 176
   typhimurium, 120—121
   typhosa, 24—25
Salmonella gastroenteritis, 103
Salmonella test, 173
Salmonellosis
   cases, number of, 25, 85, 94—95, 108—111, 120—121, 132
   clinical and epidemiological features, 13, 15
   deaths from, 90—91, 120
   etiologic agents, 7, 25, 94—97, 120—121, 149, 153—155
   outbreaks, number of, 25, 96—97, 102—103, 106—107, 120—121, 132
   physical causes, 25, 132, 149
   seasonal occurrence, 98—99
   shellfish-associated, 7
   waterborne outbreaks and all causes compared, 103
   water-contact caused, 25
Samples, drinking water, number per month, drinking water regulations, 237—241
Sampling, water systems and sources, 172—173, 175, 179—180, 217, 227—228
Sand filtration methods, see Rapid sand filtration; Slow sand filtration
Sanitary surveys, 236—241, 243—250, 252
SB broth, 199
Schistosoma
   haematobium, 6
   japonicum, 6
   mansoni, 6
Schistosome dermatitis, 6, 37
Schistosomiasis, 5—6
Schizothrix calcicola, 9
Schmutzdecke removal, 257—258
Schools, outbreak data for, 138—140, 143, 145, 149
SDWA, see Safe Drinking Water Act

Seasonal effects on slow sand filtration process, 258—259
Seasonal occurrence and variation, see also as subheading beneath specific illnesses by name
  1920—1980, 89—93, 96—101, 127—129, 133—135, 139—142, 144—147
  1971—1980, 142, 146—147
  1981—1983, 162, 165
  community water system problems, 126—129, 133, 135, 142, 147, 162
  hotels, motels, and rooming houses, data for, 139, 144
  incidence rate by month, 184—185
  individual water system problems, 127, 129, 134, 136, 142, 147, 162
  industrial facilities, data for, 140, 145
  noncommunity water system problems, 127, 129, 133—134, 136, 142, 146, 162
  parks, campgrounds, and recreational areas, data for, 139, 144
  schools, data for, 140, 145
  treatment deficiency-caused, 133, 138
  water-contact diseases, 29, 38
Secondary Drinking Water Regulations, National 44, 46, 251
Sedimentation process, 226, 260, 262, 272
Selective plating media and techniques, 198—201, 205—207
Selenium, 45, 47, 56—57, 125, 148
  intake, excessive, results of, 57
Septicemia, 32—33
Serratia marescens, 258—259
Serum agar, 207
Serum cholesterol, chlorination and, 53
Sewage-caused water-contact diseases, 24—27, 37—38
Sewage-contaminated water, isolation of pathogens and indicator organisms from, 197—199, 203—204
Sex-determined incidence rates, 185—186
Shellfish-associated illnesses, 6—7, 12, 112
Shellfish harvesting waters, isolation of coliforms from, 197—198
Shigella
  boydii, 16, 201
  dysentariae, 16, 201
  flexneri, 120—121, 201
  sonnei, 16, 114, 120—121, 201
  sp., 8, 13, 16, 26, 114—115, 120, 129, 149, 163—165, 176, 182—183
  isolation of, 201
Shigella gastroenteritis, 26
Shigellosis
  cases, number of, 26, 94—95, 107—111, 120, 132
  clinical and epidemiological features, 13, 16
  deaths from, 87, 89—91
  dye testing for, 224—225
  etiologic agents, 7, 26, 93—96, 119—121, 149, 153—155, 163—166
  isolation of *Shigella* sp., 201

outbreaks, number of, 26, 96, 102, 106—107, 119—120, 132
physical causes, 26, 132, 137, 149
seasonal occurrence, 97—98
shellfish-associated, 7
waterborne outbreaks and all causes compared, 103—104
water-contact caused, 26, 201
Silicon, 51
Silver, 45, 47
  disinfection with, 269
Simazine, 47
Single-event common-source outbreaks, 183
Skirrow's medium, 203
Slanetz-Bartley broth, 199
Slow sand filtration, 78—80, 256—260, 272
  microorganism removal capabilities, 257—260
  process description, 256—257
SM agar, 207
Sodium, 44, 47, 52—56
  average-daily dietary intake, safe, 54—55
  intake, excessive, results of, 54—56
  maximum contaminant level (MCL) for, 56
Sodium-restricted diets, 55
Softening, water, 50, 52—53, 56
Soft water, 51—53
Sources of contamination, see Contamination sources; Physical causes
Specimens, collecting, 175, 179—180
Springs
  engineering evaluation of, 224—225
  epidemiologic investigations, 173, 181, 184—185, 192
  outbreak statistical data, 128—129, 134—135, 140, 142—143, 149—150, 166—167
  sanitary survey form for, 245
Standard plate count bacteria, removal of, 257—259, 264—266
Standard plate count test, 46, 172, 180, 236—237
Standards, drinking water, see also Safe Drinking Water Act, 44—48
  regulations and surveillance, see also Regulations; Surveillance, 234—252
Standard Super-Cel® diatomaceous earth, 264—265
Staphylococci, 28
Staphylococcus aureus, 31
State regulatory agencies, responsibilities of, drinking water standards, 250—251
Statistical methods, data analysis, 187—191
  2 × 2 contingency table, 188—191
  level of statistical significance, 187
  probability determinations, 189—191
Storage facilities (water)
  epidemiologic investigations, 180—181
  outbreak statistics, 133, 136, 141, 147, 150
  sanitary survey form for, 247—248
Streptococci, 28, 198—200
  fecal, isolation of, 198—200
  Group D, 198—199
Streptococcus
  avium, 198, 200

bovis, 198, 200
equinus, 198, 201
faecalis, 198—199
faecium, 198—199
mitis, 200
salivarius, 200
Sulfate-reducing bacteria test, 173
Sulfates, 46—47, 50, 221
Super-Cel® diatomaceous earth, 264—265
Surface and groundwater water facility contamination
 barriers to disease transmission, 256, 273
 engineering aspects of outbreak investigation, 216, 218, 224—225, 228
 epidemiologic investigations, 179—181, 191—192
 isolation of pathogens and indicator organisms from, 196—210
 organic chemical contaminants, 60—65
 outbreak statistics, 80, 105, 116—117, 119—120, 122—152, 154, 166—168
 routine monitoring for pathogens, 209—210
 sanitary survey form for, 245—246
Surveillance, drinking water system, see also Sanitary surveys, 242—252
 consumer responsibilities, 242
 epidemiologic investigations, see Epidemiologic investigations
 federal agency responsibilities, 251—252
 importance of, 168, 275
 local health official responsibilities, 244—249
 outbreak statistics and, 83, 85, 168, 275
 state regulatory agency responsibilities, 250—251
 supplier responsibilities, 242—244, 249
Suspected case, defined, 175
Swimmer's ear, 28—29, 204
Swimmer's itch, 37
Swimming, illnesses from, see Water-contact diseases
Swimming pool granulomas, 29
Symptoms, summary of, 12—14
Syndrome, clinical, see Clinical syndrome
Synthetic organic chemicals, volatile, see Volatile synthetic organic chemicals

T

Taste, water, 44, 217, 223
T-2 bacteriophage, 264
TCBS agar, 205
TCE, see Trichloroethylene
Technical assistance, state regulatory agencies, 250—251
Temperature effects on filtration and disinfection processes, 258—260, 267, 269—272
Temporal associations, 175, 178, 183—184
Tests and testing, epidemiologic investigations, 172—173, 175, 179—180, 187, 191
Tetrachloroethylene, 8, 48, 64—65
Tetrathionate broth, 200

Thallium, 47
Thallous-acetate agar, 199
Thiosulfate Citrate Bile-Salt Sucrose agar, 205
Time associations, 175, 178, 183—184
Tin, 14, 19
Toluene, 47
Total coliforms, isolation and removal of, 196—197, 257—258, 260, 262, 264—266
Total dissolved solids, 46, 50
Total trihalomethanes, 44, 61—62
Toxaphene, 45, 47
Toxigenic *E. coli* diseases, see also Enterotoxigenic *E. coli,* 90—91, 95—96, 110—111, 114—115, 119, 153
Toxins, see Chemical agents; specific types by name
2,4,5-TP, 45, 47
Tracer tests, 219—220
Trachoma, 5
Transmission and distribution mains, computing contact time in, 220—221
Transmission barriers, 256—273
 disinfection, see also Disinfection, 256, 268—273
 filtration, see also Filtration, 256—268, 272
 general discussion, 256, 272—273
Traveler's diarrhea, 114—115
Treatment, evaluation of, 223, 225—227
Treatment deficiencies, water systems, see also Disinfection, inadequate; Filtration, inadequate
 drinking water standards, 44—48
 epidemiologic investigations, 172—173, 179—181, 191—192
 seasonal occurrences caused by, 133, 138
Treatment-provided barriers to disease transmission, 256—273
Treatment units, sanitary survey form for, 246—247
Trichloroethane, 47—48, 64—65
Trichloroethylene, 8, 48, 64—65
Trihalomethanes, 44, 61—62, 269
Tularemia, 94, 96, 101—102, 108—109
Turbidity, 44
 breakthrough, 263
 engineering evaluation of, 221—222, 225
 Giardia cyst turbidity data, DE filtration tests, 266—267
 regulations on, in drinking water, 237—240, 242, 249
 removal by filtration, 257, 260—261, 263—268
 tests, 180
 treatment to reduce, 222
Turbidity units, 222, 240, 257, 260—261, 263, 267—268
Two-layer enrichment procedure, 197—198
Two-layer membrane filter method, 197—198
Two-step membrane filter method, 197
Two-times-two contingency table, 188—191
Typhoid fever
 cases, number of, 24—25, 75, 94—95, 106—111, 117—118, 132
 clinical and epidemiological features, 13, 15—16
 control of, 5

deaths from, 78, 86—92, 117
  mean number, 89, 92
early cases of, reporting, 235
etiologic agents, 6—7, 24—25, 93—96, 117—118, 153—154
outbreaks, number of, 24—25, 96, 106—107, 132
physical causes, 24—25, 133, 135, 137
seasonal occurrence, 97—98
shellfish-associated, 6—7
waterborne outbreaks and all causes compared, 102
water-contact caused, 24—25

U

Ultraviolet light (UV) disinfection process, 143, 256, 269
Uranium, 47
Urease tests, 197, 202

V

Vanadium, 47
Vehicle-specific incidence rates, 186
Verification tests, for coliforms, 197—198
Vibrio
  alginolyticus, 28—29, 32—35, 205
  anguillarium, 205
  cholera, 205
    infection, 102
  cholera 01, 6, 12—14, 102, 164, 166
  mimicus, 29
  parahaemolyticus, 7, 24, 28—29, 32—34, 205
  sp.
    halophilic, 28—29, 35, 38
    isolation of, 205
  vulnificus, 29, 32—35
Vinyl chloride, 48, 64
Viral diseases and agents, see also specific diseases and agents by name
  clinical and epidemiological features, 13, 17—18
  contact time, 218—219
  engineering aspects of investigating, 216, 218—219, 221, 224, 227—228
  epidemiologic investigations, 173, 177, 182
  general discussion, 5
  incubation period, 216
  isolation of viruses, difficulty of, 196, 209
  outbreak statistics, 92, 96, 98, 106, 108—114, 118—119
  transmission barriers to, 257, 260, 264, 268, 270—271
  virus inactivation by disinfection, 268, 270—271
  virus removal by filtration, 257, 260, 264, 268
Viral gastroenteritis (nonbacterial)
  cases, number of, 118—119
  dye testing for, 224
  etiologic agents, 96, 153—155, 166—167

outbreaks, number of, 96, 118—119, 132
seasonal occurrence, 98, 100
water-contact causes, 26—28
Viral hepatitis, forms, see also specific forms by name, 118—119
Virus-like 27-nm particles, 112—113
Volatile synthetic organic chemicals, see also specific types by name, 44—45, 48, 60—61, 63—65
  minimum contaminant level (MLC) for, 64
  occurrence in groundwater, 64—65
  recommended maximum contaminant levels (RMLC) for, 44—45, 48, 64—65
Vydate, 47

W

Wadowsky and Yee's medium, 206
Water
  boiling order, 192
  corrosivity, see Corrosivity and corrosive effects, water
  hardness, see Hardness, water
  quality monitoring, 244
  safe, programs to ensure, see Regulations; Safe Drinking Water Act; Surveillance
  sampling, 172—173, 175, 179—180, 217, 227—228, 237—241
  soft, see Softening, water; Soft water
  standards, drinking water, 44—48, 234—252
Water-associated illnesses, see Water-related illnesses; specific types by name
Water-based diseases, defined, 5—6
Waterborne diseases, see also specific diseases by name
  cases, number of, see Cases, number of
  chemical diseases and agents, 44—65
  clinical and epidemiological characteristics of, 12—19
  deaths, see Deaths
  defined, 5
  drinking water-caused, see Drinking water-caused diseases
  epidemiologic investigation of, see also Epidemiologic investigations, 172—192
  etiologic agents, see Etiologic agents
  general discussions, 4—9, 74, 275—276
  occurrence and distribution, see Occurrence and distribution; Seasonal occurrence and variation
  outbreaks, see Outbreaks
  pathogen and indicator organism isolation and identification, 196—210
  physical causes, see Physical causes
  preventive regulations and surveillance, see also Regulations; Surveillance, 234—252
  transmission barriers, see also Treatment barriers, 256—273
  water-contact types and agents, see also Water-contact diseases, 24—38

Water-contact diseases, 24—38
   cases, number of, 24—26, 30—31, 33—36
   deaths from, 31, 33, 36—37
   epidemiologic investigations, 24, 26—29, 173, 178, 180—181
   etiologic agents, 24—38
   etiology, 100—102
   general discussion, 5, 8—9, 29, 37—38
   isolation of pathogens and indicator organisms, 197—198, 200—201, 203—204, 208—209
   outbreaks, number of, 24—26, 35—37, 112, 162
   physical causes, 24—28
   seasonal occurrence, 29, 38
   types, see also specific types by name, 24—38
Water-related diseases, see also Waterborne diseases; Water-contact diseases; specific diseases by name
   Bradley's classification, 4—6
   causes, miscellaneous, see also Physical causes, 6—9
Water supplier responsibilities, drinking water standards, 242—244, 249
Water supply
   direct associations with, 183, 186—187
   disinfection of, see Chlorination; Disinfection; Filtration
   distribution systems, see Distribution (water) systems
   evaluation of, 216—217, 223—227
   standards, drinking water, 44—48
   treatment deficiencies, see Treatment deficiencies
Water systems, see Community water systems; Distribution systems; Individual water systems; Noncommunity water systems
Water-vectored diseases, defined, 6
Water-washed diseases, defined, see also Water-contact diseases, 5
Weedicides, 224

Weils's disease, see also Leptospirosis, 35
Wells
   chemical contaminants in, 48—50, 52, 56—57
   disinfection of, 275
   engineering evaluation of, 216, 224—225
   epidemiologic investigations, 173, 179, 181
   outbreak statistical data, 128—133, 135—137, 139—140, 142—144, 149—150, 154, 162, 166—168
   sanitary survey form for, 245
Wound infections, 29—30, 32—35, 38

X

XLD agar, 201
Xylene, 47
Xylose lysine desoxycholate agar, 200

Y

Yersinia
   enterocolitica, 13, 16—17, 96, 106, 114—116, 167, 176
      isolation of, 201—202
   frederiksenii, 201
   intermedia, 201
   kristensenii, 201
   pseudotuberculosis, 176
   sp., 114, 164, 176, 201—202
Yersiniosis, 166—167

Z

Zinc, 14, 19, 46—47, 52—53